KV-014-327

Microcomputer Applications in Geology

COMPUTERS and GEOLOGY

a series edited by Daniel F. Merriam

1976 Quantitative Techniques for the Analysis of Sediments

1978 Recent Advances in Geomathematics

1979 Geomathematical and Petrophysical Studies in Sedimentology (edited by D. Gill & D. F. Merriam)

1981 Predictive Geology: with Emphasis on Nuclear-Waste Disposal (edited by G. de Marsily & D. F. Merriam)

Professor Merriam also is the co-Editor-in-Chief of *Computers & Geosciences* — an international journal devoted to the rapid publication of computer programs in widely used languages and their applications.

Other Related Pergamon Publications

Books

HOLLAND
Microcomputers and Their Interfacing

HOLLAND
Illustrated Dictionary of Microcomputers & Microelectronics

MARSAL
Statistics for Geoscientists

NORRIE & TURNER
Automation for Mineral Resource Development

Journals

Applied Geochemistry — NEW
Automatica
Computers & Geosciences
Computer Languages
Information Processing & Management
Information Systems
International Journal of Nuclear Geophysics — NEW
International Journal of Rock Mechanics and Mining Sciences & Geomechanics Abstracts

Microcomputer Applications in Geology

edited by

J. THOMAS HANLEY

Geologist, United States Geological Survey, Reston, Virginia

and

DANIEL F. MERRIAM

Endowment Association Distinguished Professor of the Natural Sciences and Chairman, Department of Geology, Wichita State University, Wichita, Kansas

PERGAMON PRESS

OXFORD · NEW YORK · BEIJING · FRANKFURT
SÃO PAULO · SYDNEY · TOKYO · TORONTO

U.K.	Pergamon Press, Headington Hill Hall, Oxford OX3 0BW, England
U.S.A.	Pergamon Press, Maxwell House, Fairview Park, Elmsford, New York 10523, U.S.A.
PEOPLE'S REPUBLIC OF CHINA	Pergamon Press, Qianmen Hotel, Beijing, People's Republic of China
FEDERAL REPUBLIC OF GERMANY	Pergamon Press, Hammerweg 6, D-6242 Kronberg, Federal Republic of Germany
BRAZIL	Pergamon Editora, Rua Eça de Queiros, 346, CEP 04011, São Paulo, Brazil
AUSTRALIA	Pergamon Press Australia, P.O. Box 544, Potts Point, N.S.W. 2011, Australia
JAPAN	Pergamon Press, 8th Floor, Matsuoka Central Building, 1-7-1 Nishishinjuku, Shinjuku-ku, Tokyo 160, Japan
CANADA	Pergamon Press Canada, Suite 104, 150 Consumers Road, Willowdale, Ontario M2J 1P9, Canada

First edition 1986
Reprinted 1986

Library of Congress Cataloging in Publication Data

Main entry under title:
Microcomputer applications in geology.
(Computers & geology; v. 5)
Includes index.
1. Geology — data processing. 2. Microcomputers — Programming.
I. Hanley, J. Thomas. II. Merriam, Daniel Francis. III. Series.
QE48.8.M53 1986 551'.028'5416 85–21835

British Library Cataloguing in Publication Data

Microcomputer applications in geology.—
(Computers and geology; v. 5)
1. Geology — Data processing. 2. Microcomputers.
I. Hanley, J. Thomas II. Merriam, Daniel F. III. Series
551'.028'5404 QE48.8

ISBN 0–08–031452–X

Printed in Great Britain by A. Wheaton & Co. Ltd., Exeter

Contents

List of Contributors

James D. Bliss, Mail Stop 84, 345 Middlefield Road, U.S. Geological Survey, Menlo Park, California 94025 USA

Joseph Moses Botbol, Branch of Atlantic Marine Geology, U.S. Geological Survey, Quissett Campus, Woods Hole, Massachussetts 02543 USA

Roger W. Bowen, 920 National Center, U.S. Geological Survey, Reston, Virginia 22092 USA

John C. Butler, Department of Geosciences, University of Houston, Houston, Texas 77004 USA

L. J. Chouteau, Departement de Genie mineral, Ecole Polytechnique, C.P. 6079, Succ. A. Montreal, Quebec H3C 3A7 CANADA

Michael H. Dehn, Department of Geology, Rice University, P.O. Box 1892, Houston, Texas 77251 USA

E. Deloule, Laboratoire de Geochimie et Cosmochimie, Universite de Paris, (LA 196) IPG, 4, place Jussieu, 75230 Paris CEDEX 05 FRANCE

Gerald I. Evenden, Branch of Atlantic Marine Geology, U.S. Geological Survey, Quissett Campus, Woods Hole, Massachussetts 02453 USA

Andrea G. Fabbri, Instituto di Geologia Marina, Consiglio Nazionale delle Ricerche, Via Zamboni, 65 - 140147, Bologna ITALY

J-F. Gaillard, Laboratoire de Geochimie des Eaux, Universite de Paris, (LA 196) IPG, 2, place Jussieu, 75251 Paris CEDEX 05 FRANCE

J. Thomas Hanley, 920 National Center, U.S. Geological Survey, Reston, Virginia 22092 USA

H. A. Hildebrand, Landmark Graphics Corportation, 1011 South Highway 6, Houston, Texas 77077 USA

Frank W. Jennings, Branch of Atlantic Marine Geology, U.S. Geological Survey, Quissett Campus, Woods Hole, Massachussetts 02453 USA

Michael M. Kimberley, Department of Marine, Earth and Atmospheric Sciences, Box 8208, North Carolina State University, Raleigh, North Carolina 27695-8208 USA

Stephen A. Krajewski, Geosim Corporation, 1625 Broadway, Suite 2600, Denver, Colorado 80202 USA

Marian M. Lankston, Geo-Compu-Graph, Inc., P.O. Box 1848, Fayetteville, Arkansas 72702 USA

Robert W. Lankston, Geo-Compu-Graph, Inc., P.O. Box 1848, Fayetteville, Arkansas 72702 USA

William P. Leeman, Department of Geology, Rice University, P.O. Box 1892, Houston, Texas 77251 USA

R. Mark Maslyn, Consulting Geologist, P.O. Box 502, Golden, Colorado 80402 USA

David J. Matty, Department of Geology, Rice University, P.O. Box 1892, Houston, Texas 77251 USA

J. J. Mayrand, Departement of Genie mineral, Ecole Polytechnique, C.P. 6079, Succ. A. Montreal, Quebec H3C 3A7 CANADA

Daniel F. Merriam, Department of Geology, Wichita State University, Wichita, Kansas 67208 USA

H. Roice Nelson, Jr., Landmark Graphics Corporation, 1011 South Highway 6, Houston, Texas 77077 USA

A. C. Olson, Magnum Computer Systems, Inc., P.O. Box 620038, Littleton, Colorado 80162 USA

James P. Reed, RockWare, Inc., 7195 West 30th Avenue, Denver, Colorado 80215 USA

Charles B. Reynolds, Charles B. Reynolds & Associates, P.O. Box 1004, Belen, New Mexico 87002 USA

L. Robertson, Department of Soil Survey, The Macaulay Institute for Soil Research, Craigiebuckler, Aberdeen AB9 2QJ SCOTLAND

John C. Stormer, Department of Geology, Rice University, P.O. Box 1892, Houston, Texas 77251 USA

D. J. Unwin, Department of Geography, University of Leicester, Leicester LE1 7RH GREAT BRITAIN

Availability of Software

Several of the authors in this volume have made their software available through the Computer-Oriented Geologist Society (COGS). For further information contact COGS directly at the following address:

COGS
P.O. Box 1317
Denver, CO 80201-1317

The following authors have contributed their software:

J. C. Butler,

E. Deloule and J-F. Gaillard,

C. B. Reynolds,

J. C. Stormer and others, and

D. J. Unwin.

Other software mentioned in the book are generally available from the authors. Please contact them for further information.

TRADEMARKS

Reference to products with trademarks occur at many places in the book. Where the names are not commonplace, an attempt has been made to include the company name and address.

Preface

The age of the microcomputer is here! The long anticipated computer revolution in the Earth Sciences finally arrived in the 1980s with general acceptance of the micro. Since the American Association of Petroleum Geologists convention in Calgary, where micros were on prominent display in the exhibit area, it has been obvious that micros were an essential ingredient in all aspects of the geosciences. This collection of papers is a testament to their essentiality.

After an introduction to the state-of-the-art in computing in the Earth Sciences in the early 1980s, papers extol the use of micros in geophysics, image analysis, and computer graphics. Other papers describe the versatile application of micros in a private company and in a government research group. Several papers outline how micros can be used effectively in management of data files and databases. They also explore some of the problems with data quality and file manipulation. Finally, one paper explains the use of two computer programs (expert systems) based on artificial intelligence (AI) principles in the search for mineral resources. This paper anticipates, then, a future direction in the development of micros - programs incorporating AI principles - that of "thinking" computers able to make decisions.

The group of papers were collected by the editors on a personal basis. Many of the authors were known to us as practitioners in the use of micros; other were located through friends or professional contacts; and several others had submitted papers for consideration for publication in _Computers & Geosciences._ It seemed desirable to collect a group of papers under one cover to give the profession some idea as to the extent of the microcomputer revolution in the Earth Sciences - and this volume is the result of that desire.

From this volume the reader should get some feeling for the versatility and adaptability of micros. Their portability and ease in use allows applications impossible with mainframe computers or even minis. The computing power available with micros now far outstrips that of mainframes just a few years ago. Technology (hardware) has advanced extremely rapidly in the past decade and shows no sign of abating. Perhaps the catch comes in the availability of software.

Although most users write their own software or purchase commercially available general utility programs, it is desirable to exchange or make available in some form, programs of interest to Earth scientists. This is being done in several manners. User groups, such as COGS in Denver, regularly provide an exchange through their facility. Dial-up information exchanges, such as that provided by David Crane in Dallas, serve as a clearinghouse on the availability of programs. Meetings, such as the Geochautauqua, are held annually and bring together workers in the field where they can exchange information on a personal basis and learn of developments. Of course, there is the publication medium, such as Computers & Geosciences, that publish computer programs for micros. Several of the authors of papers presented here have programs they are willing to make available to those interested.

So as the use of micros becomes more widespread, the procedures for research and operations of geoscientists will change - they will become more efficient, effective, and economical. New approaches will be possible. This publication records just some of the changes taking place now in the mid-1980s and points to some new directions. By the end of the century the geosciences should have benefitted from these changes and geoscientists will have fulfilled the prophesy of P.C. Hammer in the 1960s that

> "...those people who are thinking about what they are doing are using computers..."

D. F. Merriam
Department of Geology
Wichita State University
Wichita, Kansas 67208

Introduction

In late 1983, Dan Merriam and I were talking about the use of microcomputers in geology. Although, we personally know many geologists using microcomputers in research, we were struck by the lack of published material on the subject. The compilation of this volume is an effort to disseminate information about scientists' experiences with microcomputers.

The definition of microcomputer has changed during the past few years. Several years ago a popular computer-term dictionary defined a microcomputer by the size of its programmable memory. The boundary between a microcomputer and a minicomputer was arbitrarily set at 100K bytes. This definition was outdated quickly, along with many other aspects of microcomputers. Today, a more useful definition of a microcomputer is a desktop-size computer with a small "footprint" that functions at normal office temperature, humidity, and electrical power.

The papers in this volume deal with the application of microcomputers to geologic problems. They represent a wide range of sources (academia, government, and industry), and cover a variety of topics. They can be grouped into several general subject categories: geophysics, data management and retrieval, digitizing and graphics, and other applications.

The first two papers in the book by J.M. BOTBOL <u>(An overview of the state of computing in the Earth sciences)</u> and S.J. KRAJEWSKI <u>(Microcomputers for explorationists)</u> are introductory papers. BOTBOL presents a brief overview of the state of computing in geology. He discusses the types of computers and other selected hardware devices available to geologists, the uses of statistics and mathematics in geologic research, and databases and their place in the overall picture. In addition, he provides insight about the future in his discussions of each of these topics.

KRAJEWSKI describes microcomputers in general, including the various hardware components that make up a microcomputer and the software that is available. In addition, he offers advice on how to select a microcomputer and discusses the current uses and future trends of microcomputers in geology.

The next five papers by R.W. LANKSTON and M.M. LANKSTON <u>(The microcomputer: A low-pass filter of the effects of economic trends for small geophysical contracting companies),</u> J.J. MAYRAND and M.C. CHOUTEAU <u>(Performance evaluation of microcomputer-based interpretation system for gravity and magnetic data),</u> C.B. REYNOLDS <u>(Analysis of shallow seismic-refraction data by microcomputer),</u> H.R. NELSON and H.A. HILDEBRAND <u>(Interactive 2D- and 3D-seismic interpretation on a microcomputer),</u> and J.P. REED <u>(Computer-assisted drafting of down-hole data)</u> deal with the application of microcomputers to geophysics. LANKSTON and LANKSTON describe the experiences of a small geophysical contracting firm where a microcomputer is used to manage and analyze gravity, magnetic, radiometric, electrical, and seismic-refraction data. The microcomputer has allowed the company to change the geophysical technique they emphasize and more closely reflect the type of exploration in demand at a particular time. The authors view the microcomputer as a low-pass filter of economic and exploration trends for their company.

MAYRAND and CHOUTEAU provide the results of a study that compares several microcomputers on the basis of central-processing-unit (CPU) performance, plotter performance, memory requirements, and input/output speed. In addition, both interpreter and compiler speeds are tested. An interactive, two-dimensional, gravity and magnetic interpretation system was used as the benchmark software. The critical subroutines are included in the paper.

REYNOLDS describes two programs that are useful in analyzing shallow seismic-refraction data. The first program, LNFT, fits an inverse velocity line to a set of offset distance/time pairs for a refraction profile. The second program, ZBYTO, solves for velocity thicknesses to a maximum of 5 layers. Included in this paper is the BASIC source code, which runs on a TRS-80 microcomputer.

NELSON and HILDEBRAND describe the development of a microcomputer system for manipulating and displaying two- and three-dimensional seismic surveys. Both the hardware and the software used to construct the system are discussed. A series of photographs shows the system's capabilities for two specific examples.

REED describes a system that plots down-hole strip-log data by utilizing the microcomputer. The software uses natural-language translation to identify the correct lithologic pattern to be plotted. The system when used on a microcomputer can be transported easily to a drill-site for rapid analysis of logs.

The next two papers, R.M. MASLYN (EXPLOR and PROSPECTOR - Expert systems for oil and gas and mineral exploration) and F.W. JENNINGS, J.M. BOTBOL, and G.I. EVENDEN (A hybrid microcomputer system for geological investigations) each cover a specific area of interest. MASLYN describes two geologic expert-system programs, PROSPECTOR and EXPLOR, that have been used for oil, gas, and mineral exploration. His paper also describes the field of artificial intelligence which provides a theoretical framework for implementing expert systems on computers. An example included shows how EXPLOR was used in the Denver-Julesberg Basin.

JENNINGS, BOTBOL, and EVENDEN describe the application of a hybrid microcomputer to several research areas. The system is a flexible device that can be modified for use in many applications. These include: laboratory automation (such as analog-to-digital conversion), process control, data inventory, field data reduction, and word processing. System components are described, and a list of vendors is provided.

The following three papers by J.D. BLISS (Management of Earth-science databases and a small matter of data quality), R.W. BOWEN (Micro-Grasp, a microcomputer data system), and J.C. BUTLER (An interactive program for creating and manipulating data files using an Apple microcomputer system) deal with data management and retrieval. BLISS describes the considerations that are involved in creating a database, irrespective of the database size and host computer. He summarizes the steps that should be taken before a database is set up and the follow-up steps as the database grows.

Geologists around the world are familiar with the U.S. Geological Survey's GRASP (Geologic Retrieval And Synopsis Program). BOWEN, coauthor of GRASP, describes a scaled-down version that runs on several microcomputers. It is a command-oriented system that operates on ASCII data files in a matrix form, each row is a record and each column is a field. An example run is included with the attendant data set. The software is available from the USGS.

BUTLER has written a data-file-creation and management program. The main purpose of his paper is to describe the background, process, and pitfalls of developing a useful program rather than to present new or unique methods of file management. The program was written in BASIC for an Apple II microcomputer and it is available through the COGS public-domain software library.

The next four papers: J.C. STORMER and others (Program ROCALC: norms and other geochemical calculations for igneous petrology), M.M. KIMBERLEY (Sketching a cross section of folded terrain with a microcomputer), E. DELOULE and J-F. GAILLARD (A computation method for drawing mineral-stability diagrams on a microcomputer), and A.G. FABBRI (Promising aspects of geological image analysis) each deal with a separate topic in geology. STORMER and others describe an application of the microcomputer to petrologic calculations. The program, ROCALC, transforms rock analyses in oxide weight percent to molecular and ionic equivalents. It calculates the CIPW and other "norms" as well as ternary projections, various ratios, and other chemical indices. ROCALC is part of a library of software, under development, designed to analyze chemical data from igneous rocks. The BASIC code for ROCALC is included and is available from the COGS public-domain software library.

KIMBERLEY's program plots structural cross sections in folded terrains by interpolating bedding-plane observations along a traverse. The interpolation between outcrops is calculated by utilizing either a cubic spline or Lagrangian polynomials. The Pascal program, TRAVERSE, is included in the paper, as well as a discussion of the program in which the author suggests several options for modifying the code to fit the user's own needs. The source-code for the program, which runs on an Apple II, is included.

DELOULE and GAILLARD describe a new computational method for constructing mineral-stability diagrams. This method is incorporated into microcomputer software that draws activity diagrams depicting chemical equilibria among minerals and aqueous solutions. The programs were written in MBASIC and FORTRAN for a Victor 9000 microcomputer. They are now available for the IBM PC from COGS. The code for the Victor 9000 is included at the end of the paper.

FABBRI summarizes the use of microcomputers in image analysis. He also discusses recent developments in hardware and software that are of importance to image analysis and all of the geosciences. New applications in economic and environmental geology, the study of textures in polycrystalline aggregates, and interpreting remote-sensing data are examined.

The last three papers: A.C. OLSON (Mapping applications for the microcomputer), D.J. UNWIN (DIGITS: a simple digitizer system for collecting and processing spatial data using an Apple II), and L. ROBERTSON (Multiarea measurement from maps and soil thin sections using a microcomputer and a graphics tablet) deal with digitizing and displaying map information. OLSON describes a system of interactive programs designed to aid geologists and engineers in using a microcomputer for volumetric calculations and mapping applications. In addition, he discusses the special problems posed by conversion of a large, existing mapping program from the mainframe environment to microcomputer operation.

UNWIN describes software that will digitize location information. This type of information is essential to many types of geologic research and is usually difficult for an individual researcher with limited funds to obtain. UNWIN has provided a breakthrough in this bottleneck. The software is included for an Apple II used with an Apple Graphics Tablet.

ROBERTSON describes a system for digitizing irregular shapes from maps and photomicrographs. The software is included along with modifications that are necessary to the manufacturer's software. The program runs on an Apple II and uses the Apple Graphics Tablet.

In closing, I thank all the contributors for their help and cooperation. My sincere hope is that these papers will encourage the neophyte to use microcomputers and the experienced user to find more and diverse uses. Finally, I thank my wife, Terry, for her great help and encouragement in preparing this book.

J. T. Hanley
U. S. Geological Survey
920 National Center
Reston, VA 22092

An Overview of the State of Computing in the Earth Sciences

J. M. Botbol

U.S. Geological Survey

ABSTRACT

Computers (especially microcomputers), mathematics, and statistics now are bona fide inclusions on the list of essential geologic disciplines, and the present technological "generation gap" will be reduced with time. University geology curricula eventually will include appropriate courses in computing, mathematics and statistics, and information science, and the efficacy of advanced methods will be tested and resolved.

The most significant problems facing the geomathematical community today relate to computer-resource and database administration. Conventional and traditional Earth-science project and site-management projects and site-management procedures are inadequate to support modern computer-supported geologic operations. Major emphasis must be placed on the planning of sympathetic computer and database-management procedures appropriate to the unique aspects of the geological sciences.

INTRODUCTION

Geology, a multidisciplinary science, has incorporated computers, statistics, and machine-readable databases for many years. Since the acceptance of computers, both statistical applications and database activities have grown enormously. The recent availability of low-cost microcomputer and minicomputer systems has affected significantly the geologic sciences, causing an almost exponential growth in computing activity. Elaborate mathematical and statistical techniques, once considered impractical and too expensive, now are considered routine. Complex cartographic operations are performed as parts of program packages that readily are available.

With respect to computer hardware and applications in computing, geology currently is undergoing a transition from traditional approaches to new-wave computer technology. Unfortunately, much of the new technology arrived after most of the current working geologists graduated from university. The most recent technology has not been incorporated yet into geology departments. Consequently, although state-of-the-art technology in computing and statistics is being used increasingly in the profession, the user community is small.

The new technology is not without liabilities. Any single computing system (and, therefore, the programs executed on it) generally interfaces with several scientists and more than one map or project. Although the concept of managing computing resources has become an essential component of a modern geological laboratory, it nevertheless requires geologists to budget for support services in a discipline that is alien to most of them. In addition, many geologists have a justifiable reluctance to use the easily, and in most situations, almost instantly gained result of statistical procedures which they can neither derive nor verify by familiar analytical methodology.

COMPUTER HARDWARE

Almost every type of computing device is used in geology. Hardware can be divided into three types: central processors (CPU's), mass data-storage devices, and other peripheral equipment.

Central Processors

Basically, five types of CPU's presently are in use: microcomputers, minicomputers, mid-sized computers, large mainframes, and super-fast giant mainframes.

Microcomputers. --Microcomputers are used both as process controllers and as stand-alone computing devices. For process control in geological laboratories, they are combined with a wide variety of analytical and field data-collection devices that usually generate voltage proportional to whatever attribute they measure. These analytical devices include seismometers, gravimeters, magnetometers, sediment analyzers, location-measuring devices, spectrometers, and diffractometers. The computer is used to convert sensor measurement to more appropriate units of measure. Depending on the sophistication of the program, the computer may generate either printed or graphical output and store the collected data on tape or disk. These computers usually are dependent on the sensing device to which they are attached, and the source programs that drive the devices in these "turnkey" systems invariably are inscrutable, unavailable, expensive, and proprietary. In process control, micros also are used as drivers for graphics-display devices. Used for output of increasingly popular color and animation, the micros efficiently manipulate local images and cause minimal impact on the host computer providing the data.

Microcomputers generally are used as stand-alone computing devices because of unavailability of alternative larger computer support, negative impact on a larger support computer, or prohibitive costs of a larger support computer. The most recent developments in microcomputer technology include a 32-bit word virtual-memory chip supported by a sophisticated operating system. Performance of this configuration parallels that of many minicomputers and some of the mid-sized computers.

However, consider a scientific project that initially requires a microcomputer to be used as a sophisticated calculator and for limited program development. Eventually, the use pattern evolves to a level of activity that includes database storage and retrieval, graphical output, and an increasing number of users. The function of the micro that was intended originally for program development has grown to emulate a larger system requiring its own staff. The danger is an eventual overcommitment of the micro.

Many geologists in financially offensive computing environments have turned to micros for program development, limited statistical applications, and limited local database applications. Some laboratories use micros as intelligent remote-access terminals having local data-reduction and editing capability.

Minicomputers. - Most minicomputer applications in geology are single site-support systems or elaborate process controllers. As single site-support systems, minicomputers emulate larger, faster, and more expensive systems. As process controllers, the minis either have been replaced by newer, more powerful micros, or have had their process-control function thoroughly integrated with data reduction and other data-processing tasks.

Mid-sized Computers. - The "midi's" are presently the most popular of the agency or institutional distributed computer systems. In almost all respects, they emulate their larger and faster counterparts, but at a considerably lower purchase price. Communications networks and applications and systems software are available widely, and they require a small operating staff. Mid-sized computer systems seem to be replacing previously existing minicomputer single-site systems.

Large Mainframes. - Large mainframes have been the traditional institutional and agency major computing resource. Most modern systems have time-sharing capability and host numerous concurrent activities. Their large size and speed make them particularly useful for database activities. Sophisticated operating systems tend to minimize user/machine dialogue. Many geologic applications on the host computer are limited because of high purchase price and operating costs, and proper scientific evolution thereby is constrained.

Super-fast, Extra-large Mainframes. - Extra-large mainframes are rare, expensive, and absolutely essential to large, compute-bound data treatment. Large geophysical models that simulate Earth systems and complex numerical analyses on large numeric arrays are among the typical uses for these giants. Except for those few who are fortunate enough and sufficiently well endowed to own their own system, all access to them is by communications network, mail, and personal visits and is always by prearrangement. Fast-raw computing speed is the essential functional capability.

Storage

As increasing amounts of data are gathered, databases become progressively larger. Magnetic tape is the principal medium for data storage. Disks are becoming increasingly more compact and faster; however, their high cost and relative fragility make them impractical for other than active online storage in computer-room environments. In hazardous or rugged environments, disk fragility can be overcome by use of plug-for-plug disk-compatible core. Core cost is higher significantly than disk cost. Traditional media such as punched cards and paper tape rapidly are falling into disuse.

Many geophysical data are being collected directly in digital form to maximize recording of data dynamic range. Magnetic tape is the major storage medium. Present experiments in storage of these data convert digital data to video or photographic analog form in order to economize on storage space.

The major mass data-storage mechanism of the near future is the laser-written disk. A highly focused laser permanently burns bit patterns into a disk approximately the size of an LP (long-playing) record. One disk can store the equivalent of 20 to 40 conventional magnetic tapes.

Experiments in the early 1960's have shown that energy levels of outer electrons of individual atoms of certain solids can be modified externally. The different energy levels then are used to represent binary data. Because the number of atoms/mole equals the number of bits/mole which is Avogadro's number (6.023×10^{23} atoms/mole), this distant-future technology will revolutionize completely mass data storage.

Peripheral Equipment

Graphic devices are the peripheral devices most significant in geologic applications. Of these, the graphics cathode-ray tube (CRT) is the most popular peripheral output device. CRT's are used principally as previewing devices for both numeric and graphical output, and it is rare to have a computer without some type of CRT. Compared with various types of plotters, the CRT is less expensive, has adequate resolution, and, most importantly, serves as both a conventional terminal and a graphical display device.

Graphics input has been impacted significantly by the use of the flying spot scanner, which rapidly and accurately encodes existing graphic documents. Other graphics input devices presently in wide use are conventional digitizing tables and rotating drum microdensitometers. These are software dependent, and there has been little significant recent change in the traditional methods and software.

User Community

Almost all computer-oriented geologists have little, if any, formal computer training. For the most part, they use applications programs written by others and usually have sufficient knowledge of their host system to operate these in-house applications programs and possibly to write other minor utility or applications programs. Most such programs are functional but suffer from lack of proper program structure and documentation, and they are useful only to the geologist (and his team) who wrote the program.

Junior geologists compose most of the hardware users. Current practices almost universally include an alarming degree of delegating computing responsibility to junior staff. Not only does delegation further the generation gap, it forces junior staff to learn computing from a utilitarian standpoint only.

Future Outlook

The future for computing systems and operating system software looks exceptionally bright. Current system capabilities far outweigh the capacity of most geologists to use them. Increasingly lower equipment costs make computer hardware within the reach of most geologic laboratories.

Few educational institutions, however, include an adequate computing background in their geology curricula. Most computer courses taken by geologists are oriented exclusively toward programming, whereas in an information-science or computer-science background, programming would be only one part of the necessary course of study.

STATISTICS AND MATHEMATICS

Geological statistics and mathematics applications are divided into three distinct categories: special-purpose programs, program packages, and modeling. Usually this grouping reflects organizational, computing resource, and specific project limitations rather than partitions due to statistical method utilization.

Special-Purpose Programs

Almost every geological organization has unique aspects in its method of data handling and the way in which information flows from the point of capture to the ultimate residence of the information. Furthermore, each organization stresses statistical applications unique to its specific needs and computing capacity. For these reasons, many statistical and mathematical applications programs are written specifically for the treatment of data on one computer in one organization. These programs generally take maximum advantage of local hardware features and compute only those statistics immediately relevant to local analysis.

An example is the treatment of marine seismic data collected while the ship is underway. These data can be treated any time from the moment of capture to some point in time long after the cruise has been completed. Only the design and capacity of the host computer (both at sea and in the laboratory) determine the extent and nature of the data treatment. Of the many data manipulations possible, only those that are useful, possible with the local computer, and preferred by local staff will be used. Local computer constraints may require special overlaying procedures for handling large arrays, and locally developed, proprietary, curve-fitting algorithms may be incorporated into the local program package.

Many aspects of the special-purpose programs preclude machine portability and interorganizational collaboration because of exceptionally high degrees of machine dependency, proprietary formulations, and unique or exclusive data input/output characteristics. Special-purpose programs constitute a significant part of mathematical and statistical data treatment in geological applications. Unfortunately, these programs are among the most poorly documented and usually are not available for general use. Their use in the generation of results of investigations is referred to in passing and the credibility of the results must be taken on faith.

Geophysical consulting companies, ore-reserve estimation companies, and laboratory analytical equipment companies are marketing special-purpose programs together with computing hardware systems. The programs are designed specifically to be implemented on specific machines, and the systems are referred to as "turnkey" systems, many of which are sold with object code only. Geologists who wish to share results and verify their analyses with colleagues in other organizations know that the proprietary and machine-dependent aspects of these systems prohibit a facile exchange.

Program Packages

This discussion is limited to those program packages that include a variety of standard, verifiable, treatment procedures and that are machine portable to the extent that they can be installed on most computers that support ANSII FORTRAN. They are available from private vendors or institutions and range in price from gratis to tens of thousands of dollars. Although the commercial packages have proprietary aspects, they are used so widely that interorganizational collaboration is not impeded significantly. Examples of these packages include BMD, STATPAC, SAS, SPSS, SIMSAG, IMSL, BLUPAC, and COVPAC.

Geological data sets pose problems for some commercial packages. These problems relate spatially to dependent data that are not distributed evenly, exceptionally large number arrays, and qualified data values such as "greater than" or "less than" geochemical determinations. Packages written specifically for geologic data have minimal problems in these areas, however.

Modeling

"Modeling" here refers to any geological study in which numerical analysis is used to simulate a natural or economic system. These studies occur in almost every aspect of geology and are becoming increasingly popular because of the widespread use of computers and the rapid growth of geologic databases. Detailed discussion of specific procedures is beyond the scope of this report; therefore, a small sample of the many procedures will be presented.

In economic geology, grade- and tonnage-distribution models are standard fare. Economic values of minerals within a deposit are plotted singly or in concert in order to determine optimum property development. Kriging techniques have become popular in interpolation of ore reserves, and almost all major copper porphyries have been modeled by these methods. Mineral exploration engages many multivariate statistical modeling procedures. Techniques such as artificial intelligence (for example, the Prospector program package) and pattern recognition (such as, the Characteristic Analysis program package) are used when areas in question are compared quantitatively with precomputed occurrence models. Factor analytic methods are used widely to describe resource exploration models quantitatively. Analysis of variance, correlation analyses, and discriminant analysis are used to portray the behavior of mineral-resource data so that useful relationships can be isolated, identified, amplified, and utilized. The economic development of past and present petroleum resources has been modeled and used as an index for prediction of development of future resources. Confidence limits on probabilistic assessment of undiscovered mineral resources have been improved because of these modeling studies.

Marine geology utilizes complex mathematical methods to simulate the physical and chemical behavior of the oceans. Predictions regarding mass movement of sediments are used for beach-erosion studies, offshore-well placement, modeling of channel filling, and modeling of pollution dispersement. Such predictions are based upon studies of enormous quantities of time- and space-dependent data. The eroding capacity of water movement is increasingly more predictable because of wave modeling studies. Suboceanic geology and mineral-resource estimates are becoming defined better owing to the development of seismic stratigraphy models.

Modeling is not a newcomer in geophysics. Gravity studies, magnetic studies, and seismic-data analyses regularly use mathematical modeling techniques to determine the best alternative geological explanation of measured observations.

The modeling story continues in geochemistry, remote sensing, sedimentology, and other aspects of geology. Of major importance is the increasing sophistication of most applications, which is correlated directly with the recent increased availability of computing resources, particularly low-cost storage, and the proliferation of interactive graphics-display devices. Recent advances in computer graphics have had enormous influence on the use of modeling procedures. For example, the laborious, time-intensive, batch plotter graphics that were once necessary are being replaced by instantaneous CRT images. These images may be in color; in some computer systems, the images are animated.

User Community

Almost all geologists use at least some statistics. Inasmuch as statistical and modeling applications are limited only by available computing resources, the more usable the computing resource, the higher the degree of use. Many geologists in geophysics and mineral economics regularly use statistics and modeling as integral parts of data analysis, and local computer support is sympathetic to these needs. However, for other geologic disciplines in many institutions, government agencies, and companies, adequate mission-oriented statistics can be expected, but modeling is a luxury reserved for specific explicit situations. These limitations are due entirely to computer costs imposed on geologists by local management.

A generation gap in the use of computer facilities is obvious. Younger geologists, who have some background in computing, accept statistics, mathematics, and modeling as essential tools of the trade unlike the older generation of geologists, who, for whatever reasons, continue to ignore or reject this technology.

Future Outlook

Advanced statistics and modeling procedures now have arrived at the point where, without wider utilization to test their current effectiveness, any future refinements would be counterproductive. Present-day computer hardware certainly is adequate for implementing all the procedures now existing. The near future will be focused predictably on the adoption of existing mathematical, statistical, and modeling procedures by larger segments of the geologic community.

DATABASES

To most geologists, databases are the dullest, most boring, and most bothersome aspects of their work. Consequently, little attention is paid to the details of geologic data storage and retrieval. Project funding characteristically avoids such matters, and database maintenance usually falls to the hands of computer centers, outside data-collection agencies or companies, or a few local geologists who have the dedication and keen perception to realize the true value of data. For purposes of this report, geological databases are considered from the standpoint of degree of user activity and, therefore, are classified as archival or active databases.

Archival and Active Databases

In the geologic community, many databases can be considered archives. Data are stored typically on magnetic tape and can occupy thousands of reels of tape. Even data that are used only once, usually when they are collected, fall into this category. The unique aspects of archives are that they are not kept online, updates are simple concatenation, and retrieval activity is low or nonexistent.

An extreme example of a geological archival database is the U.S. Geological Survey's marine multichannel-seismic database. Thousands of reels of tape containing original, unreduced seismic data are stored in a warehouse. Each year, hundreds of reels of tape are added to the collection. Reduced versions of these data also are on tape and, after they are used, they too are added to the archive. On rare occasions, an area that has been studied is reinvestigated, and relevant data are extracted from the files.

Active databases are those that have sufficient storage and retrieval activity to justify online maintenance. Activity obviously differs from site to site but, in general, these databases exist at almost every geological laboratory and are the focal point of most data storage and retrieval. Examples of active databases include digitized coastlines and political boundaries, geochemical analyses, sedimentary attributes, geologic ages, and mine production.

Most databases are active for at least some period of their existence. This is when they constitute a data subset relevant to a given project. When the host project is completed, the data become inactive and are migrated offline.

Interestingly, most projects that expend funds for database support, expend those funds only for the active aspects of the database. That is, only the immediately necessary and obvious computer expenses are accommodated. Funding for the data superset or archival database rarely is considered, if at all. If active data structures prove to be inadequate and require renovation that extends to archives, the time, money, and man-hours required to effect a change may be underestimated by the active project. In these situations, either a new database file structure is invoked, thereby terminating any link to the archival superset, or the database-management group must cover the entire archival and active database-reconstruction costs by inflating computer service charges.

Data Retrieval

Most database activity in geological laboratories is simple retrieval. Although many organizations support elaborate database-management systems, geologists are interested primarily in those data that have been stored and that are related to their immediate projects. Consequently, they are concerned with limited and familiar files on a retrieval basis only. The powerful storage and multiple file structure capabilities of database-management systems usually are wasted.

Some geological laboratories use the local system editor or quickly written one-time utility programs to retrieve desired subsets. This type of retrieval is characteristic of organizations having no definite database or computer support group and reflects a situation where computer-readable data have been aggregated with no thought given to future data utility. One of the major drawbacks to such limited use is that only the most simple combinations of subsets are ever retrieved by the average user. Variable output formats and complex logical combinations of variables can be implemented effectively only with programs designed to perform these tasks.

User Community

With the increasing availability of computer support via remote terminals, geologists are becoming more aware of the utility of databases. At the current time, there is an overall surge of activity in the compilation of sensible, useful, and usually well-organized databases. Like it or not, most geologists involved in data capture are involved in database compilation. Almost all analytical laboratories are feeding their data to at least one database.

The number of retrievers of data, however, seems to be limited. Interproject data syntheses are compiled probably by those who collect the data, not by those who request the syntheses. In other words, geologists are concerned with their own data.

Multidisciplinary data synthesis is a growing activity and, for the most part, it is confined presently to resource appraisal and massive editorial compilations. These users are indeed a small part of the geologic community.

Future Outlook

No doubt exists that geological databases will continue to grow. File structures will remain project and data dependent, thus requiring interproject data syntheses to be performed by increasingly knowledgeable users.

Many attempts have been made to aggregate huge master databases. Most of these have failed because of site and project dependencies. With respect to these databases, the tool that seems most viable for the future is the database index in which the detailed descriptions and references to relevant data files are compiled instead of the actual data.

The most discouraging aspect for the future outlook on databases is the failure of most educational institutions to incorporate database-related courses in their curricula. Courses in mathematics, computing, and statistics exist, but as yet only computer scientists or information scientists have any formal training in information science.

Microcomputers for Explorationists

S. A. Krajewski

GEOSIM Corporation

ABSTRACT

Microcomputers are playing an increasingly important role in geological exploration of all types. They can increase significantly the interpretative efficiency and productivity of explorationists. This paper reviews the hardware that makes up a microcomputer system and the software that allows the user to do a variety of tasks. Guidelines are presented that will help with purchasing a complete system. The current and future uses of microcomputers in an exploration office are discussed.

INTRODUCTION

A variety of microcomputer hardware systems now are available for explorationists to use in their search for hydrocarbons. These systems are desktop in size, user friendly, capable of rapidly performing complex series of calculations, able to produce sophisticated graphic printouts of geologic and geophysical data, and affordable to either company or independent explorationists. More importantly, use of these systems takes much of the mystery out of computer data analysis by allowing the user to interactively work with the system and the data, and to construct and fine-tune geological and geophysical interpretations and models for exploration prospects.

MICROCOMPUTER HARDWARE

A microcomputer can be defined as a small computer that uses a microprocessor (a single electronic chip) as its central processing unit (CPU), and will fit easily on the top of a desk.

Currently, there are about 1,000 companies marketing microcomputers. These companies include the more familiar vendor names such as IBM, Apple, Hewlett-Packard, Radio Shack, COMPAQ, Sperry, and Columbia. Obviously, a large number of smaller companies and new companies are attempting to establish a name for themselves in a volatile market.

Microcomputers sold by these vendors differ greatly in configuration, capability, and price. Configuration refers to the assembled hardware components of the microcomputer, and generally, configuration can include any or all of the following components:

- Processor or CPU (Central Processing Unit),
- Memory,
- Disk,
- Keyboard,
- Printer and display, and
- Power supply.

The CPU is the brain of the microcomputer and consists of an arithmetic unit and a control unit. The control unit is the part that carries out all of the program instructions. The arithmetic unit is the heart of the microcomputer, where the data actually are processed. You may think of the latter as being a complicated piece of equipment but it is not. The arithmetic unit simply adds two numbers together. The following are how each basic arithmetic operation is handled by the processor: subtraction - it makes the second number negative and adds; multiplication - is simply repeated addition; division - is repeated subtraction; and comparison - results from subtraction. From these operations other, more complex operations are derived. Thus microcomputer programs are made up of millions of steps based on addition. The machine makes up for limited ability by performing millions of operations per second.

Memory is made up of thousands of microscopic transistors. It is generally of two types: RAM and ROM. RAM stands for random-access memory and is the memory where programs and data reside. It is volatile memory which and thus when the electricity is shut off, the instructions in RAM disappear. ROM stands for read-only memory and programs or data can reside there. The difference is that this is nonvolatile memory; whatever is in ROM when the electricity is shut off will be there when the microcomputer is turned back on. Also, ROM is memory that can be read from and unlike RAM, you cannot write anything onto it. ROM generally is used for operating systems, "bootstap" programs, different character fonts for foreign languages, and different programming languages. Because it is expensive to produce ROM, it is used only for operations that are basic to the microcomputer or are needed in the computer's functioning.

Disk storage is the main form of mass storage of data and programs used in microcomputers. Tapes play a minor role as storage medium for microcomputers. Disk units store data on a spinning disk whose surface is coated with a magnetic material. Data or programs can be read or written over repeatedly by the read/write head. This head is attached to an arm that moves back and forth over the disk surface to read the data that are recorded on concentric tracks on the disk. The disk controller controls the head position and transmits the data or programs between RAM and the disk.

Keyboards provide input for the microcomputer. They generally are set up the same as a typewriter and are an integral part of the human-computer interface. Output from a microcomputer usually is through the display. When the output is required in more permanent form, it can be directed to a printer. Printers are either dot matrix, where the characters are printed as a matrix of dots, or letter quality, where the characters are of typewriter quality. In addition, plotters are available for outputting publication-quality graphics.

Finally, the power supply is the device that supplies electricity to the microcomputer in the form that is required by the machine. These devices generally transform the electricity so that the microcomputer can use it and provide some form of protection from loss of power or power surges.

Microcomputer configurations can be specified by the user at the time of purchase, or components can be added-on, substituted, or removed after the purchase as the user's needs change.

Capability refers to the way that the microcomputer handles data, that is, whether it works with one or two characters at a time (respectively an 8- or

16-bit machine), how fast the machine operates (machine power or speed is measured in millions of instructions per second, or MIPS) and the amount of working memory (RAM) available for completing operations.

The price of microcomputer hardware ranges from approximately $2,000 for a basic configuration to well over $10,000. Price will differ depending on the desired configuration and on the vendor supplying the microcomputer.

MICROCOMPUTER SOFTWARE

As noted here, there are about a thousand companies marketing hardware, and about 10,000 companies marketing up to 100,000 software programs. Just as with hardware, software programs differ greatly in the efficiency and language used to process data and the tasks performed, capability (the amount of data that can be processed), and price.

Simply stated, a software program consists of a set of programmed commands that instruct the CPU on how to process the data, and in turn, which circuits or switches in the CPU should be open or closed. The program is written in logical problem-solving steps just as if the problem were to be solved without using a computer and it can be written in a variety of languages (BASIC, Pascal, FORTRAN, etc.).

Software can be purchased in the form of written code that the user must enter into the microcomputer each time that it is used, or in the form of machine-readable code on floppy disks. The latter are placed into the micro's storage devices for self-booting (reading) after the computer's power is turned on. These programs are referred to as turnkey software.

The greatest number of software programs are available for the more usual business applications. These include:

1. word processing - software that manipulates text;

2. graphics - software that uses the CPU to generate and display pictures, line drawings, special characters, etc. on either the monitor or printer;

3. database management - software that organizes data on the storage devices (floppy or hard disks) so that the data can be manipulated without having to reenter the data each time it is needed;

4. spreadsheet - software that uses a programmable matrix (rows and columns) for financial modeling and other repetitive calculation needs; and

5. communications - software that allows transfer of data between two computers by telephone.

In addition, a large number of geologic-applications software programs now are available for use with the microcomputers.

Software that is menudriven is the easiest to learn and use for new computer users. This software presents the user with a list of selections as to what the user can do within the program. Once the intricacies of the software are mastered, menus become sources of frustration because it takes time to step through them.

Software programs are either copyprotected or not copyprotected. With copy-protected software, the user cannot make copies of the programs, therefore, the vendor should supply both a prime and backup disk containing the program. Software that is not copyprotected allows the user to make copies of the disk.

Software prices can range from the price of a disk (usually for freeware software) to about $10,000 depending on the program's capabilities. A high price is not correlative always with extensive capability. In addition to the purchase price of the software, vendors usually charge maintenance and upgrading fees.

WHERE TO START

One of the ways to obtain information about microcomputer hardware and software is through local users groups. A variety of these groups exist in cities throughout the United States. They usually are organized around or dedicated to the use of a particular machine or manufacturer, such as IBM and Apple among others. Interested newcomers can attend the monthly meeting of users groups and tap a large pool of knowledge and experiences at no cost. In addition, a users group normally will have an extensive library of public-domain software that members can purchase for the price of a disk and test before spending their own money for more expensive applications programs.

In addition to users groups, information can be obtained through computer application committees of local technical societies. For example, in Denver, the Rocky Mountain Association of Geologists, the Denver Geophysical Society, the Colorado Ground Water Society, the Association of Engineering Geologists and several other societies all have standing microcomputer or computer applications committees that meet on a monthly basis and sponsor a variety of activities for their members.

Another source of information is the Computer-Oriented Geological Society (COGS). COGS is a professional organization which actively encourages the application of computers (and more specifically, microcomputers) to natural-resource exploration and development. Its primary purpose is to act as an information clearinghouse. It is dedicated to providing a forum for self-help and discussion of common problems so that its members can locate low-cost solutions to their problems and benefit from other member's experiences.

Since its founding in December of 1982, the organization has grown to a membership of about 720, has established sections in two other cities, and has sections in the formative stages in seven additional cities. The COGS membership is diverse and ranges from geologists who merely are curious about the use of computers to expert programmers who write and market commercial geologic software.

COGS accomplishes its objectives through the following activities:

1. A monthly newsletter.

2. Monthly technical meetings. Past topics have included: computer-aided mapping, interactive seismic modeling, log analysis, trend-surface analysis, and geologic database analysis.

3. Publication of a catalog listing and describing about 300 available geologic programs. The catalog groups the programs by 19 end-use categories, such as log analysis, seismic modeling, economics, and by type of computer needed to run the program.

4. Distribution of public-domain geologic programs. Three disks currently are available with about 30 programs. The disks include programs for everything from cash-flow analyses to crossplot log analysis to decline-curve analysis routines to contouring.

5. Publication of a directory listing members and their interests.

6. Operation of a public computer bulletin board which contains past newsletters, the catalog listing, programs on the public-domain disks, and space for posting notices.

7. Cosponsoring conferences on computer applications in geology.

Additional information about COGS can be obtained by writing COGS, P.O. Box 1317, Denver, CO 80201.

Information on state-of-the-art geologic applications also can be obtained from the proceedings of microcomputer-based technical conferences that have been held during the past two years. Two such conferences, GeoTech '83 and '84 were held in Denver and at least four similar conferences will be held in Denver, Houston, Dallas, and Calgary in 1985. These conferences consist of technical sessions, in which application papers are presented, workshops on specialized topics, and exhibit areas where attendees can obtain information from hardware and software vendors, and consultants offering microcomputer services.

Finally, as interest in the use of microcomputers continues to grow, articles in the form of case-history analysis and other types of studies are starting to appear in a variety of technical journals. This trend will continue as explorationists gain more confidence in their abilities to use effectively microcomputers in their daily project activities.

MICROCOMPUTER USE TRENDS

Recently, several surveys have been completed to characterize microcomputer use trends in geology. The December 1984 issue of American Association of Petroleum Geologist's (AAPG) EXPLORER summarized the results from of a 43 question survey conducted in mid-1984 by Tedd F. Sperling. The survey was sent randomly to more than 2,000 oil and gas exploration companies; and, 231 (11.5%) of the questionnaires were returned.

The survey results indicated that less than one-half (46%) of the responding companies use microcomputers, and that 36.4% of the personnel within the companies use micros. Those using microcomputers categorized themselves as: managers (62.9%), geologists (30.0%), geophysicists (5.0%), and geoscience related (2.1%). The age of the respondents was: 21-30 (16.5%), 31-40 (36.9%), 41-50 (19.4%), 51-60 (20.8%), and 60+ (6.4%). The tasks being completed with microcomputers are: well database management (12.6%), reservoir analysis (11.0%), word processing (10.6%), lease database management (9.9%), accounting (9.6%), geology programs (8.8%), contouring (8.7%), spreadsheets (8.7%), graphs and pie charts (6.1%), geophysics programs (5.0%), seismic modeling (4.7%), and seismogram construction (4.3%). The types of microcomputers in use are: IBM PC, PCjr, XT, and PC 3270 (41.1%); Apple II, III, Macintosh, Lisa (23.0%); Hewlett Packard (6.1%); TRS 80 (6.1%); Texas Instruments (3.0%); COMPAQ (1.8%); DEC 100 (1.2%); and other (17.7%). Of the IBM microcomputers being used, the IBM PC was the most popular (27.2%). Similar data were obtained in user surveys conducted by COGS and by the Computer Applications Committee of the AAPG.

These data indicate that within a 2 to 3 year period, microcomputers, and in particular IBM microcomputers, have become an important tool for assisting explorationists in their search for hydrocarbons. Also of interest was that a large percentage of senior-level explorationists are using microcomputers (46.8% of the users are > 41 years old) in the workplace - microcomputers are not tools used primarily by recent graduates.

Another indirect indicator of microcomputer-use trends can be obtained by summarizing the software programs listed in the COGS catalog of known geologic software. The programs can be summarized as follows:

Type of Program	IBM	Apple	Other
	Microcomputer Type		
Data management	E	D	E
Drilling engineering	E	E	E
Economic analysis	A	D	E
Gravity and magnetics	C	C	E
Image processing	E	C	E
Lease records	D	E	E
Log analysis	D	D	E
Mapping/contouring	B	C	E
Mining economics	E	E	E
Mud logging	E	E	E
Paleontology	E	D	E
Petroleum accounting	A	E	E
Reservoir analysis	C	D	E
Seismic analysis	C	A	E
Trend analysis	D	E	E

IBM = IBM PC & XT

Apple = Apple II

Other = 14 types

A = 11+ programs

B = 8 to 10 programs

C = 5 to 7 programs

D = 2 to 4 programs

E = 0 to 1 program

These data also indicate that both a greater diversity and a larger number of software programs are available for the IBM microcomputer than for any other type and undoubtedly, this trend will continue.

WHAT TO EXPECT

If you are considering the acquisition of a microcomputer, some general advice is in order. Before you purchase anything, develop some familiarity with the subject. As noted earlier, information can be obtained from users groups, COGS, libraries, magazines, and friends. Once you have a basic familiarity with microcomputers, carefully define the job or jobs you want to do with the microcomputer. Next, identify and evaluate the software that will do those tasks that you (not the vendor) want to do. Finally, select and purchase the hardware that runs the software you want. The purchase should be based on both current and future (projected) needs.

Unfortunately, this common sense approach seldom is used. Instead hardware is purchased first either when "a deal you can't refuse" comes along, you act on the apparent sound advice of a friend or relative, or the high-pressure sales pitch wins out. Remember, the microcomputer is an exploration tool that will assist you in your business. Its acquisition should be evaluated as any other business decision or venture and sound judgment should prevail.

If you are new to using a computer, expect to spend a lot of time becoming familiar with the operation of both the software and hardware. You will learn that the systems are not as "user friendly" as the sales literature and salesman led you to believe.

The reason for this is simple. Software and hardware have to be used in exact ways, for example, data have to be entered in a precise way and operational commands have to be consistent with those defined in the operating system. What seems as a small variation to you usually will cause the microcomputer to abort processing of the command or data. As a result, user frustration quickly sets in unless you are persistent and prepared to spend many long hours developing computer-compatible thought processes. Be prepared to spend some time getting to know your system. Remember, it took more than a day for you to learn how to walk. Because successful microcomputer use is predicated on logical data-analysis procedures, this learning process will be difficult especially for those explorationists who are not organized in their work habits.

Once the basic computer literacy skills have been mastered, other sources of frustration will arise in differing degrees. These will include:

- poorly written manuals;
- inadequate or unavailable technical support;
- software or hardware that does not work as it should, for example, the vendor forgot to tell you that what you need "still has some bugs that are being worked on", or that "the modified software will be available next week" (this type of software frequently is referred to as "vaporware" because it may not materialize);
- complex data entry procedures;
- costly upgrade and maintenance agreements;
- parameter specification or rescaling restrictions (inability to get exactly what you want from the system);
- limited capacity to analyze data (to date micros cannot handle independently analyzing large databases, and cannot complete quickly extremely complex mathematical calculations);
- company pecking orders that dictate when and who can use the system;
- management indecisiveness about software and hardware acquisitions;
- limited amounts of time to learn how to use fully in-house systems;
- unavailability of funds to expand software and hardware systems;
- decision-makers and in-house technical experts who feel threatened by the new technology; and
- invariably, the system going down at noon and nine cash-flow-model variations are needed for a management meeting at 2:00 PM.

In spite of the anticipated frustration, the acquisition of a microcomputer is a sound business decision, and worth the investment of both time and money. Acquisition of a desktop system will give the explorationist full control over the data-analysis procedure, and, if the workstation is configured appropriately, if the software capabilities are understood properly, and if the exploration database is prepared logically and valid, then microcomputers will improve significantly and cost-effectively efficiency within the explorationist's office.

FUTURE TRENDS

The use of microcomputers by explorationists has increased dramatically during the past two years, and this trend undoubtedly will continue. Hardware will continue to decrease in size, decrease in cost, and increase in capability. An example of this change can be demonstrated with the IBM PC. It has an 8088

Intel microprocessing chip in the CPU and is rated at between .20-.25 MIPS (it can handle 200,000-250,000 instructions per second). In addition, it can work on up to 640,000 bytes (characters) of information in RAM memory, and up to 40 megabytes of information can be stored and accessed on a hard disk. This $5,000 desktop microcomputer is the technologic equivalent of a large, room-sized minicomputer that rented for more than $5,000 per month 15 years ago.

As the use of microcomputers continues to increase, a dramatic "shaking out" within the software industry will occur. Software that is not "user friendly" or capable of performing sophisticated data-analysis tasks will disappear, as will companies that are not responsive to user needs. In addition, software prices will continue to decrease, new and improved algorithms will be developed for processing and analyzing geologic data, and geologic modeling will become more commonplace and move from one and two-dimensional modeling to interactive three-dimensional modeling. Software uses also will shift from general applications, such as word processing, database management, spreadsheets, etc., to more technically oriented applications such as statistics, trend analysis, correlation and classification, and modeling/simulation thereby further quantifying geologic concepts and processes. More software in the form of "freeware" will become available through organizations such as COGS and from explorationists who write their own programs. Thus, a lower price alternative will be available and cut into the marketplace of the commercial software developer. Finally, software using "expert system" programming algorithms (artificial intelligence) will be available on the market in an attempt to make the geologic data-analysis process more efficient.

Explorationists will continue to increase their computer competency and realize that microcomputers have certain limitations. They will begin to develop a renewed interest in the use of mini or mainframe computer systems that can process rapidly and analyze large quantities of data. Transition back to the larger systems will have been facilitated by the development of a "computer-comfort factor" from use of the microcomputers.

Even with these technologic advances, the fundamental problems of using microcomputers to process and analyze exploration data probably will remain. Microcomputers will not analyze data more efficiently in sparse data areas; it will be time-consuming (and expensive) to organize and enter data into the microcomputer; the microcomputer will process and analyze the entered data according to the limitations of the algorithms developed by the software programmer and the programming language; the microcomputer only will output what the user tells it to; and, the microcomputer only will produce output that is as good as the information being fed into the micro (in other words, garbage in, garbage out).

SUMMARY

Microcomputers are playing an increasingly important role in the explorationist's office just as other data-analysis tools have in the past, and, these tools can increase significantly the interpretative efficiency and productivity of professional explorationist. It is unlikely, however, that microcomputers will replace explorationists and their judgment expertise in the search for oil and gas because the microcomputer is controlled fully by the user. As a result, oil and gas reserves will continue to be located "in the minds of men and women" and not "in the microprocessors of microcomputers."

The Microcomputer: A Low-pass Filter of the Effects of Economic Trends for Small Geophysical Contracting Companies

R. W. Lankston and M. M. Lankston

Geo-Compu-Graph, Inc.

ABSTRACT

The Apple II microcomputer became the first widely used microcomputer in the geosciences in late 1978 or early 1979 because of its portability, its low cost, and its ability to provide color CRT graphics. These features made the microcomputer attractive to small geophysical contractors involved in remote operations. The history of Geo-Compu-Graph, Inc.'s use of the microcomputer since 1979 is an illustration of how the versatility of the microcomputer has allowed the small firm to weather boom and recessional times in the minerals, engineering, and petroleum industries and to be competitive during the entire period. The microcomputer has been employed for all phases of processing and graphical display of gravity and magnetic, radiometric, electrical, and seismic refraction and reflection data as well as such tasks as advertising and publishing. The microcomputer, therefore, has served as a low-pass filter to smooth out the peaks (and pokes) of recent economic cycles thereby allowing the company to not only stay in business but actually to advance the area in geophysics particularly in areas such as high-resolution seismic-refraction methods and engineering seismic-reflection methods. The proven capability and reliability of the microcomputer, coupled with greater acceptance of such systems and greater availability of software, suggest that the next generation of 32-bit microcomputers will become even more widespread, in a shorter period of time, than the older, but viable 8-bit machines.

INTRODUCTION

Late 1978 and early 1979 saw the convergence of several events in the geosciences. The seeds for the petroleum boom of 1981-82 were sprouting, the uranium exploration boom was at its peak, and the microcomputer revolution was beginning with the emergence of such machines as the Apple II and the TRS-80. Boom times in the petroleum and minerals industries usually invite many geoscientists to leave their secure positions with the major companies in order to strike out on their own as contractors or as independent exploration companies. Whether these upstart enterprises survive past the end of the boom is a function of many variables. However, those firms that formed in 1978-79 that incorporated the microcomputer into their routine operations had the best chance of surviving. This paper illustrates how one geophysical contractor, Geo-Compu-Graph, Inc., was able to establish a new business and to maintain it through the boom and recessional times since 1979. The microcomputer was the central tool that filtered the high-amplitude swings of the economic cycles so that the new geophysical company could sustain an orderly growth during the first five critical years of its existence.

NECESSITY AND INVENTION

In 1978-79, the uranium boom was at its peak; both industrial and government geologists and geophysicists were engaged in all manner of data acquisition, and evaluating areas from the size of single mining claims to large regions of the country for their uranium potential. Because of its low cost and ease of operation, the portable gamma-ray spectrometer is one of the most popular instruments for these evaluations.

Gamma-ray spectrometer data have to be corrected for the effects of Compton scattering. Although, some of the instruments that were used during the uranium boom made such corrections internally, many others required the user to make the corrections with the aid of a calculator. The corrections were not difficult to make, and programmable pocket calculators usually were employed. However, even with the programmable calculators, the corrections were time consuming, and keystroke errors did occur. After the data were corrected for the Compton scattering and converted from count rates to potassium, uranium, and thorium concentrations, ratios of the elemental concentrations were calculated, also not a difficult task, but time consuming. The final task of the data-reduction process was displaying the data in the form of multichannel profiles. This phase of the work required the greatest amount of attention because it was from these profiles that decisions regarding the exploration effort were made. The data reduction and plotting needed to be completed at the end of each day so that decisions could be made regarding the next day's field work. Many sites, large and small, were evaluated using the slow, manual method as outlined.

The microcomputer, however, offered the field person the opportunity of expediting all aspects of the reduction, plotting, and archiving of the gamma-ray spectrometer data. The first aspect of the microcomputer that made it attractive for data reduction was its relative portability. A portable color television set (there were no color monitors at that time), a dual floppy disk system, a small printer, and the processor itself could be transported easily and set up at a field office, usually a motel room. The second aspect of the microcomputer that made it attractive was its relatively low cost. At a cost of about $3000, the microcomputer system was about one-half the cost of the gamma-ray spectrometer.

With a simple data input, edit, and print routine, the field geologist or geophysicist could enter, check, and edit all of a day's data in a matter of minutes and make a neat tabulation of the raw data. With another relatively simple program, all of the data could be corrected for the Compton scattering, the ratios could be calculated, and the data could be stored on diskette and the corrected data could be tabulated. Finally, the color CRT (cathode-ray tube) graphics capability of micros allowed the multichannel profiles to be displayed in a matter of seconds instead of the hours that were required with manual plotting.

The philosophy that the microcomputer would be used for processing and displaying the gamma-ray spectrometer data at field sites was a viable one. Software, termed the COMSTRIP package, thus was developed, providing a suite of programs to complete the reduction, display, and archiving of the gamma-ray spectrometer data. The hardcopy was accomplished through the process known as a "graphics dump" where the dot image on the CRT screen was transferred, dot for dot, to a dot-matrix printer. An illuminated point on the screen became a printed point on the paper. Figure 1 is a multichannel profile generated by the output program in the COMSTRIP package.

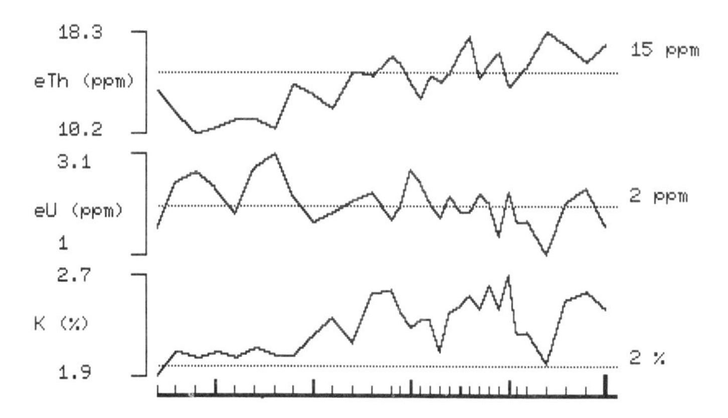

Figure 1. Graphics dump of multichannel display of KUT data. Tic marks along horizontal axis represent station locations.

FILTERING THE FIRST ECONOMIC TROUGH

With the rapid decline of the uranium industry after the Three Mile Island problem in mid-1979, it was necessary to look for other areas for microcomputer applications. The philosophy of portable geophysical data reduction and graphing was expanded from gamma-ray spectrometer data to other geophysical data sets. Magnetic data lent themselves to the philosophy of rapid data reduction and display in the field. Moreover, with a relatively straightforward program, anomalies could be modeled and interpreted preliminarily at the field office. The program, MAGMOD, was written employing equations presented by Gay (1966), which were presented initially for generating a set of master curves that were used in the field for comparing to observed anomalies and making initial interpretations of the magnetic data. With only a slight change in operational philosophy, the pre-microcomputer technique was adapted so that the portability of the master curves was maintained and more flexibility in modeling complex features was achieved simultaneously. Figure 2 is a graphics dump of the CRT output of one model. On the color CRT, the dikes are outlined in green, the observed anomaly is large, white plus signs, and the calculated anomaly is a continuous white line.

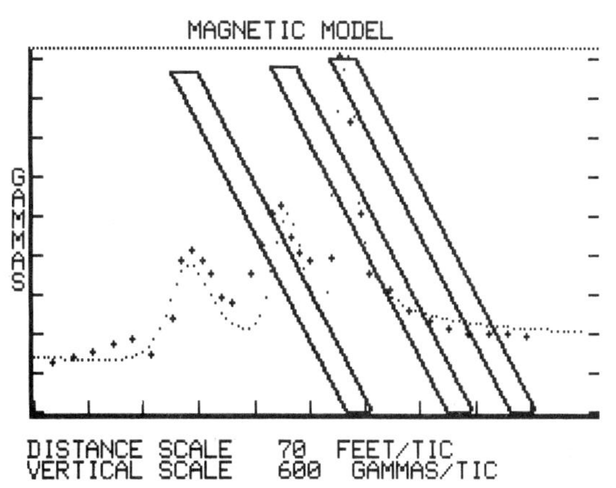

Figure 2. Graphics dump of MAGMOD display. Calculated anomaly is small dots; observed anomaly is large + symbols. Dikes are outlined by solid lines. (Observed data from Telford and others, 1976, p. 767.)

Because some geophysical data are better displayed in map form than in profile form, another program in the expanding geophysical applications package was one that used color-coded pixels on the color CRT screen to depict where data had been acquired and to give a quick look at the amplitude and distribution of the anomalies. COLORMAP, as the program was termed, was useful for many types of geophysical and geochemical data display. COLORMAP could be run each evening to update the data file, to see if any anomalies had been mapped, and to monitor the progress of the data-acquisition effort. As with many of the tasks the microcomputer was asked to do in the early days, COLORMAP was not sophisticated, but it made the field explorationist's work more efficient. One of the most important aspects of COMSTRIP, MAGMOD, and COLORMAP programs was that the field geologist or geophysicist was able to make decisions in the field that, in the days prior to the introduction of the microcomputer, might not have been made until the field crew had moved off of the site. By being able to preview data and make decisions based on the data, the exploration projects could be more efficient and more productive with the overall result that exploration was less costly. A suite of elementary programs for processing and displaying many types of geophysical and geochemical data with the microcomputer at field sites also was developed. The hardware/software system was dubbed the GO-ANYWARE system.

RIDING THE NEXT WAVE

The CRT graphics display and the dot-matrix screen dump were satisfactory initially for data output. Data input, however, was more difficult. Manually keying the data into the computer had its limitations, particularly if the input data were in the form of a graphical display and not in a data table such as would be generated during spectrometer, gravimeter, or magnetometer surveys. This was the situation interpreting seismic reflection data that were recorded with what were known popularly as "engineering" seismographs; hand-portable, 12-channel, signal-enhancement units. The microcomputer allowed the engineering seismic-reflection method to become a viable tool that is being employed today for mapping groundwater, shallow mineral targets, and hazardous waste sites.

In its earliest form, the engineering seismic-reflection method made use of x-squared t-squared analysis. This is a classic, graphical method (Dobrin, 1976) of deriving subsurface depths and interval velocities from the reflection arrivals. The method requires making a graph on which is plotted the square of the arrival time versus the square of the respective source to geophone offset distance. The velocity and depth information is derived from the straight line that best fits the data and the intercept of that line with the time squared (vertical) axis. Although the task is time-consuming when done manually, it becomes trivial with the aid of the microcomputer and CRT graphics capability. The geophysicist needs only to key in the interpreted arrival times and the source to geophone distances. The machine proceeds to perform the squaring operations, plot the graph on the screen, perform a least-squares linear regression and calculate the velocity and the depth from the slope and y-intercept. Through the use of the microcomputer, the time necessary for interpreting the reflection data decreased to less than 25% of that required using the manual method. Figure 3 is a graphics dump of the CRT display of the reflection-analysis program.

Being a new technique, the engineering seismic-reflection method was difficult to sell. Engineers, hydrologists, and exploration geologists were skeptical about the method because the field data were not in a form they could understand readily, and the final cross section (Fig. 4) had no obvious tie to the field records. The solution was to input the entire seismic trace into the computer and to process it the way the petroleum industry does, although on a reduced scale. This would reduce the skepticism because the engineer or geologist could see the field data, the processed data, and then the geological interpretation. The steps between each stage would be smaller and, consequently, easier to grasp. Constructing the seismic section first required transmitting the seismic data to the computer. In 1979, few instruments were available for this task, and those that were available were expensive. For about $1000, about 20% of the cost of a field, digital, seismic-data recorder, and x-y digitizer could be integrated with the microcomputer system. The digitizer allowed the seismic traces on the paper field records to be digitized and sent to the computer. The digitizer had another advantage. It was not limited to the input of seismic data. It could be used for the input of any data that was in profile form, and it could be used to input x, y, and z (amplitude) values from maps of gravity, magnetic, or geochemical data.

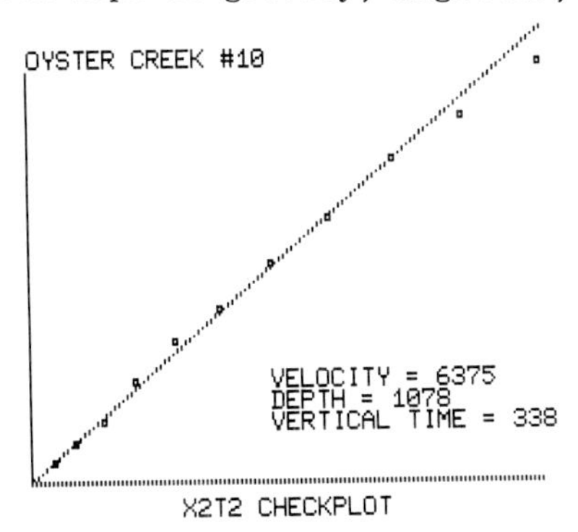

Figure 3. Graphics dump of X^2T^2. Squares represent input arrival times and offset distances. Diagonal line shows quality of least squares linear regression line. Velocity is in ft/sec; depth is in ft; time is in msec.

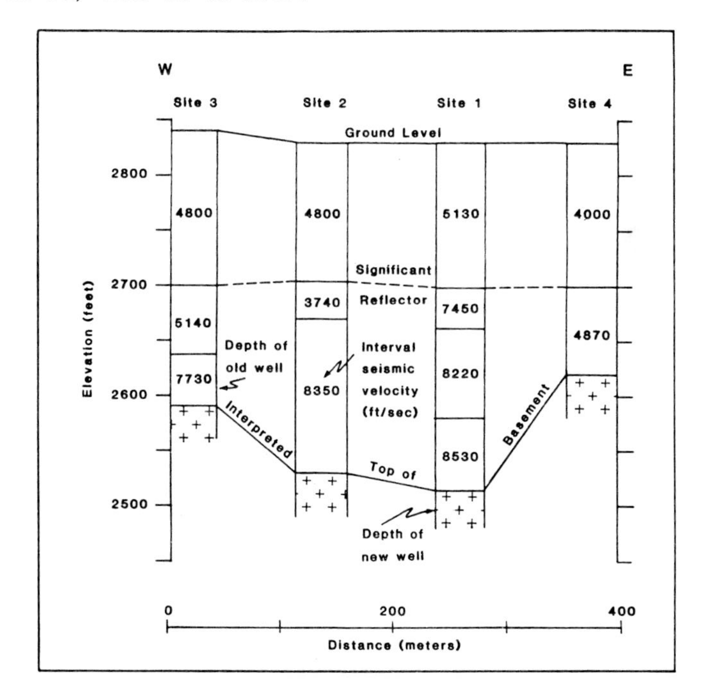

Figure 4. Interpretation cross section based on correlation of four reflection records (after Lankston and Lankston, 1983a).

In June 1980, the GO-ANYWARE system was used on a field site in the Cascade Mountains to process engineering seismic-reflection data collected during a groundwater exploration project. Each seismic trace was hand digitized and analyzed within 24 hours after it was recorded. Although the data input and processing procedures had been applied previously to other engineering seismic-reflection data, the Cascades project was the first in-the-field implementation. Figure 5A is a portion of the seismic section that was developed finally. The dot-matrix image was first generated on the CRT screen and then dumped to the printer. In order to obtain the needed vertical (time) resolution, the data were processed twice, once for the earlier part of the section and once for the later part of the section. The two parts then were spliced together. The two parts fit well together although the printer used only a friction paper-feed mechanism. The interpretation of the processed seismic data (Fig. 5B) indicated that the depth to the bedrock was 660 (±30) feet. The engineering firm that was studying the aquifer subsequently drilled into bedrock at 666 feet. The close agreement between the drilling and seismic work certainly was a result of being able to process some of the data at the field office thereby ensuring that optimum field data-acquisition parameters were being employed.

The engineering seismic-reflection data-processing package is significant in the evolution of the use of the microcomputer in geophysical applications for several reasons. First, it made use of both graphical data input and graphical data output. Second, it was implemented on a microcomputer and could be executed in a meaningful time frame at a field office. This is significant because of the sophistication of many of the algorithms used in correcting the seismic data prior to generating the seismic section. Many believed that seismic-reflection data processing was a mainframe computer task. As was demonstrated with the GO-ANYWARE reflection package and numerous other packages that were developed later, complex tasks do not require necessarily the resources of a large machine. The important consideration is whether the microcomputer can perform the task time and cost effectively. The application of the seismic-reflection method for engineering studies never would have advanced if petroleum-industry scale and cost of operations were all that was available.

The engineering seismic-reflection package also was significant because of its organization. The software system had to be arranged in modules with each one performing one of the many tasks that must be completed before the final section is plotted, because the programs required for the seismic-reflection data processing are more sophisticated than those discussed previously. The first task, and therefore the first program module in the system, performs the data input via the digitizer. The large number of data traces in a typical survey requires that some method for keeping track of the data must exist. In addition, one of the programs in the system must perform the task of identifying those data traces that will be processed or displayed together.

In late 1980, a new method was developed for interpreting seismic-refraction data. The generalized reciprocal method (GRM) (Palmer, 1980) is a powerful interpretation aid and is suited ideally to implementation on a microcomputer. A popular refraction data interpretation in widespread use in 1980 was one developed by the U.S. Geological Survey (Scott, 1977). This technique had limitations of not being able to handle lateral velocity changes within a horizon, of requiring increasing velocity with depth, and of requiring the resources of a minicomputer, or larger system, for execution. Moreover, the program was written in FORTRAN. All of these factors worked against the program ever running on a microcomputer.

GRM employs a series of graphs prepared by a relatively simple equation and from which the interpreter obtains parameters about the subsurface including lateral velocity changes. The parameters are brought together in one final equation from which the interpreted geologic cross section is plotted by the GREMLIN package. It is a modular package, written in BASIC, with each module performing one of the GRM tasks. The first task is the input of the first break arrival times into the computer. This could be done manually from the keyboard. However, the digitizer is suited to transferring the times directly from the paper field seismic record to the diskette storage system.

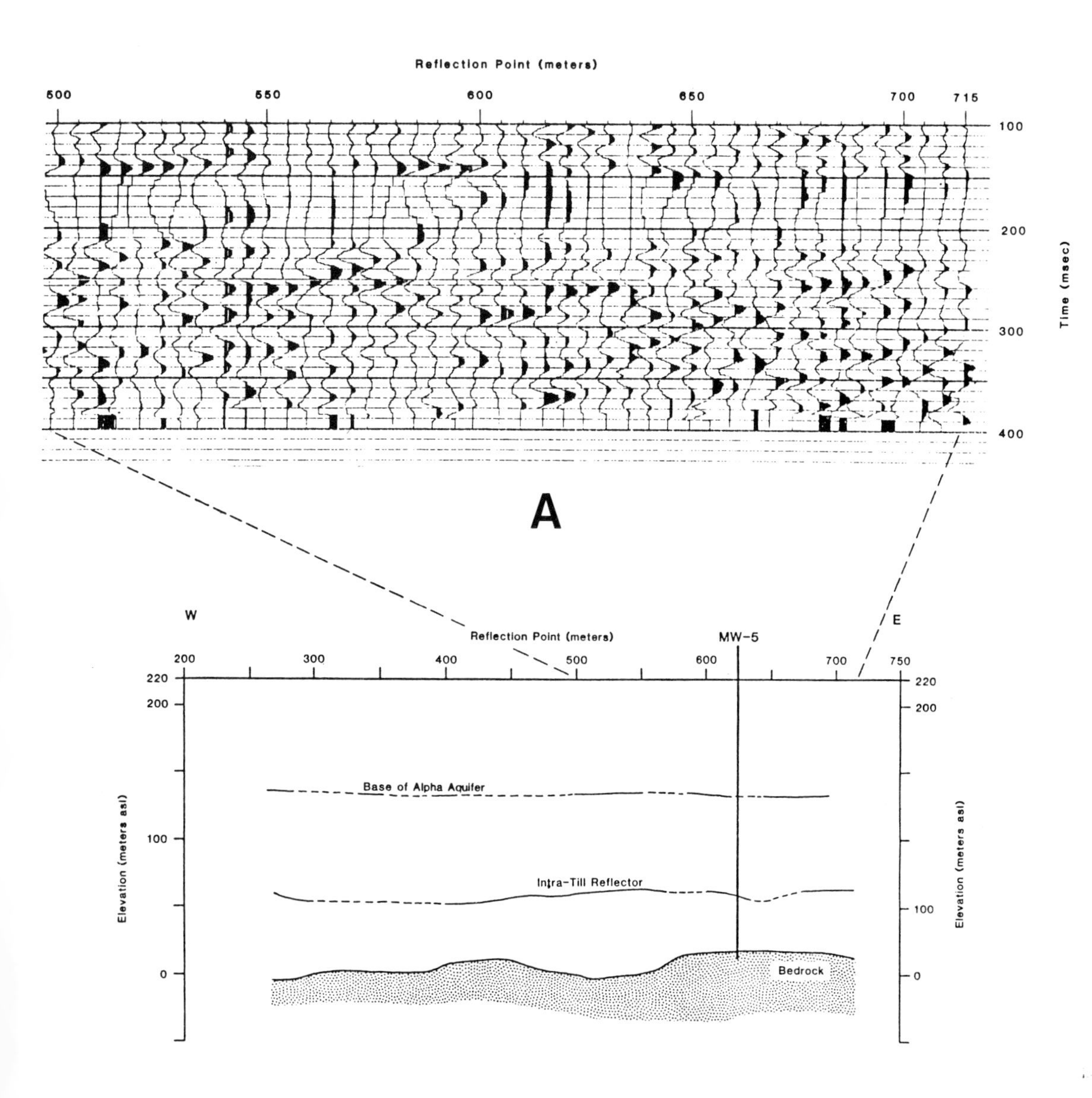

Figure 5. A, portion of seismic section generated using CRT graphics display and B, interpretation of processed seismic data. Seismic section (A) was generated in panels on CRT. Splice between top and bottom halves can be seen at 250 msec. Gaps between individual CRT screens are obvious at 540 and 605 m.

Each of the subsequent stages in the GRM processing of the data requires the production of some form of graph. These could be programmed easily for display on the CRT. However, this is one situation where the graphs are used more easily if they are on paper. The graphics screen dump used in other programs was not satisfactory because of the difficulty in changing scaling factors on the graphs. A pen and ink plotter was integrated with the microcomputer system.

The pen and ink plotter, similar to the digitizer, is usable with many forms of data and was incorporated into the gravity and magnetic software packages, reflection package, and geochemical data statistics and display programs. Figures 6 and 7 show the pen and ink plots of the refraction data and the interpretation of the data, respectively. The programs in the GREMLIN package follow the same philosophy that was adopted with the COMSTRIP programs: simple computational schemes, graphical data display, and portability of the entire system to field sites. By being modular similar to the engineering seismic-reflection programs, the microcomputer is able to perform a complex data-processing sequence leading to the final interpretation of refraction data by breaking it down into its logical steps.

Because of the flexibility of the microcomputer, it could be applied to virtually any exploration effort at hand. For example, during the years 1981 to 1983, gravity data for petroleum and groundwater exploration were collected, processed, and interpreted. Individual projects differed with some being completed entirely in the central office and some involving both field and central-office data processing. The microcomputer, however, was the only computer employed for any of the data processing, modeling, or graphical and tabular display. The flexibility of the GO-ANYWARE package is illustrated by the interest during the last two or three years of the petroleum industry in obtaining more detailed information about the near-surface section so that static corrections could be determined more precisely for application to reflection data. The previously ignored first-break data contain much of the desired information. However, few in the petroleum industry were able to match the data processing and interpretation power of the GRM implemented on a microcomputer. An implementation such as the GREMLIN package is as comfortable with 192 trace split-spread records from a petroleum-style survey as it is with data collected with a single-channel seismograph.

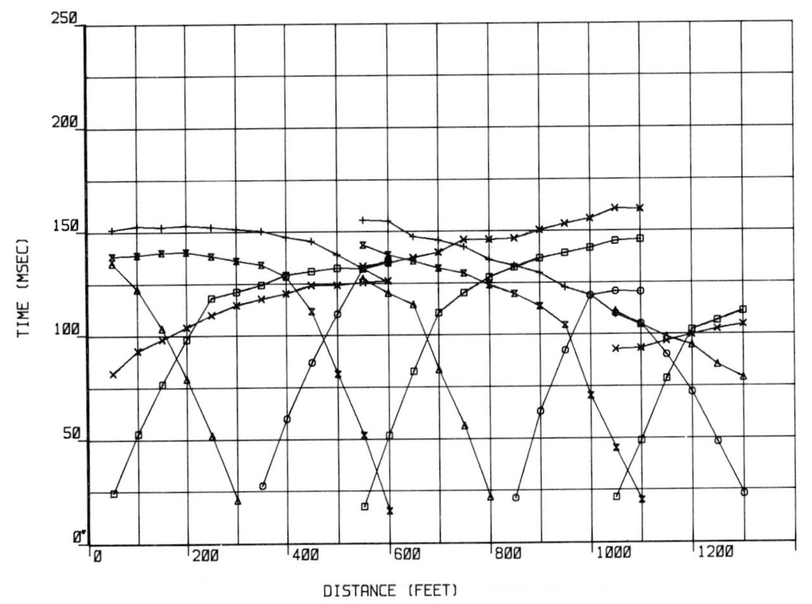

Figure 6. Pen and ink plot of time-distance data.

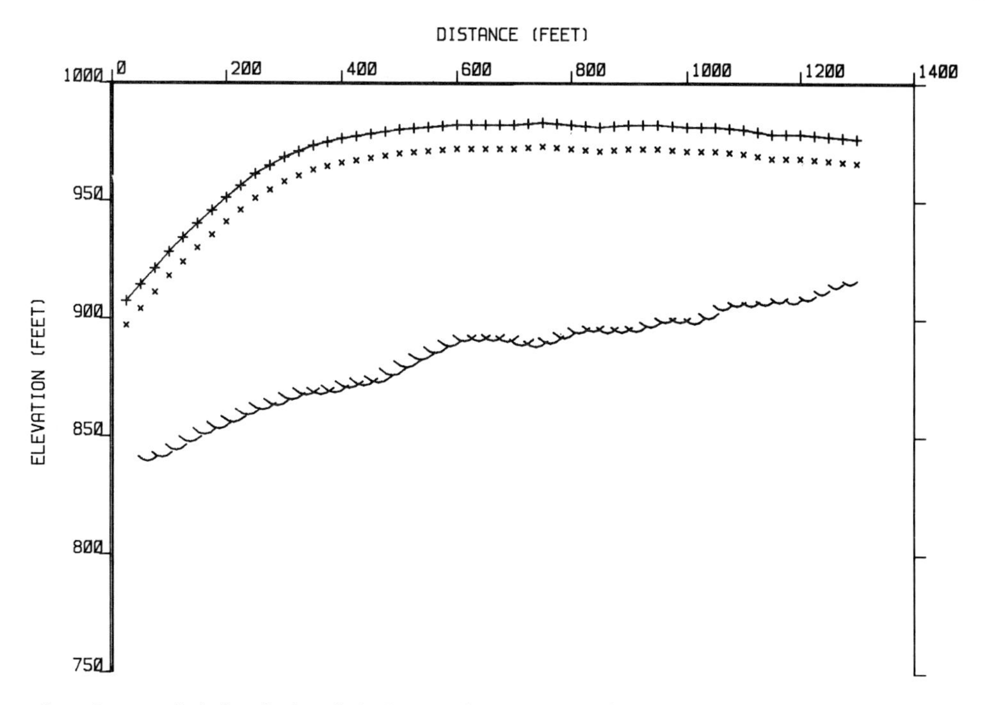

Figure 7. Pen and ink plot of GREMLIN interpretation.

STEADY STATE REACHED

By 1982 several, relatively sophisticated geophysical data-processing and display packages had been developed. These were applicable to most of the popular geophysical techniques. The Apple II-based GO-ANYWARE system had been implemented at many installations. Modularity of the hardware and software allowed the correct computer tools to be assembled for the task at hand. BASIC compilers had been introduced so that data processing throughput was increased by as much as five times. More sophisticated modeling and processing routines were developed. Talwani modeling (Talwani, Worzel, and Landisman, 1959) was used on gravity and magnetic data with the results displayed on the pen and ink plotter (Fig. 8); velocity filtering and deconvolution were incorporated into the seismic-reflection software package (Fig. 9). Figure 8, although reduced from the originally plotted size, shows that the computer/plotter combination was able to accomplish many routine drafting tasks such as applying explanations and titleblocks to illustrations. The seismic records in Figure 9 were computer plotted, but the annotation was added manually.

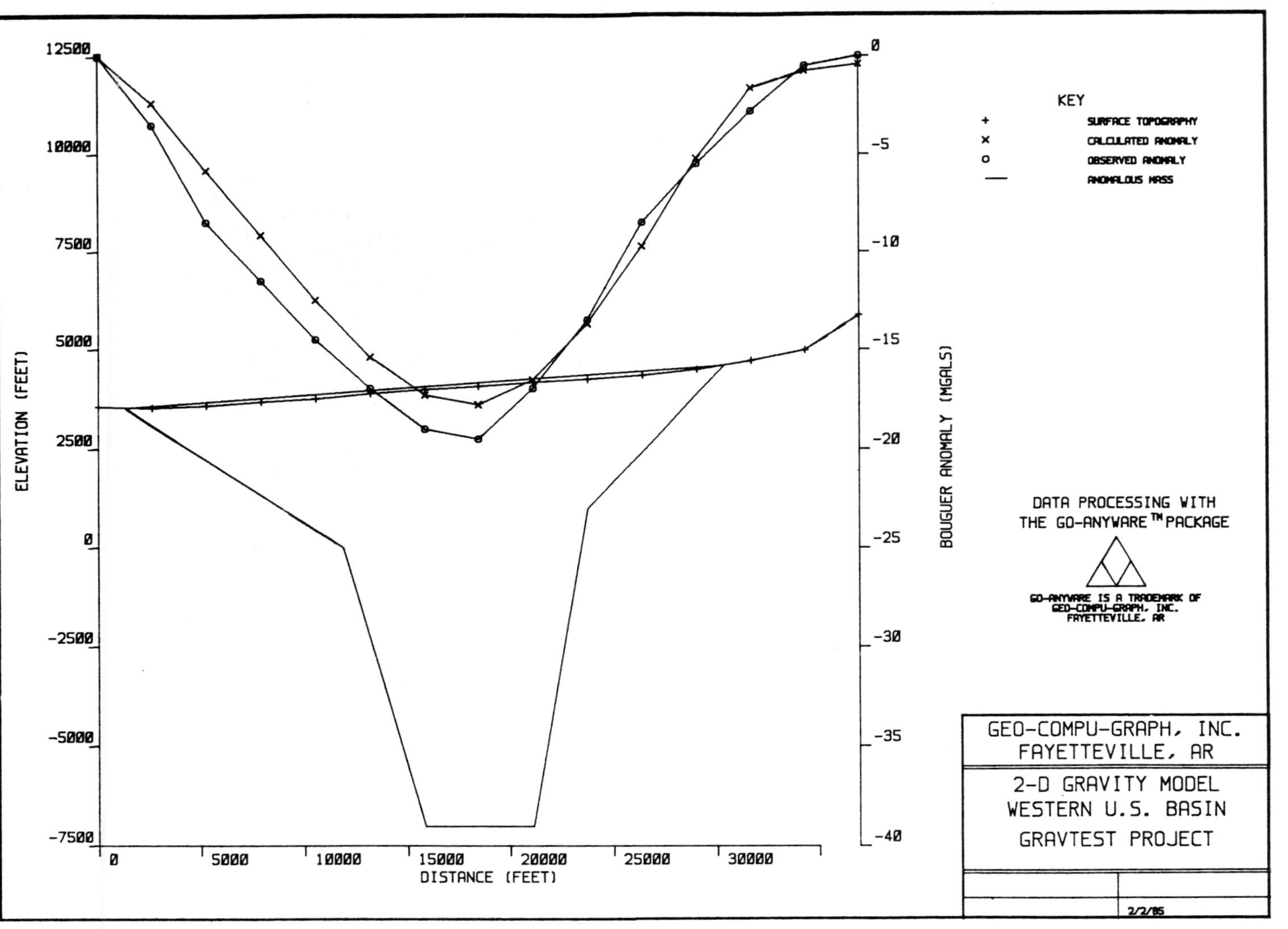

Figure 8. Report-grade plot of Talwani-style gravity model.

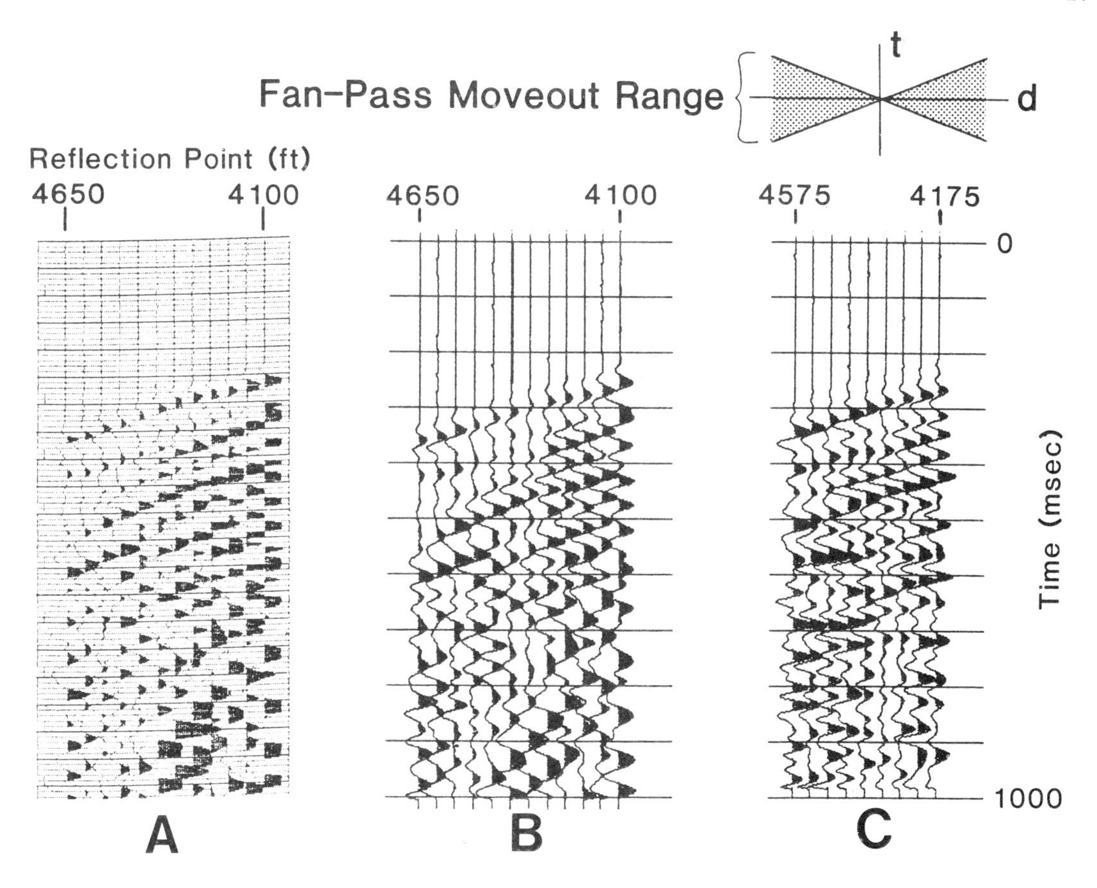

Figure 9. A, field record, B, computer plotted field record, and C, velocity filtered record. Surface waves, clearly visible in field data, are removed by velocity filtering (after Lankston and Lankston, 1983b).

The microcomputer was so flexible as to be used for tasks other than geoscientific data processing and display. Each contract project undertaken required a final report. Word processing, therefore, became almost as great a use for the machine as the numerical use. The word-processing power also was used to generate quarterly newsletters to keep clients informed of new field techniques and new data-processing capabilities.

As a primary developer of the engineering seismic-reflection method, our geophysicists wrote a monograph describing the method. The first and second editions, published in 1981 and 1982 respectively, were typescripts prepared entirely on the microcomputer's word processor. In 1983, through another step forward in technology, the word-processor data files that contained the third edition of the monograph (Lankston and Lankston, 1983a) were transmitted via a telephone link directly to the typesetter's computer. Galley proofs of the entire monograph were available within hours instead of days, and the number of mistakes in the text were minimal compared to the number expected if the monograph had been rekeyed completely by the typesetter.

MOVING AHEAD

By late 1982, the microcomputer world was changing rapidly. Many new machines were being introduced that offered more portability, better CRT graphics, or faster processing speed than the Apple II. The microcomputer was accepted widely at all levels in the geosciences by those who saw a powerful tool that would be less costly and more flexible than time-sharing services or inhouse mainframes. Today, the microcomputer is seen in geoscience companies' business offices, in the professionals' and technicians' offices or laboratories, and routinely in the field. Manufacturers of portable geophysical equipment, initially slow to accept the concept of field data processing, have designed data-acquisition instruments with computer interfaces so that data can be dumped from an instrument's memory into the computer's memory in a matter of seconds. The magnetometer of 1979 that required manual transcription and keying of data into the computer now as been replaced by units that can transmit station data at a rate of 30 stations per second. Portable tape-recorder systems can transmit seismic data exactly as digitized in the field into the computer in about one-half the time required for hand digitizing and at many times the precision.

The newer, faster, more portable microcomputers certainly do not indicate the older microcomputers are obsolete. The GO-ANYWARE system evolved into the GO-ANYWARE workstation, and it is a viable tool with proven history and, therefore, relatively few bugs left to discover. Turnkey operations with systems such as the GO-ANYWARE workstation, even by inexperienced personnel, are usual, and some companies offer systems for rent for short-term field or office use.

Through the eyes of the 1985 geoscientist who has become comfortable with the presence of the microcomputer, many of the evolutionary steps outlined here may seem trivial. However, each step represented a significant advance and one that was accompanied by days or weeks of frustration and considerable cost either in expenditure for hardware or software labor, or both.

Early users worked with 8-bit microcomputers, but now the next generation of 32-bit microprocessors are available. Fitting into a package about the same size and costing only slightly more than the one for an Apple II, the larger and faster 32-bit machines are expected to improve the art of field processing and interpretation of geophysical data and to provide more flexibility to the smaller geophysical service company for minimizing the effects of economic swings.

The GO-ANYWARE 16 workstation (Fig. 10), as the 32-bit version is being termed, will be compatible upwardly with the many 8- and 16-bit microcomputers that currently are on the market. It will support the same functions as its 8-bit sibling, but larger data files (such as would be needed in CDP seismic-reflection data processing) will be accommodated more easily. Moreover, the upward compatibility will make more software available to run on one machine.

Figure 10. GO-ANYWARE 16 Geophysical Workstation. In addition to micro-computer and dot-matrix printer shown, workstation may be configured with pen and ink plotter, digitizer, hard-disk system, and 8-bit and 16-bit emulators.

The next generation of 32-bit-based microcomputers will suffer from the same limitation as the 8-bit machines. That is, it will be able only to move as far as one's imagination will push it. With the performance demonstrated by the 8-bit machines, is the limit of the 32-bit machines in field and office geophysical operations even imaginable today? Certainly, in the same sense that the programmable calculator had been accepted widely in 1979 and the 8-bit microcomputer provided the next technological buffer against the swings of economic cycles, the users of the 32-bit machines will be able to low-pass filter the economic peaks and pokes of the next five years.

REFERENCES

Dobrin, M.B., 1976, An introduction to geophysical prospecting: McGraw-Hill Book Co., New York, 630 p.

Gay, S.P., 1966, Standard curves for interpretation of magnetic anomalies over long tabular bodies: Mining Geophysics, v. II, Soc. Explor. Geophys., p. 512-548.

Lankston, R.W., and Lankston, M.M., 1983a, An introduction to the utilization of the shallow or engineering seismic reflection method: Geo-Compu-Graph, Inc., Fayetteville, Arkansas, 39 p.

Lankston, R.W., and Lankston, M.M., 1983b, Shallow seismic reflection experiments on the Monte Cristo Claims, Nevada: Expanded abstracts of the Technical Program, Soc. Explor. Geophys., Tulsa, p. 183-185.

Palmer, D., 1980, The generalized reciprocal method of interpreting seismic refraction data: Soc. Explor. Geophys., Tulsa, 104 p.

Scott, J., 1977, SIPB-a seismic refraction inverse modeling program for batch computer systems: U.S. Geol. Survey Open File Report 77-366, 108 p.

Talwani, M., Worzel, J., and Landisman, M., 1959, Rapid gravity computations for two-dimensional bodies with application to the Mendocino submarine fracture zone: Jour. Geophys. Res., v. 64, no. 1, p. 49-59.

Telford, W., Geldart, L., Sheriff, R., and Keyes, D., 1976, Applied geophysics: Cambridge Press, New York, 860 p.

Performance Evaluation of a Microcomputer-based Interpretation System for Gravity and Magnetic Data

J. J. Mayrand and M. C. Chouteau

Ecole Polytechnique, Montreal

ABSTRACT

Microcomputer software has a definite appeal for interpretation of geophysical data in the field. Our performance evaluation of an interactive two-dimensional interpretation system indicates wide throughput variations for available microcomputers. The computing time required for the calculation of a typical gravity-anomaly curve ranges from 6 minutes to 20 seconds depending on the microcomputer configuration used. A typical magnetic-anomaly curve requires a 13 minute to 40 second delay. Execution times increase for more complex interpretation models and may be reduced by a factor ranging from 3 to 16 by using a BASIC compiler instead of a BASIC interpreter. The topology validation of user-defined models has proven to require minimal time compared to calculations and thus has been considered a useful addition to the system. Hard-copy printouts of results are obtained in 3 to 10 minutes. The random-access memory requirements and the input/output speed are no limitation to the system performance. Because of their low-calculation speed, several of the smaller microcomputers currently available are not suited for the efficient interactive interpretation of two-dimensional models.

INTRODUCTION

A number of computer programs that permit two-dimensional quantitative interpretation of gravity and magnetic data are available. Their execution performance ranges from a fraction of seconds of computer time for the mainframe-based computation of a typical anomaly curve (Rudman and Blakely, 1983), up to a couple of hours for pocket calculator programs such as those presented by Vorce and Pearson (1982a) and Vorce and Pearson (1982b).

Of particular interest to the mining industry are microcomputer programs that made possible the efficient interpretation of data directly in the field. Reeves and MacLeod (1983) have stated that an interpretation program running on an IBM Personal Computer can compute an anomaly curve in a few seconds and that much of what they have programmed "...could be adapted without great loss of effectiveness to hardware costing much less" (Reeves and MacLeod, 1983, p. 23). The purpose of this paper is to assess in a more systematic way the last part of this statement.

The first step in this study was to package an interactive system running on a microcomputer which could handle the two-dimensional interpretation of both gravity and magnetic data. In order to obtain the performance evaluation of a computer system suited for industrial use, the study sought to obtain a system that could provide full user guidance through a video displayed documentation as well as complete validation of all data entered by the user and provide an acceptable computational performance.

The second part of this study assessed the software- and hardware-related limitations to the system's performance.

ALGORITHMS AND SYSTEM USED

Talwani, Worzel, and Landisman (1959) introduced simple formulae enabling the rapid computation of an anomaly by using polygons as an approximation to complex two-dimensional geologic bodies. As shown on Figure 1, the shape of the polygons are assumed to be representative of the geologic cross section of the anomalous body studied. The method assumes an infinite extension of the geologic formations along strike. The mathematical formulae used in our system to compute an anomaly are presented in Appendix 1.

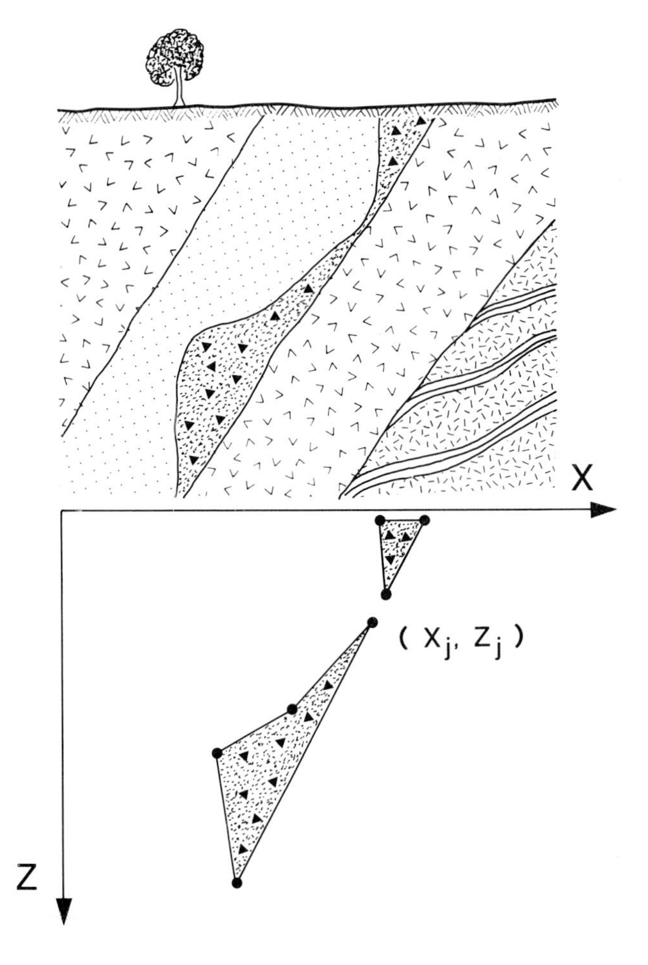

Figure 1. Approximation of geologic cross section of mineralized body with use of polygons.

The system is written in BASIC, and runs on a TRS-80 Model III microcomputer using Appart's Inc. NEWDOS and BASIC interpreter. The system comprises six programs totaling about 4000 lines of coding, 2800 of which are executable statements. All promptings and internal program comments are in French.

The system allows the user to file, edit, and obtain hard-copy printouts of field data and of model parameters. Up to five different eight-sided polygons can be used to define an anomalous body. Gravity, vertical, and total magnetic field anomalies can be computed.

Graphic printouts of results (8 1/2 x 11 inches) are produced with a TRS-80 Multi-Pen Plotter.

PERFORMANCE EVALUATION

Four possible limiting factors to the system performance have been considered. These are the central processing unit (CPU) performance, plotter performance, memory requirements, and input/output (I/O) speed. The first two factors have turned out to be more important than the last two. The four factors will be considered separately.

The CPU Performance

The CPU performance depends on the type of benchmark program being executed and on the microcomputer used. Both considerations are examined in more details hereafter.

Four of the system subroutines have proven to be potentially CPU-bounded. These are:

(a) the subroutine performing the gravity-anomaly calculations. The benchmark derived from this subroutine is referred to as TEST #1.

(b) the double-precision version of the subroutine performing the magnetic-anomaly calculations. Double-precision arithmetic has been deemed necessary for this subroutine in view of the formula instability for limiting situations. The benchmark derived from it is referred to as TEST #2.

(c) the subroutine which validates the definition of polygons. As shown on Figure 2, the user can provide inadvertantly the computer with a list of summits that do not define properly the polygon. A subroutine which validates the polygon thus is a useful addition to the system. The listing of his subroutine is given in Appendix 2. The benchmark derived from it is referred to as TEST #3.

(d) the subroutine which determines the rotation (clockwise or counterclockwise) used to list the polygon summits. Equation (2) and (3) (Appendix 1) both require that summits be scanned clockwise. A listing of this subroutine is presented in Appendix 3. The benchmark derived from it is referred to as TEST #4.

The four benchmark programs described previously were executed on most of the following six different microcomputer configurations:

(a) a TRS-80 Model III using the TRSDOS version 1.3 operating system and a BASIC interpreter. This hardware/software configuration os referred to as the TRS-INTERPRETER.

(b) an Apple IIe using the Apple Computer Inc. DOS 3.3 operating system and a BASIC interpreter. This configuration is referred to as the APPLE-INTERPRETER. TEST #2 was not executed on this configuration because double-precision arithmetic is not available on it.

(c) an IBM Personal Computer using the PC DOS version 2.0 operating system and the Microsoft BASICA interpreter. This configuration is referred to as the IBM-INTERPRETER.

(d) a TRS-80 Model III using the RSBASIC version 2.4 BASIC compiler of Ryan-McFarland Corp. This configuration is referred to as the TRS-COMPILER. TEST #1 and TEST #2 were not executed on this configuration due to BASIC syntax incompatibilities with the compiler.

(e) an Apple IIe using the BASIC Applesoft Compiler version 2.01. This configuration is referred to as the APPLE-COMPILER.

(f) an IBM Personal Computer using the Microsoft Inc. BASIC compiler version 1.01. This configuration is referred to as the IBM-COMPILER.

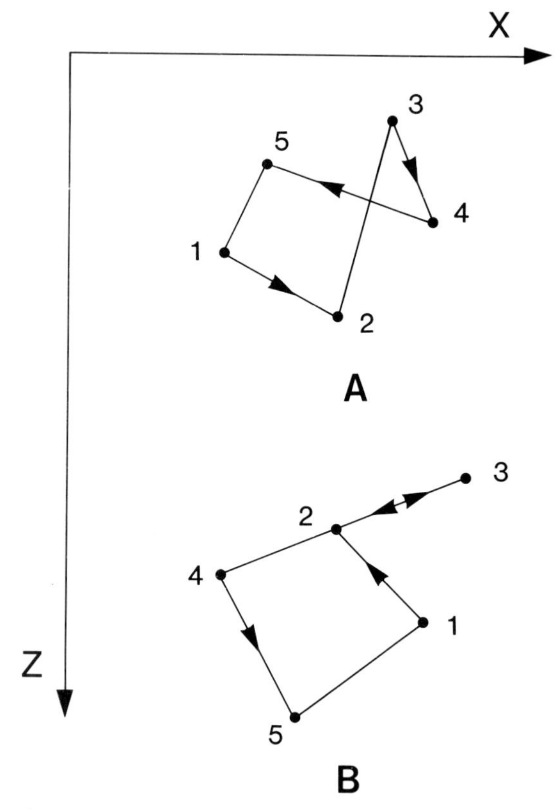

Figure 2. Topology of polygons. Polygon (a) is faulty as two sides intersect each other, whereas polygon (b) should be rejected because of two overlying and probably ill-defined polygon sides.

The computer time required to run each of the four benchmark programs on the various microcomputer configurations is presented schematically on Figure 3. These execution times were established for what was deemed to be a typical interpretation model, namely, a two polygon model for which the anomaly curves were calculated at 51 points. One polygon had three summits and the other had four.

A study of the data on Figure 3 shows that there is a significant difference in the effectiveness of the system when operated on different microcomputers. We feel that the computation of an anomaly curve probably should not exceed one to two minutes. As for longer delays, the proportion of idle time to productive time becomes excessive. Based on this criterion, several of the microcomputer configurations tested in this work would not be suited for the two-dimensional interpretation of potential fields. Likewise, the reduction of execution time by a factor of about 3 to 16 through the use of a compiler instead of an interpreter is striking. Given the price range of $225 to $400 for such compilers, it is obvious that productivity gains would permit to recuperate the purchase price in a few weeks.

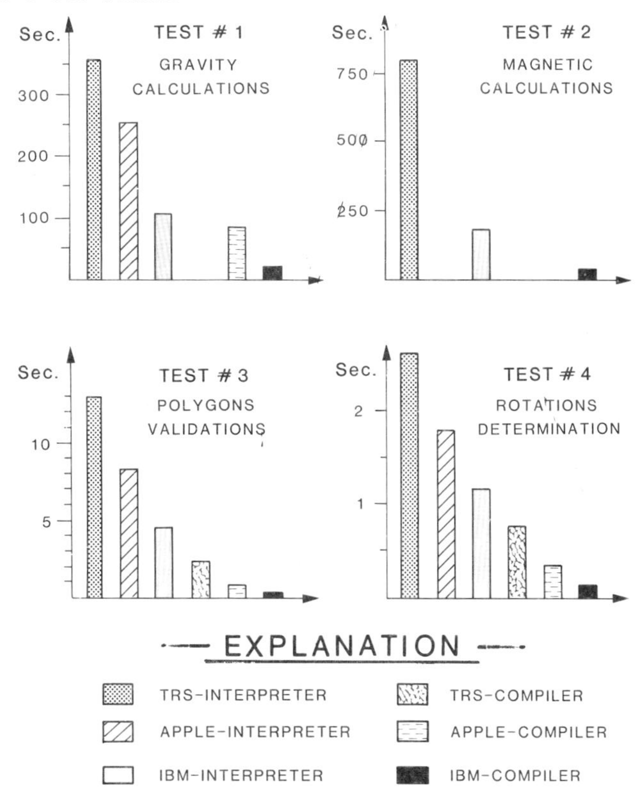

Figure 3. Comparative execution time (in seconds) for various microcomputer configurations used.

Both the polygon validation (TEST #3) and the rotation determination (TEST #4) subroutines have execution times less than that of the anomaly-calculation routines, so that whatever configuration is used, the validation features are useful additions to an interpretation system. As indicated by the comparison of TEST #2 and TEST#1 results, double-precision arithmetic may not be handled satisfactorily by a microcomputer even though single-precision arithmetic is.

The execution time of an interpretation run is determined largely by either of the first two subroutines from which TEST #1 and TEST #2 derive. If (Ti) is the total execution time required by a given microcomputer for handling our so-called typical model, the total time (T) needed by the same micro to compute the N-points anomaly curve of an M-sided polygon then is approximately equal to:

$$T = Ti \times N \times M \times 3 \times 10^{-3}.$$

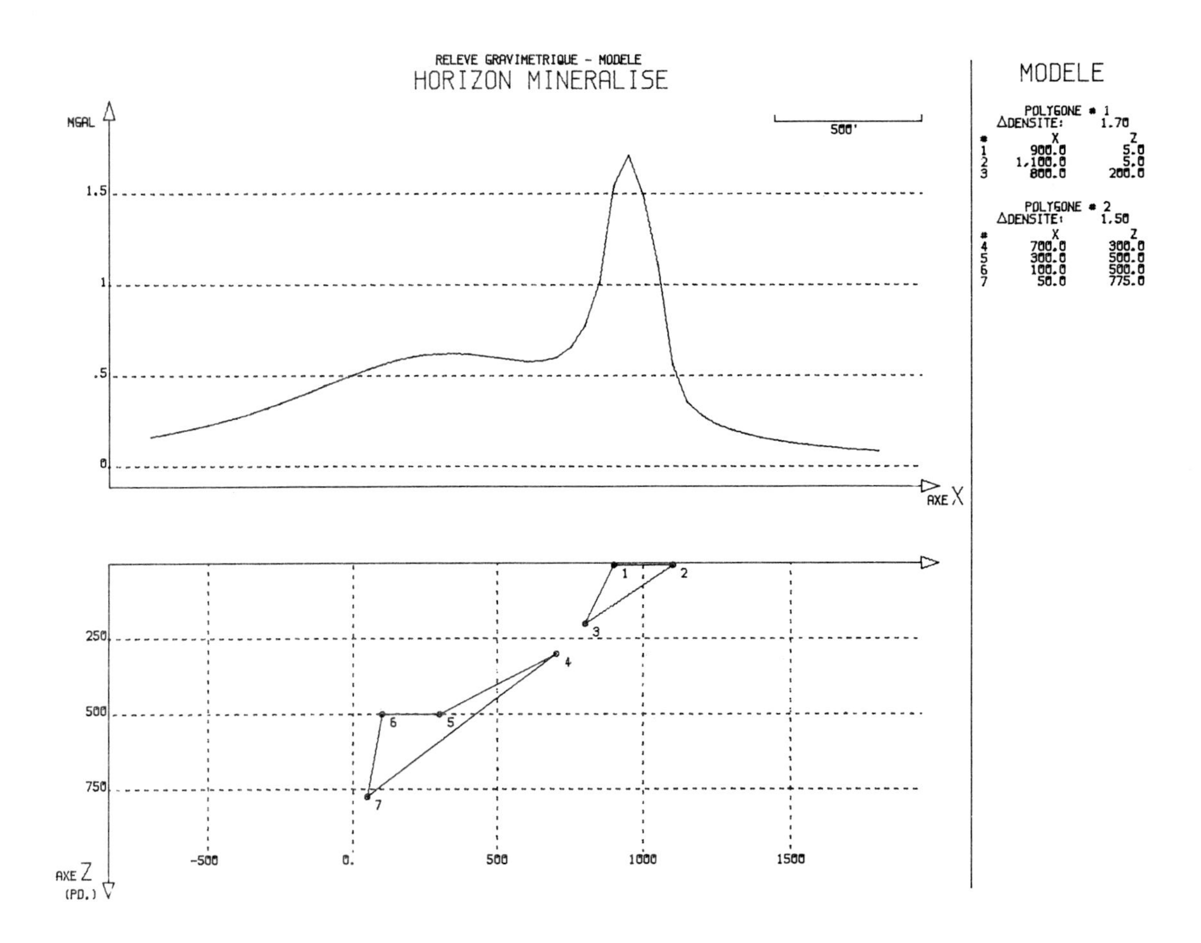

Figure 4. Gravity-anomaly curve drawn with a TRS-80 Multi-Pen Plotter in about four minutes.

Plotter Performance

A sample of the TRS-80 Multi-Pen Plotter printout of a gravity-anomaly curve is shown in Figure 4. The plotting time required for such a drawing is about four minutes. The plotting of field data and of more complex polygons, however, may raise that figure to ten minutes. On the basis of technical specifications, we understand the Epson MS-80 dot-matrix printouts, similar to those presented by Reeves and MacLeod (1983), could be produced in six minutes, however elaborate the models are. We have confirmed that the more recent Epson FX-80 requires a flat three minutes delay to produce an 8 1/2 X 11 inch graphic printout.

The evaluation of other plotting devices would permit valuable comparisons. Despite the limited data at our disposal, we consider that: plotting devices that can be carried in the field have an acceptable throughput for the printing of final versions of interpretation curves. Preliminary interpretation results, however, would be displayed more efficiently on the microcomputer screen. Likewise, although pen plotters usually produce better quality graphics than dot-matrix printers, they require machine-dependent software which may duplicate the software needed to produce screen graphics.

Random Access Memory Requirements

The system has been segmented in six programs which chain to each other as required by the options selected by the user. The memory requirements of the programs range from 5K to 28K. Such requirements make the programs easily adaptable to most microcomputers on the market and do not constitute constraints on the system performance.

A substantial part of the memory requirements comes from the video-displayed documentation. All promptings and instruction texts are provided to the screen through the execution of BASIC statements so that no access to a message file is needed. Despite such a memory-consuming approach and a maximum program size of 28K, the documentation has been determined to be adequate to guide unfamiliar users through the whole interpretation process with no other help than a one-page description of how to load the diskettes.

Input/Output Speed

The I/O speed has been determined to be no material limitation to the system performance.

The loading delay of programs from the diskettes to the microcomputer memory ranges from four to six seconds and accounts for a negligible part of a typical working session.

The I/O of data produces no user detectable delays except when a packed version approach to diskette files is taken. The reading or writing of all parameters of an interpretation model then requires eight seconds. That delay can be avoided by realizing a less efficient packing of data on the diskette.

CONCLUSIONS

The evaluation of the interpretation system described in this work and its execution on six different microcomputer configurations permit the following conclusions:

(a) the effectiveness of a two-dimensional potential fields interpretation system can be reduced significantly by operating it on the smaller, less-capable microcomputers available on the market.

(b) due to markedly improved calculation performance, the use of a BASIC compiler rather than a BASIC interpreter is justified economically.

(c) the validation of the interpretation model polygons requires minimal computing time compared to the actual anomaly computation and improved the reliability of results.

(d) when required, double-precision arithmetic requires special evaluation of the CPU performance.

(e) portable plotters are appropriate tools for final printouts of interpretation results, but less suited than video displays for the rapid plotting of preliminary interpretation curves.

(f) the random-access memory requirements are no limitation to a system even when extensive documentation texts are supported by the microcomputer memory.

(g) the input/output speed poses no important limitation to the system performance.

ACKNOWLEDGMENTS

This work was part of a student research project (UPIR) carried out from October 1983 to April 1984 at Ecole Polytechnique of Montreal.

We thank Mr. Robert Bazinet for his help in microcomputer communication links and Mr. Gaston Pouliot for reviewing the manuscript. We also acknowledge the financial support of the Direction de la Recherche de l'Ecole Polytechnique de Montreal.

REFERENCES

Grant, F.S., and West, G.F., 1965, Interpretation theory in applied geophysics: McGraw Hill Book Co., New York, 584 p.

Reeves, C.V., and MacLeod, I.N., 1983, Modeling of potential field anomalies - some applications for the microcomputer: First Break, v. 1, no. 8, p. 18-24.

Rudman, A.J., and Blakely, R.F., 1983, Computer calculation of two-dimensional gravity fields: Indiana Geol. Survey Occasional Paper 40, 22 p.

Talwani, M., Worzel, J.L., and Landisman, M., 1959, Rapid gravity computations for two-dimensional bodies with application to the Mendocino submarine fracture zone: Jour. Geophys. Res., v. 64, no. 1, p. 49-59.

Vorce, K.A., and Pearson, W.C., 1982a, A TI-59 calculator program for determining the gravity anomaly of a 2-D geologic body: Computers & Geosciences, v. 8, no. 3-4, p. 335-339.

Vorce, K.A., and Pearson, W.C., 1982b, A TI-59 pocket calculator magnetic modeling program: Computers & Geosciences, v. 8, no. 3-4, p. 349-354.

APPENDIX 1

ANOMALY CALCULATION FORMULAE

Gravity anomalies are obtained separately for each polygon with equation (2) which is equivalent to equation (10-7) of Grant and West (1965).

$$A(0,0) = 2G\Delta\rho \sum_{j=1}^{n} \left(\frac{X_j Z_{j+1} - X_{j+1} Z_j}{(X_{j+1} - X_j)^2 + (Z_{j+1} - Z_j)^2} \right) \times \left\{ \frac{(Z_{j+1} - Z_j)}{2} \ln \left(\frac{X_{j+1}^2 + Z_{j+1}^2}{X_j^2 + Z_j^2} \right) + (X_{j+1} - X_j) \left[tg^{-1} \left(\frac{X_{j+1}}{Z_{j+1}} \right) - tg^{-1} \left(\frac{X_j}{Z_j} \right) \right] \right\} \quad (2)$$

$A(0,0)$:	gravity anomaly at point (0,0)
G:	gravitational constant
$\Delta\rho$:	density contrast
n:	number of polygon sides
X_j, Z_j:	coordinates of the polygon summits; the positive side of the Z axis is downward
Σ:	the summation is performed clockwise over the polygon summits.

Vertical magnetic field anomalies are obtained separately for each polygon with equation (3) taken from Grant and West (1965) with a correction for an apparent misprint of sign in the original.

$$A(0,0) = C_1 \sum_{j=1}^{n} \frac{1}{(1 + a_j^2)} \times$$

$$\left\{ C_2 \ \ln \sqrt{\frac{(1 + a_j^2)Z_{j+1}^2 + 2a_jb_jZ_{j+1} + b_j^2}{(1 + a_j^2)Z_j^2 + 2a_jb_jZ_j + b_j^2}} \right.$$

$$+ C_3 \left[tg^{-1} \left(\frac{(1 + a_j^2)Z_{j+1} + a_j}{b_j} \right) \right.$$

$$\left. \left. - tg^{-1} \left(\frac{(1 + a_j^2)Z_j + a_j}{b_j} \right) \right] \right\} \qquad (3)$$

$A(0,0)$:	magnetic anomaly at point (0,0)
n:	number of polygon sides
C_1:	$2kH_0 \sqrt{1 - \cos^2\lambda \cos^2 i}$
C_2:	$a_j\sin\beta - \cos\beta$
C_3:	$\sin\beta + a_j\cos\beta$
k:	magnetic susceptibility contrast
H_0:	total magnetic field of the earth
i:	magnetic inclination
λ:	magnetic declination of the strike of the body
β:	$tg^{-1}(tg\ i / \sin\lambda)$
a_j:	$(X_{j+1} - X_j) / (Z_{j+1} - Z_j)$
b_j:	$(X_jZ_{j+1} - X_{j+1}Z_j) / (Z_{j+1} - Z_j)$
X_j, Z_j:	coordinates of the polygon summits; the positive side of the Z axis is downward.
Σ:	the summation is performed clockwise over the polygon summits.

For the total field, the constants C_1, C_2, and C_3 become:

C_1:	$2kH_0(1 - \cos^2\lambda\cos^2 i)$
C_2:	$-a_j\cos2\beta - \sin2\beta$
C_3:	$a_j\sin2\beta - \cos2\beta$

APPENDIX 2

POLYGON VALIDATION SUBROUTINE

The polygon definition is presented as a list of summit identification numbers available in variable PS% (*,*).

The validations performed are:

(a) all summits referenced in the list have been defined;

(b) at least three summits have been used to define the polygon;

(c) the same summit is not used twice in the list; and

(d) two sides of the polygon do not intersect each other.

```
10 REM
20 REM  --------------------  SOUS-PROGRAMME
30 REM  -----  VALIDATION D'UN POLYGONE
40 REM
50 REM  -----  PARAMETRES:
60 REM  -----      INTRANTS:
70 REM                PO%:        NUMERO DU POLYGONE VALIDE
80 REM                PS%(5,8):   IDENTIFIE LES 8 SOMMETS DEFINISSANT
90 REM                            CHACUN DES CINQ POLYGONES
100 REM                SX!(40):    COORDONNEE EN 'X' DES QUARANTE
110 REM                            SOMMETS POSSIBLES DES POLYGONES
120 REM                SZ!(40):    COORDONNEE EN 'Z' DES QUARANTE
130 REM                            SOMMETS POSSIBLES DES POLYGONES
140 REM                NS%:        NOMBRE TOTAL DE SOMMETS DEFINIS
150 REM                PC$(5):     CODE INDIQUANT SI LES 5 POLYGONES ONT
160 REM                            ONT ETE DEFINIS ('O') OU NON ('N')
170 REM  -----      EXTRANTS:
180 REM                VA%:        '0' SI POLYGONE 'OK'
190 REM                            PAS EGAL A '0' SI LE POLYGONE EST MAL DEFINI
200 REM
210 VA% = 0
220 IF PC$(PO%) = "N" THEN RETURN
230 REM
240 REM  -----  S'ASSURER QUE TOUS LES SOMMETS ONT ETE DEFINIS
250 REM
260 SO% = 0
270 FOR W% = 1 TO 8
280    W1% = PS%(PO%,W%)
290    IF W1% = 0 THEN GOTO 330
300    IF W1% < 0 OR W1% > NS% THEN VA% = 1
310    SO% = SO% + 1
320    NEXT W%
330 IF VA% <> 0 THEN RETURN
340 REM
350 REM  -----  S'ASSURER QUE LE POLYGONE A AU MOINS 3 SOMMETS
360 REM
370 IF SO% < 3 THEN VA% = 2
380 IF VA% <> 0 THEN RETURN
390 REM
400 REM  -----  LE MEME POINT N'A PAS ETE UTILISE DEUX FOIS DANS LA LISTE
410 REM
420 FOR W1% = 1 TO SO% - 1
430    FOR W2% = W1% + 1 TO SO%
440       IF PS%(PO%,W1%) = PS%(PO%,W2%) THEN VA% = 3
450       IF VA% <> 0 THEN RETURN
460       NEXT W2%
470    NEXT W1%
```

```
480 REM
490 REM  -----  VERIFIER LA TOPOLOGIE DU POLYGONE
500 REM
510 REM  -----  BOUCLES POUR PASSER EN REVUE TOUTES LES
520 REM  -----  PAIRES POSSIBLES DE COTES DU POLYGONE
530 REM
540 FOR W1% = 1 TO SO% - 1
550    FOR W2% = W1% + 1  TO SO%
560       X1! = SX!(PS%(PO%,W1%))
570       Z1! = SZ!(PS%(PO%,W1%))
580       X2! = SX!(PS%(PO%,W1%+1))
590       Z2! = SZ!(PS%(PO%,W1%+1))
600       X3! = SX!(PS%(PO%,W2%))
610       Z3! = SZ!(PS%(PO%,W2%))
620       W% = W2% + 1
630       IF W% > SO% THEN W% = 1
640       X4! = SX!(PS%(PO%,W%))
650       Z4! = SZ!(PS%(PO%,W%))
660       GOSUB 1110
670 REM
680 REM  -----  CALCUL DES PENTES DES DEUX COTES
690 REM
700 W! = X2! - X1!
710 IF ABS(W!) < 1E-6 THEN M1! = 1E9 ELSE M1! = (Z2! - Z1!) / W!
720 W! = X4! - X3!
730 IF ABS(W!) < 1E-6 THEN M2! = 1E9 ELSE M2! = (Z4! - Z3!) / W!
740 REM
750 REM  -----  CALCUL DES INTERCEPTES DES DEUX COTES
760 REM
770 B1! = 1E9
780 B2! = 1E9
790 IF M1! <> 1E9 THEN B1! = Z1! - M1!*X1!
800 IF M2! <> 1E9 THEN B2! = Z3! - M2!*X3!
810 REM
820 REM  -----  ANALYSE DES POINTS D'INTERSECTION DES DEUX COTES
830 REM
840 IF ABS(M2!-M1!) < 1E-4 AND ABS(B2!-B1!) < 1E-4 AND (B1! <> 1E9 OR (B1! = 1E9
 AND ABS(X3!-X1!) < 1E-6)) THEN GOSUB 1240
850 IF VA% <> 0 THEN RETURN
860 IF ABS(M2! - M1!) < 1E-4 THEN GOTO 1040
870 REM
880 REM  -----  CALCUL DES COORDONNEES DE L'INTERCEPTE
890 REM
900 IF M1! = 1E9 THEN IX! = X1!
910 IF M1! = 1E9 THEN IZ! = M2!*IX! + B2!
920 IF M2! = 1E9 THEN IX! = X3!
930 IF M2! = 1E9 THEN IZ! = M1!*IX! + B1!
940 IF M1! <> 1E9 AND M2! <> 1E9 THEN IX! = (B2! - B1!) / (M1! - M2!)
950 IF M1! <> 1E9 AND M2! <> 1E9 THEN IZ! = M1!*IX! + B1!
```

```
960 REM
970 REM  -----  VALIDATION DU POINT D'INTERSECTION
980 REM
990 IF W1% = W2% - 1 OR (W1% = 1 AND W2% = SO%) THEN GOTO 1040
1000 IF IX! - F1! >-1E-4 AND F2! - IX! >-1E-4 AND IX! - F3! >-1E-4 AND F4! - IX!
 >-1E-4 AND IZ!-F5! > -1E-4 AND F6!-IZ! > -1E-4 AND IZ!-F7! > -1E-4 AND F8!-IZ!
> -1E-4 THEN VA% = 4
1010 REM
1020 REM
1030 REM
1040       NEXT W2%
1050    NEXT W1%
1060 RETURN


1070 REM
1080 REM  --------------------  SOUS-PROGRAMME
1090 REM  -----  CALCUL DE MAXIMUM ET DE MINIMUM
1100 REM
1110 IF X1!  < X2! THEN F1! = X1! ELSE F1! = X2!
1120 IF X1! >= X2! THEN F2! = X1! ELSE F2! = X2!
1130 IF X3!  < X4! THEN F3! = X3! ELSE F3! = X4!
1140 IF X3! >= X4! THEN F4! = X3! ELSE F4! = X4!
1150 IF Z1!  < Z2! THEN F5! = Z1! ELSE F5! = Z2!
1160 IF Z1! >= Z2! THEN F6! = Z1! ELSE F6! = Z2!
1170 IF Z3!  < Z4! THEN F7! = Z3! ELSE F7! = Z4!
1180 IF Z3! >= Z4! THEN F8! = Z3! ELSE F8! = Z4!
1190 RETURN


1200 REM
1210 REM  --------------------  SOUS-PROGRAMME
1220 REM  -----  VALIDATION SPECIALE DE L'INTERCEPTE
1230 REM
1240 W5! = F1!
1250 IF F3! > F1! THEN W5! = F3!
1260 W6! = F2!
1270 IF F4! < F2! THEN W6! = F4!
1280 V1% = 0
1290 IF W6! - W5! > -1E-4 THEN V1% = 4
1300 W5! = F5!
1310 IF F7! > F5! THEN W5! = F7!
1320 W6! = F6!
1330 IF F8! < F6! THEN W6! = F8!
1340 V2% = 0
1350 IF W6! - W5! > -1E-4 THEN V2% = 4
1360 REM
1370 REM
1380 REM
1390 IF V1% = 4 AND V2% = 4 THEN VA% = 4
1400 RETURN
```

APPENDIX 3

ROTATION DETERMINATION SUBROUTINE

Equation (2) and (3) will yield a reversed sign anomaly value if the polygons summits are scanned counterclockwise in the summation.

The subroutines presented here determine the rotation sense used for the definition of a polygon. It returns a (+1) or (-1) factor when the rotation seems clockwise or counterwise on a diagram similar to Figure 1. This factor then can be used to correct the sign of the anomaly value. The user thus is relieved of the mandatory clockwise definition of polygons.

The mathematics underlying this determination use Stokes Theorem. Using the vectorial field F(x,z) = (-1/2z,1/2x), we have:

$$S = \iint_S dxdz$$

$$= \iint_S \left(\tfrac{1}{2} - \left(-\tfrac{1}{2}\right)\right) dxdz$$

$$= \iint_S (\nabla \times \bar{F}).dS = \oint \bar{F}.dr$$

$$= \oint \left(\left(-\tfrac{1}{2}z\right)dx + \left(\tfrac{1}{2}x\right)dz\right)$$

$$= \tfrac{1}{2}\oint (-zdx + xdz) \tag{4}$$

where: S; area enclosed by the polygon.

r; counterclockwise path along the polygon edges.

The polygon area computed with equation (4) will be positive if the polygon has been defined counterclockwise on an X-Z diagram where the positive side of the Z axis is upward. The rotation implicit in the user's definition of the polygon can be determined by calculating the polygon area with equation (4).

```
10 REM  --------------------  SOUS-PROGRAMME
20 REM  -----  DETERMINE LE SENS DE ROTATION UTILISE POUR
30 REM  -----  DEFINIR LE POLYGONE
40 REM
50 REM  -----  LA SURFACE DU POLYGONE EST CALCULEE AU MOYEN
60 REM  -----  D'UNE INTEGRALE DE LIGNE LE LONG DE LA
70 REM  -----  FRONTIERE DU POLYGONE.  EN VERTU DU THEOREME
80 REM  -----  DE STOKES, LA SURFACE AINSI CALCULEE SERA
90 REM  -----  POSITIVE SI LE POLYGONE EST DEFINI DANS LE
100 REM  -----  SENS ANTI-HORAIRE ET NEGATIVE AUTREMENT.
110 REM
120 REM  -----  PARAMETRES:
130 REM  -----      INTRANTS:
140 REM                PO%:        NUMERO DU POLYGONE DONT ON CALCULE
150 REM                            LA SURFACE
160 REM                PS%(5,8):   IDENTIFIE LES 8 SOMMETS DEFINISSANT
170 REM                            CHACUN DES CINQ POLYGONES
180 REM                SX!(40):    COORDONNEE EN 'X' DES QUARANTE
190 REM                            SOMMETS POSSIBLES DES POLYGONES
200 REM                SZ!(40):    COORDONNEE EN 'Z' DES QUARANTE
210 REM                            SOMMETS POSSIBLES DES POLYGONES
220 REM  -----      EXTRANTS:
230 REM                DP!(5):     PREND LA VALEUR DE '+1' OU '-1'
240 REM                            SELON LE SENS DE ROTATION UTILISE
250 REM                            POUR DEFINIR CHACUN DES 5 POLYGONES.
260 S3! = 0.
270 DP!(PO%) = +1
280 REM
290 REM  -----  BOUCLE POUR BALAYER LES SOMMETS DU POLYGONE
300 REM
310       FOR SO% = 1 TO 8
320          IF PS%(PO%,SO%) = 0 THEN GOTO 490
330          X1! = SX!(PS%(PO%,SO%))
340          Z1! = SZ!(PS%(PO%,SO%))
350          W% = SO% + 1
360          IF W% > 8 THEN W% = 1
370          IF PS%(PO%,W%) = 0 THEN W% = 1
380          X2! = SX!(PS%(PO%,W%))
390          Z2! = SZ!(PS%(PO%,W%))
400 REM
410 REM  -----  CALCUL DE LA SURFACE
420 REM
430          S3! = S3! - 0.5 * (Z1!*(X2!-X1!) - (Z2!-Z1!)*X1!)
440          NEXT SO%
450 REM
460 REM  -----  AJUSTE LE FACTEUR EN FONCTION DU SIGNE DE LA
470 REM  -----  SURFACE
480 REM
490 IF S3! < 0. THEN DP!(PO%) = -1.
500 RETURN
```

Analysis of Shallow Seismic-Refraction Data by Microcomputer

C. B. Reynolds

Charles B. Reynolds & Associates

ABSTRACT

Two microcomputer programs have proven to be especially useful to us in analysis of shallow seismic-refraction data for geologic engineering purposes. LNFT, which fits an inverse velocity line by least squares to a set of offset distance/time pairs for a refraction profile, yields refractor velocity, zero intercept time and quality of line fit. ZBYTO uses the results of LNFT to solve for velocity layer thickness by the zero-intercept method for situations of two to five layers.

INTRODUCTION

Shallow seismic-refraction data are used in several types of geologic investigations. The geologic information derived from shallow refraction probes is used in geologic engineering investigations in connection with excavation and foundation design, especially if there are problems with shallow rock or unstable soil conditions. Shallow-refraction studies are employed as part of programs for exploration and development of mineral deposits and groundwater. They also are utilized to provide the shallow velocity information needed to correct seismic-reflection data to a reference or datum plane.

In analysis of shallow-refraction profiles two microcomputer programs, LNFT.BAS and ZBYTO.BAS, have been utilized.

EXAMPLE I - A SIMPLE GEOLOGIC MODEL

Figure 1A shows a typical refraction-probe layout and simple geologic or velocity model. The seismic energy source, which usually is a sledgehammer struck on a metal plate on the ground, a heavy weight dropped to the ground, or a small explosive charge, is located at position A. Part of the seismic energy travels along a ray path such that it enters a deeper, higher velocity layer, is refracted along its upper surface and rises again to reach in turn each of a series of seismic receivers or geophones (positions B, C, D, E, F, or G as shown by Figure 1A). The angle by which the seismic ray enters the lower layer is a function of the velocities of the upper and lower layers as related by Snell's Law. A reverse recording of the probe or profile is desirable; in this situation, after recording the refraction probe with the source at A, a geophone would be placed at A, the geophone at G replaced by the source and a second recording taken. The reciprocal times of the two profiles, that is, the travel time from A to G and the travel time from G to A, should be equal or nearly so.

A time-distance plot displaying the first arrival times at the geophones for the example model of Figure 1A is shown in Figure 1B. The arrival times at geophones B, C, and D fall on a straight line through the origin; this is because the direct wave through the upper layer reaches these geophones before the wave refracted from the second or deeper layer. The inverse slope of this line, that is, distance/time, is the velocity of the first (shallower) layer. The arrival times at geophones E, F, and G also plot on a straight line, but this line intersects the zero distance (vertical) ordinate at some time after time-zero. This zero-distance time-intercept later will be referred to as the zero intercept. The inverse slope of this second line is the velocity of the second (deeper) layer. If the refractor (second layer) were not horizontal as in Figure 1A but dipping, the refractor velocity thus measured would not be correct. The mean of the refractor velocities determined from the two reversed recordings described previously will be close to the true refractor velocity.

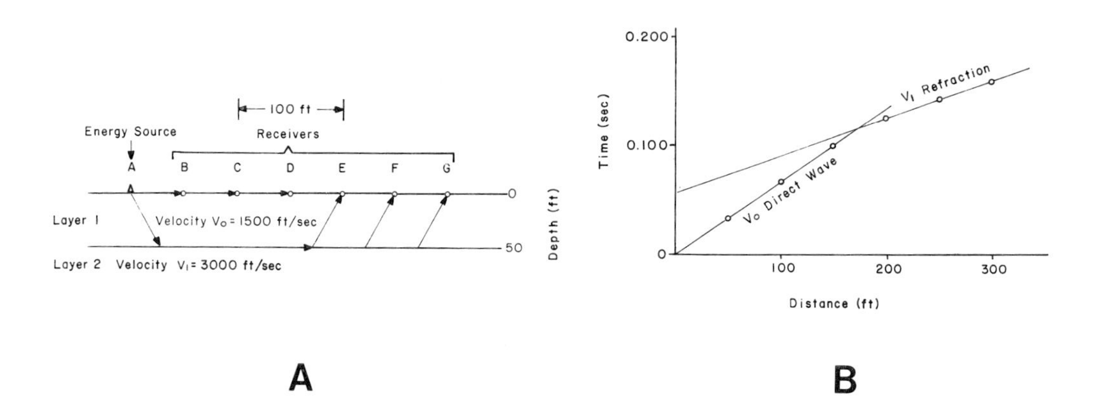

Figure 1. A, Typical shallow-refraction layout with simple geologic model; B, time-distance plot for model.

In actual practice the refractor usually is not perfectly planar, so that there is some departure of the time-distance points from the ideal straight line. There also are other reasons which may cause some scatter of time-distance points, such as local variations in the velocity of the upper layer. Consequently, the fitting by eye of a straight line to a set of time-distance points may be subjective enough that two interpreters may draw significantly different inverse-velocity lines through the same group of points.

The main purpose of the program LNFT.BAS is to remove this subjectivity and uncertainty by fitting a straight line to a series of time-distance points by the method of least squares. There also is a considerable improvement in efficiency in that the time-distance plot need not be made; the interpreter goes directly from the paper records to the microcomputer. LNFT.BAS yields an indication of the quality of the line-fit and the zero-distance time-intercept for the line.

Figure 2A shows the computer output for the times and distances of the simple geologic velocity model of Figure 1A. Figure 2B shows the computer output of the program ZBYTO.BAS, used to solve for the thickness of the first layer of the model of Figure 1A. The difference between the solution, 48.5 ft, and the actual thickness in the model, 50 ft, results from rounding-off of the travel times. The differences between the velocities calculated by LNFT.BAS for the two layers and those of the model also result from this rounding-off. The quality of line-fit is given as -1 in both situations (Fig. 2A), which is perfect to three decimal places. Our experience suggests that a reasonable lower limit for the quality of line-fit is about -0.95.

```
LEAST SQUARES LINE FIT TO REFRACTION DATA

GEOLOGIC MODEL OF FIGURE 1 - V0 FIRST ARRIVALS

VELOCITY OF LINE FIT TO DATA IS  1493
ZERO INTERCEPT OF LINE IS  0
QUALITY OF LINE FIT IS -1
NUMBER OF DISTANCE-TIME PAIRS USED IS  3

DISTANCE-TIME PAIRS ARE:

 50 @ .033   100 @ .067   150 @ .1

DISTANCES IN FEET
```

```
LEAST SQUARES LINE FIT TO REFRACTION DATA

GEOLOGIC MODEL OF FIGURE 1 - V1 FIRST ARRIVALS

VELOCITY OF LINE FIT TO DATA IS  2941
ZERO INTERCEPT OF LINE IS  .056
QUALITY OF LINE FIT IS -1
NUMBER OF DISTANCE-TIME PAIRS USED IS  3

DISTANCE-TIME PAIRS ARE:

 200 @ .124   250 @ .141   300 @ .158

DISTANCES IN FEET
```

A

```
REFRACTION DEPTH SOLUTION BY ZBYT0/BAS

TWO-LAYER CASE: GEOLOGIC MODEL OF FIGURE 1 - V0 THICKNESS
T0= .056          V0= 1493          V1= 2941
THICKNESS OF FIRST LAYER (Z0)= 48.5 FT          VELOCITIES IN FT/SEC
```

B

Figure 2. A, Least-squares line-fit to refraction data; B, refraction depth solution by ZBYTO.BAS.

EXAMPLE II - A FIELD RECORD

Figure 3A shows a typical shallow-refraction record. The six receivers were at distances of 50, 100, 150, 200, 250, and 300 ft from the source (a dropped weight). The same record is shown by Figure 3B after being picked (first troughs marked) and the picks converted to times and corrected to first arrival times; null lines have been drawn and sets of arrivals representing different velocities recognized. The three nearest (top) traces seem to indicate one velocity and the three farthest (bottom) traces represent another, faster velocity. This information is next prepared for the microcomputer and LNFT.BAS as:

Profile X1W - V0 data

Program change: 500 DATA 50,.045,100,.075,150,.103

Responses:
Indentification: Profile X1W Las Milpas Area
Valencia Co NM - V0

Number of pairs: 3

Profile X1W - V1 data

Program change: 500 DATA 200,.129,250,.151,300,.175

Responses:
Indentification: Profile X1W Las Milpas Area
Valencia Co NM - V1

Number of pairs: 3

After the computer is put into BASIC and LNFT.BAS loaded, the first "500 DATA" statement is entered and the command RUN given. The computer will ask (via the monitor) for case identification and number of distance-time pairs. After these are entered, the results will be output on the monitor and the printer. This process is repeated for the second set of distance-time pairs. Note that in the "500 DATA" statements the distance is given first and the time is second for each pair.

The printer output from LNFT.BAS for this example (Fig. 3) is reproduced by Figure 4A. Note that the zero intercept for the line fitted to the first three distance-time pairs (V0 line) is not zero, but 0.016 sec. This indicates that this layer is not at the surface, but, in fact, is a second layer beneath a thin surface layer. In the area of this example the indicated thin surface layer has been determined (by using short receiver intervals and buried geophones) to have a mean velocity of about 900 ft/sec.

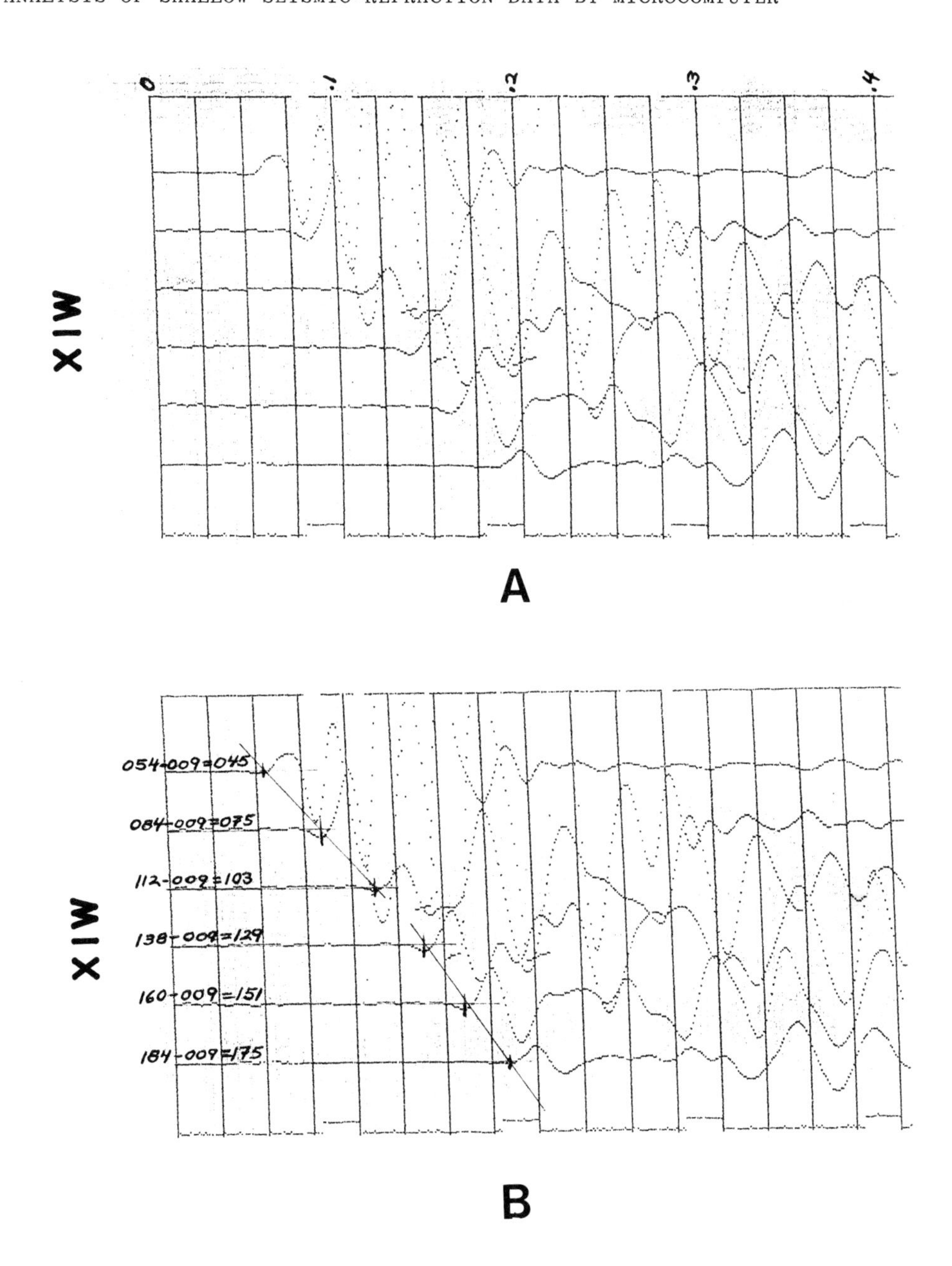

Figure 3. A, Typical shallow-refraction record; B, first troughs picked, marked, and corrected.

```
LEAST SQUARES LINE FIT TO REFRACTION DATA

PROFILE X1W LAS MILPAS AREA VALENCIA CO NM - V0

VELOCITY OF LINE FIT TO DATA IS  1724
ZERO INTERCEPT OF LINE IS  .016
QUALITY OF LINE FIT IS -1
NUMBER OF DISTANCE-TIME PAIRS USED IS  3

DISTANCE-TIME PAIRS ARE:

 50 @ .045  100 @ .075  150 @ .103

DISTANCES IN FEET
```

```
LEAST SQUARES LINE FIT TO REFRACTION DATA

PROFILE X1W LAS MILPAS AREA VALENCIA CO NM - V1

VELOCITY OF LINE FIT TO DATA IS  2174
ZERO INTERCEPT OF LINE IS  .037
QUALITY OF LINE FIT IS -.999
NUMBER OF DISTANCE-TIME PAIRS USED IS  3

DISTANCE-TIME PAIRS ARE:

 200 @ .129  250 @ .151  300 @ .175

DISTANCES IN FEET
```

A

```
REFRACTION DEPTH SOLUTION BY ZBYTO/BAS

TWO-LAYER CASE: PROFILE X1W LAS MILPAS AREA - V0' THICKNESS
T0= .016          V0= 900          V1= 1724
THICKNESS OF FIRST LAYER (Z0)= 8.4 FT          VELOCITIES IN FT/SEC
```

```
REFRACTION THICKNESS SOLUTION BY ZBYTO/BAS

THREE-LAYER CASE:                  PROFILE X1W LAS MILPAS AREA - V0 THICKNESS
T0= .037          Z0= 8.4          V0= 900          V1= 1724          V2= 2174
THICKNESS OF SECOND LAYER (Z1)= 28.3 FT             VELOCITIES IN FT/SEC
```

B

Figure 4. A, Least-squares line-fit to refraction data; B, refraction depth solution by ZBYTO.BAS.

ZBYTO.BAS computes refractor depth (or more accurately, the thickness of the layer above a refractor) by the zero-distance time-intercept method. This method is explained in standard textbooks on the subject of exploration geophysics (such as Dobrin, 1976). With the computer in BASIC, ZBYTO.BAS is loaded and the command RUN given. The computer asks:

NUMBER OF LAYERS (2 TO 5)?

In this situation where the thickness of the surface layer is to be determined, the input is 2. The computer will next ask:

CASE IDENTIFICATION?

Here we enter "Profile X1W Las Milpas Area - V0 Thickness." Next the computer asks:

ZERO INTERCEPT TIME?

We enter 0.016, after which the computer asks:

VELOCITY OF FIRST LAYER?

The entry here is 900. The computer next asks:

VELOCITY OF SECOND LAYER?

Here we enter 1724. The computer now will output to the monitor the calculated thickness of the first layer as

Z0=8.4.

A more complete result will be written on the printer as shown by the upper part of Figure 4B.

To solve for the thickness of the second layer (Z1), the first question (HOW MANY LAYERS?) is answered by entering 3. The computer will ask for the zero intercept time (0.037), the thickness of the first layer (8.4), and the velocities of the first, second, and third layers (900, 1724, and 2174). the monitor then will display the thickness of the second layer as:

Z1=28.3.

A more complete hardcopy result will be output to the printer (lower part of Figure 4B).

The velocity model calculated for the example situation of Figure 3 is shown graphically in Figure 5.

REFERENCE

Dobrin, M.B., 1976, Introduction to geophysical prospecting: McGraw-Hill Book Co., New York, 630 p.

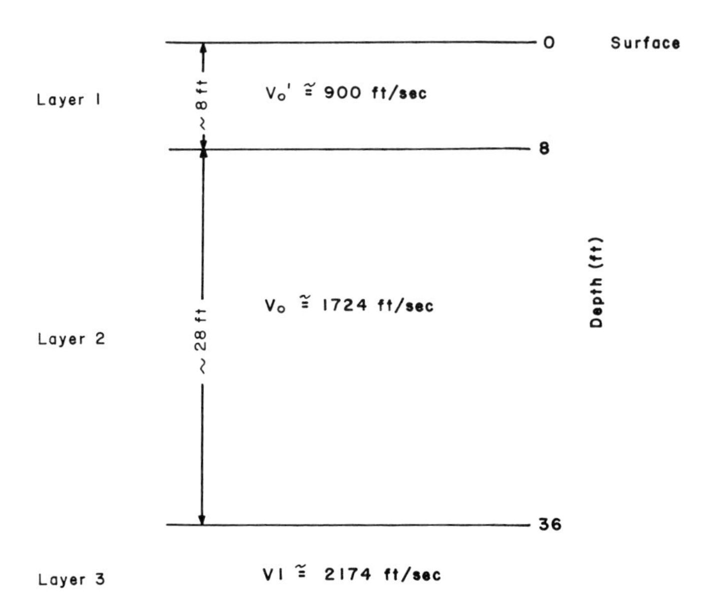

Figure 5. Geologic (velocity) model resulting from analysis of shallow seismic-refraction profile X1W.

APPENDIX

Program Listings

```
PROGRAM BY CHARLES B. REYNOLDS AND ASSOCIATES, P.O.BOX 1004, BELEN, NM, 87002
10 REM PROGRAM LNFT.BAS TO FIT A STRAIGHT LINE BY LEAST
20 REM SQUARES TO A SET OF DISTANCE AND TIME PAIRS FROM
30 REM REFRACTION DATA
100 DIM X(100),T(100)
110 CLS:PRINT"PROGRAM LNFT.BAS":PRINT
120 PRINT"DATA LINES BEGINNING WITH 500 SHOULD GIVE DISTANCE"
130 PRINT"AND TIME PAIRS WITH DISTANCE FIRST IN EACH PAIR"
140 PRINT:INPUT"CASE IDENTIFICATION ";IDENT$
150 INPUT"NUMBER OF DISTANCE-TIME PAIRS ";A%
160 FOR N%=1 TO A%:READ X(N%),T(N%):NEXT
170 FOR N%=1 TO A%
180    XADD=XADD+X(N%):TADD=TADD+T(N%):XTADD=XTADD+X(N%)*T(N%)
190    XSQADD=XSQADD+X(N%)^2:TSQADD=TSQADD+T(N%)^2
200 NEXT
210 M=(XTADD-(XADD*TADD)/A%)/(XSQADD-(XADD^2)/A%)
220 TZERO=(FIX(1000*(TADD-M*XADD)/A%+.5))/1000:VFT=FIX(1/M+.5)
230 RSQ=FIX(1000*M*(XADD*TADD/A%-XTADD)/(TSQADD-(TADD^2)/A%)-.5)/1000
240 PRINT:PRINT IDENT$
250 PRINT"VELOCITY IS ";VFT
260 PRINT"ZERO INTERCEPT IS ";TZERO
270 PRINT"FIT QUALITY IS ";RSQ
280 LPRINT"LEAST SQUARES LINE FIT TO REFRACTION DATA"
290 LPRINT CHR$(138):LPRINT IDENT$
300 LPRINT CHR$(138):LPRINT"VELOCITY OF LINE FIT TO DATA IS ";VFT
310 LPRINT"ZERO INTERCEPT OF LINE IS ";TZERO
320 LPRINT"QUALITY OF LINE FIT IS ";RSQ
330 LPRINT"NUMBER OF DISTANCE-TIME PAIRS USED IS ";A%
340 LPRINT CHR$(138):LPRINT"DISTANCE-TIME PAIRS ARE: "
350 FOR N%=1 TO A%:LPRINT X(N%);" @ "T(N%);
360 NEXT
370 LPRINT CHR$(138):LPRINT CHR$(138):LPRINT"DISTANCES IN FEET"
380 FOR N%=1 TO 5:LPRINT CHR$(138):NEXT
500 DATA 300,.097,350,.106,400,.108,450,.117,500,.120,550,.130,600,.137
999 END
```

```
PROGRAM BY CHARLES B. REYNOLDS AND ASSOCIATES, P.O.BOX 1004, BELEN, NM, 87002
10 REM PROGRAM ZBYTO.BAS TO CALCULATE THICKNESSES OF
20 REM LAYERS (2 TO 5) FROM REFRACTION DATA USING
30 REM ZERO INTERCEPT TIMES
100 CLS:PRINT"PROGRAM ZBYTO.BAS TO CALCULATE REFRACTOR DEPTHS"
110 PRINT:INPUT"NUMBER OF LAYERS (2 TO 5) ";A%
120 IF A%<2 OR A%>5 GOTO 9998
130 ON A% GOTO 9998,200,300,400,500,9998
200 CLS:PRINT"TWO-LAYER REFRACTION CASE":PRINT
205 INPUT"CASE IDENTIFICATION ";IDENT$
210 INPUT"ZERO INTERCEPT TIME ";TO
220 INPUT"VELOCITY OF FIRST LAYER ";VO
230 INPUT"VELOCITY OF SECOND LAYER ";V1
240 ZO=(TO/2)*(VO*V1/SQR(V1^2-VO^2)):ZO=FIX(10*(ZO+.05))/10
250 PRINT:PRINT"ZO= ";ZO;" FT"
260 LPRINT"REFRACTION DEPTH SOLUTION BY ZBYTO.BAS":LPRINT CHR$(138)
270 LPRINT"TWO-LAYER CASE:", IDENT$
280 LPRINT"TO= ";TO,"VO= ";VO,"V1= ";V1:LPRINT"THICKNESS OF FIRST LAYER (ZO)= ";
ZO;" FT","VELOCITIES IN FT/SEC"
290 FOR N%=1 TO 5:LPRINT CHR$(138):NEXT:GOTO 9999
300 REM THREE-LAYER CASE SOLUTION
305 CLS:PRINT"THREE-LAYER REFRACTION CASE":PRINT
307 INPUT"CASE IDENTIFICATION ";IDENT$
310 INPUT"ZERO INTERCEPT TIME ";TO
315 INPUT"FIRST (SHALLOWEST) LAYER THICKNESS ";ZO
320 INPUT"VELOCITY OF FIRST (SHALLOWEST) LAYER ";VO
325 INPUT"VELOCITY OF SECOND LAYER ";V1
330 INPUT"VELOCITY OF THIRD (DEEPEST) LAYER ";V2
335 Z1=V2*V1/(2*SQR(V2^2-V1^2))*(TO-2*ZO*SQR(V2^2-VO^2)/(VO*V2))
340 Z1=FIX(10*(Z1+.05))/10:PRINT:PRINT"Z1= ";Z1;" FT"
345 LPRINT"REFRACTION THICKNESS SOLUTION BY ZBYTO.BAS"
350 LPRINT CHR$(138):LPRINT"THREE-LAYER CASE:",IDENT$
360 LPRINT"TO= ";TO,"ZO= ";ZO,"VO= ";VO,"V1= ";V1,"V2= ";V2
365 LPRINT"THICKNESS OF SECOND LAYER (Z1)= ";Z1;" FT","VELOCITIES IN FT/SEC"
370 LPRINT"VELOCITIES IN FT/SEC"
375 FOR N%=1 TO 5:LPRINT CHR$(138):NEXT
395 GOTO 9999
400 REM FOUR LAYER CASE SOLUTION
405 CLS:PRINT"FOUR LAYER REFRACTION CASE":PRINT
410 INPUT"CASE IDENTIFICATION ";IDENT$
415 INPUT"ZERO TIME INTERCEPT OF DEEPEST LAYER ";TO
420 INPUT"THICKNESS OF FIRST (SHALLOWEST) LAYER ";ZO
425 INPUT"THICKNESS OF SECOND LAYER ";Z1
430 INPUT"VELOCITY OF FIRST (SHALLOWEST) LAYER ";VO
435 INPUT"VELOCITY OF SECOND LAYER ";V1
440 INPUT"VELOCITY OF THIRD LAYER ";V2
445 INPUT"VELOCITY OF FOURTH (DEEPEST) LAYER";V3
450 B=V3*V2/(2*SQR(V3^2-V2^2)):C=2*ZO*SQR(V3^2-VO^2)/(VO*V3)
460 D=2*Z1*SQR(V3^2-V1^2)/(V1*V3):Z2=B*(TO-C-D):Z2=FIX(10*(Z2+.05))/10:
465 PRINT:PRINT"Z2= ";Z2;" FT":LPRINT"REFRACTION THICKNESS SOLUTION BY ZBYTO.BAS
"
```

```
470 LPRINT CHR$(138):LPRINT"FOUR-LAYER CASE:",IDENT$
475 LPRINT"TO= ";TO,"ZO= ";ZO,"Z1= ";Z1:LPRINT"VO= ";VO,"V1= ";V1,"V2= ";V2,"V3=
 ";V3
480 LPRINT"THICKNESS OF THIRD LAYER (Z2)= ";Z2;" FT","VELOCITIES IN FT/SEC"
485 FOR N%=1 TO 5:LPRINT CHR$(138):NEXT
495 GOTO 9999
500 REM FIVE LAYER CASE SOLUTION
505 CLS:PRINT"FIVE-LAYER REFRACTION CASE":PRINT
510 INPUT"CASE IDENTIFICATION ";IDENT$
515 INPUT"ZERO INTERCEPT TIME FOR DEEPEST LAYER ";TO
520 INPUT"THICKNESS OF FIRST (SHALLOWEST) LAYER ";ZO
525 INPUT"THICKNESS OF SECOND LAYER ";Z1
530 INPUT"THICKNESS OF THIRD (DEEPEST) LAYER ";Z2
535 INPUT"VELOCITY OF FIRST LAYER ";VO
540 INPUT"VELOCITY OF SECOND LAYER ";V1
545 INPUT"VELOCITY OF THIRD LAYER ";V2
550 INPUT"VELOCITY OF FOURTH LAYER ";V3
555 INPUT"VELOCITY OF FIFTH LAYER ";V4
560 B=V4*V3/(2*SQR(V4^2-V3^2)):C=2*ZO*SQR(V4^2-VO^2)/(V4*VO)
565 D=2*Z1*SQR(V4^2-V1^2)/(V1*V4):E=2*Z2*SQR(V4^2-V2^2)/(V4*V2)
570 Z3=B*(TO-C-D-E):Z3=FIX(10*(Z3+.005))/10:PRINT:PRINT"Z3= ";Z3;" FT"
575 LPRINT"REFRACTION THICKNESS SOLUTION BY ZBYTO.BAS":LPRINT CHR$(138)
580 LPRINT"FIVE-LAYER CASE:",IDENT$
585 LPRINT"TO= ";TO,"ZO= ";ZO,"Z1= ";Z1,"Z2= ";Z2
590 LPRINT"VO= ";VO,"V1= ";V1,"V2= ";V2,"V3= ";V3,"V4= ";V4
595 LPRINT"THICKNESS OF FOURTH LAYER (Z3)= ";Z3;" FT","VELOCITIES IN FT/SEC"
597 FOR N%=1 TO 5:LPRINT CHR$(138):NEXT
599 GOTO 9999
9998 PRINT:PRINT"UNACCEPTABLE NUMBER OF LAYERS; TRY AGAIN"
9999 END
```

Interactive 2D and 3D Seismic Interpretation on a Microcomputer

H. Roice Nelson, Jr. and H. A. Hildebrand

Landmark Graphics Corporation

ABSTRACT

The development of sophisticated turnkey solutions for two substantial geoscience problems are described. Because of recent developments in microcomputer CPU power, digital storage, graphics processors, and other key hardware peripherals, it now is possible to build a system for stand-alone interactive interpretation of two-dimensional (2D) and three-dimensional (3D) seismic surveys. The hardware and software tools used to build a solution to these problems are described. The bulk of the paper is a set of illustrations showing the steps in doing an interactive 2D and an interactive 3D interpretation. The data used are from offshore Texas (2D) and from the North Sea (3D).

INTRODUCTION

Turnkey microcomputer applications are available that can help solve oil and gas exploration problems. During the last decade, microcomputer hardware has evolved to rival the power of mini and even mainframe computers of years past. In addition, users are becoming more computer literate and applications are being developed that solve problems in the user's language. The results are cost effective, powerful tools for simple, user-friendly solutions using complex data structures.

Interactive 2D (two-dimensional) and 3D (three-dimensional) seismic interpretation are examples of these new tools. A system for computer-aided interpretation developed by Landmark Graphics Corporation will be the basis for this chapter. First is a brief description of the hardware capabilities. Following this is a brief description of a software philosophy designed to solve user problems. The body of the paper consists of 44 illustrations that illustrate a few key steps in 2D and 3D interpretation projects.

HARDWARE

The interpretation station used for this work is designed to look similar to a desk for use in an office or a computer-room environment. A large desktop workspace provides room for paper seismic sections, maps, well logs, etc. One and two monitor configurations are available, although the two-monitor system is the most popular because the base map and other data are available concurrently. There is a small tower on the left for the computer and differing numbers of larger expansion cabinets on the right for disk storage and tape input.

The CPU (central-processing unit) is the Intel 80286 microprocessor. This is the same chip used in the new IBM AT microcomputer. The Intel 80287 floating-point accelerator is standard, along with 1M bytes of main memory. As operating system improvements develop, the full 16M bytes of main memory supported by the 80286 will be available. In addition, an optional 1 Megaflop array processor can be installed. The CPU boards are installed in a 21-slot multibus. The system uses the MS-DOS operating system, and the applications described were developed using standard ANSII-77 FORTRAN. Other languages, such as C and Pascal are also available.

Key peripherals include the graphics processor, the console subsystem, the tape subsystem, and the disk subsystem. The graphics processor supports one or two high-resolution monitors (1280 X 1024 pixels). These monitors can be 30 Hertz interlaced or 60 Hertz noninterlaced. The console subsystem consists of an alphanumeric console for text listings and system work, and a 5M byte removable Winchester disk for downloading program updates, horizon files, and other data transfers. Modems also are available for remote transfer of small files. Large files are shared between interpretation stations and hosts using the Ethernet local area network (LAN) at rates up to 1M bit per second. This allows sharing of disk and tape resources for sites with multiple system installations.

The tape and disk system can be part of a stand-alone station, or located at one node of the LAN as shared resources. Tape subsystems can be 1600 only or 6250/1600 bits per inch (bpi), and handle standard 9-track SEG-Y seismic tapes, UKOOA navigation tapes, etc. Typical disk configurations are 440M or 880M bytes. However, up to 16 separate 440M bytes disks, or 7.0G byte (gigabytes or 10**9 bytes) can be put on a single interpretation station. Data files can span multiple logical and physical drives and be up to 4.0G bytes each. This is comparable to the capabilities of computers in the mega-mini class.

As shown by the hardware options described, a microcomputer system is not limited necessarily in computer power. In fact, the expansion and networking options available allow development of extremely powerful systems. This is true especially if the capabilities of the microcomputer are packaged with appropriate peripherals to provide resources needed for specific applications, such as seismic interpretation (Nelson, Hildebrand, and Mouton, 1983).

SOFTWARE PHILOSOPHY

With the availability of these relatively inexpensive, yet powerful hardware systems, it becomes feasible to build applications and place microcomputer-driven solutions in an explorationist's office. However, because many explorationists are not computer literate, it is important to develop easy-to-use software for specific applications. Systems need to provide user-based tools to solve tasks at hand, in a manner similar to the way the job would be accomplished manually. In addition, this must be done cost-effectively.

There are two basic approaches being used to provide these solutions in the geosciences industry. The most widespread approach is for limited resources to be used to develop a quick solution for specific problems; such as creating well synthetics, doing 2D ray-trace modeling, etc. Although these systems aid in a particular niche, they solve mostly nonsubstantial problems, and usually are not tied easily into an integrated exploration system. Although there are continuing improvements in hardware capabilities, this approach has a limited potential for providing general exploration solutions.

A better approach is to define a substantial geoscience application and attack it by developing major software tools that can be used to build solutions to other geoscience applications.

The tools underlying the 2D and 3D interpretation examples presented here are based on generic and independent software subsystems that are used as building blocks for major applications. Examples include software subsystems for graphics, menu managers, the array processor, viewport managers, seismic data access, large file I/O (input/output), seismic display, horizon data access, message control, and tape handling.

Once the tools were developed for interactive 3D interpretation, the basic interactive 2D interpretation package used as an example here was developed in merely three months. Of course, this basic package will be enhanced, improved, and expanded upon for years. The development of many other geoscience applications holds a similar relationship once the basic tools are developed. Handling well logs and synthetic traces can be compared to handling wiggle seismic traces. Signal-processing algorithms to post-stack-process 3D seismic traces can be used to process 2D sections, well logs, or numerically derived model synthetics. Velocity information can be treated similar to 2D or 3D seismic data for lithology interpretation. Reservoir models are based on 3D boundaries similar to seismic horizon files. Of course, the same tools can be used by companies for development of proprietary supplements to existing application packages.

One other important aspect of the microcomputer revolution in the geosciences is that microcomputer systems have an economy-of-scale not possible with host computers (Nelson, 1984b). End-user access to microcomputers is available more widely today than access to the host computers was in the past. Instead of users inputting batch jobs to a few large computers, users now can try different ideas on many computers. Although these applications typically have been less sophisticated, more applications are being developed, and those that are worthwhile can be integrated as third-party software packages on an integrated system. This implies development of a system with the largest possible number of solutions available in the shortest possible time frame.

2D SEISMIC INTERPRETATION

Two-dimensional seismic surveys are a type of data widely used for petroleum exploration. The basic use of this data is to extrapolate geologic information from 1D (one-dimensional) well information along a 2D seismic line or series of lines to build up a 3D representation of the subsurface.

Traditional interpretation techniques involve a tremendous amount of clerical overhead time; time that is spent in data retrieval setting up processing jobs, waiting for reprocessing, posting horizons from sections to maps, etc. Dedicated systems with turnkey applications can not only cut down on this overhead time, but can provide previously unavailable application tools for detailed prospect analysis (Nelson, 1984a). These new tools include color softcopy seismic display, computer-manager section retrieval, post-stack seismic processing, scaling and horizon flattening, automatic horizon picking, and mapping.

One of the best advantages of interactive interpretation techniques is the ability to try different options. For example, it might be advantageous to close a loop of data going around to the right and then look at a connecting loop going around to the left. The ability to determine which color combination brings out the geology best for an individual interpreter can make a big difference in unraveling subtle geology. Testing the validity of a horizon or a fault pick by placing it in different locations or at various dip angles, and checking the different options on crossing or parallel lines can make a tremendous improvement in interpretation accuracy. Option testing is particularly useful in seismic stratigraphic interpretation (Simson and Nelson, 1985). This includes the ability to identify seismic facies, separate seismic sequences, and look at attributes of the seismic data such as instantaneous phase, frequency, and reflection strength.

The survey used to illustrate interactive 2D interpretation is a Grant/NORPAC spec survey from offshore Texas. It is not possible to describe completely a sophisticated application in a brief article, therefore the bulk of the allowable space is used to show and describe snapshots taken from doing an interactive 2D interpretation. Figures 1 to 26 illustrate interactive 2D interpretation using the Grant/NORPAC 2D survey.

3D SEISMIC INTERPRETATION

Three-dimensional seismic surveys expand a 2D seismic line over an area to provide a volume of seismic data representing the subsurface. Interactive techniques are a must in the interpretation of 3D seismic surveys (Cole and Nelson, 1985). This is due to the amount of information associated with a 3D survey. These surveys typically will have from 150 to 500 in-line sections, from 250 to 600 cross-line sections and about 1000 time-slice sections. In addition, there are numerous arbitrary lines or user tracks. Attempting to keep track of this many paper sections is a nightmare. This is why all of the sophisticated interactive interpretation systems started by building solutions for interactive 3D interpretation.

The fact that processed 3D surveys form a regular three-dimensional lattice of points can be used by a computer with sufficient resources to keep track relatively easily of all the traces, samples, and horizons. Thus the computer can sort traces of the data volume that are in-line, cross-line, or along some arbitrary user-specified direction, such as when connecting wells or defining the dip direction within a fault block. The computer also can create and keep track of time-slice or paleosections (sections showing amplitude distributions across or parallel to a 3D horizon), and combination time-slice and time-series sections.

However, it is not sufficient always for the computer to be able to retrieve data from disk within a fraction of a minute. One of the important characteristics of 3D surveys is that the data are well enough sampled that individual horizons can be followed on movies of parallel sections. This has been demonstrated widely through most of the last decade with 16mm time-slice movies. More recently authors have written about the advantages of using digital movies as exploration tools, particularly with 3D seismic survey (Ottolini, Sword, and Claerbout, 1984). Not only can more data be looked at more quickly when it only takes a fraction of a second between displays, but the interpreters can keep information in their mind from previous sections, note differences, and draw geologic conclusions that are not possible looking at a single section. This system is unique in allowing movies to be displayed from large Winchester disk files. An example of geologic significance of animation is to make a movie of sections pivoting around a well location. This allows evaluation of critical dip, the relationship of faulting and the well, etc.

Another important capability is to be able to make a composite of several pieces of data on a single display for simultaneous visual evaluation. It is easy to fold paper sections and get several related items altogether for evaluation (even if it is difficult to locate the paper and put it back in place so it can be located next time). Interactive systems need to provide the same capabilities with comparable ease, in order to better solve the same problems. As shown in the 3D interpretation example, this is being accomplished.

One interesting point is the emphasis users have on displaying wiggle-trace displays, until they start working an interactive interpretation project. It has been shown that in many instances it is easier to recognize reflections on variable-density displays than it is on wiggle-variable-area sections (Nelson, 1984). It also is a fact that interpreters tend to first start on an interactive system with wiggle or black and white sections, but shortly migrate to using colored variable-density displays.

The regular 3D lattice provides a nice framework for softcopy horizon picking and map display. Given picks on one section, these can be carried to a parallel section and converged to the appropriate peak, trough, or zero-crossing using the automatic picking mode. This makes picking horizons with reasonable continuity fast and accurate. Picks are posted automatically to a horizon file, and at any point of the interpretation the horizon can be evaluated in map views. Because 3D data are gridded already, the horizon forms a surface that can be displayed in several ways. In addition, horizon surfaces can be subtracted from isochrons, multiplied for depth conversion, have amplitudes extracted along or parallel to them, smoothed, and many other operations.

To illustrate and expand briefly on these concepts, displays from the interactive 3D interpretation of a small, North Sea 3D survey will be used. This comprises Figures 27 through 44.

SUMMARY

Turnkey applications are available on microcomputers that can help solve significant oil and gas exploration problems such as interactive 2D and 3D interpretation. Proper packaging of presently available microcomputer-controlled hardware can provide tools that can be taken to individual users cost-effectively. There will be a proliferation of this type of technology as geoscience computer users become more computer literate and applications mature and are developed in the user's language.

REFERENCES

Cole, R.A., and Nelson, H.R., Jr., 1985, Interactive computer graphics and the interpretation of three-dimensional seismic surveys: 54th Annual International SEG, Expanded Abstracts of the Technical Program, Atlanta, Georgia, in press.

Fegin, F.J., 1981, Seismic data display and reflection preceptibility: Geophysics, v. 46, no. 2, p. 106-120.

Nelson, H.R., Jr., Hildebrand, H.A., and Mouton, J.O., 1983, Color softcopy, animation and interactive user interface in interpretation station design: Presentation at 53rd Ann. Intern. SEG Convention, Expanded Abstracts of the Technical Program, Las Vegas, Nevada.

Nelson, H.R., Jr., 1984a, Micro-based graphics analyses overcoming seismic obstacles: Am. Oil and Gas Reporter, p. 42-45.

Nelson, H.R., Jr., 1984b, Seismic interpretation and microcomputer economies-of-scale: Presentation at the 46th Meeting of the European Assn. of Expl. Geoph., Technical Programme and Abstracts of Papers, London, England, p. 78-79.

Ottolini, R., Sword, C., and Claerbout, J.F., 1984, On-line movies of reflection seismic data with description of a movie machine: Geophysics, v. 49, no. 2, p. 195-202.

Simson, S.F., and Nelson, H.R., Jr., 1985, Seismic stratigraphy moves towards interactive analysis: World Oil, in press.

Figure 1. Survey map for spec 2D survey offshore Texas (data used in Fig. 1-26 is courtesy of Grant/NORPAC). Loops, parallel sections, zig-zag sections, or any other combination of sections within 1200-trace softcopy display limitation can be picked interactively from map for simultaneous display.

Figure 2. Variable density display of five sections closing loop. Traces in panel on far right are same as traces in middle of panel on left, only displayed in opposite direction. Linear black-to-white (peaks-to-troughs) color table was used for display.

Figure 3. Figure 2 was adjusted instantly to highlight strongest peaks and troughs. User moves digitizing puck to adjust color table, exponentially expanding zero-crossing color.

Figure 4. Figure 3 was adjusted instantly to give AGC (automatic gain control) appearance. Peaks and trough colors are moved towards zero-crossing color, increasing contrast.

Figure 5. Unlimited color selections are available to user. Here, sections are given higher frequency appearance by assigning two identical spectra, one from peak to zero crossing and one from zero crossing to trough. Note that because data were 2D migrated steeply dipping deeper layers do not tie.

Figure 6. Figure 5 has been zoomed using pixel replication. Pixel replication takes individual display points (picture elements or pixels) and expands them from 1 to 4 (2 pixels x 2 pixels) to 9 to 16, etc. Although 16 zoom steps are possible, seismic and map data become blocky with more than two or three zoom steps.

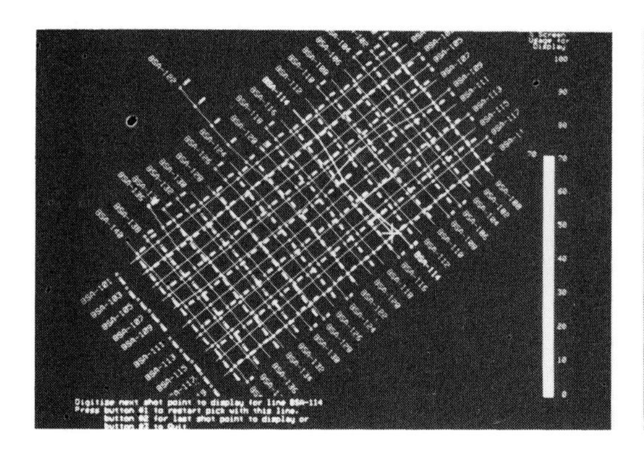

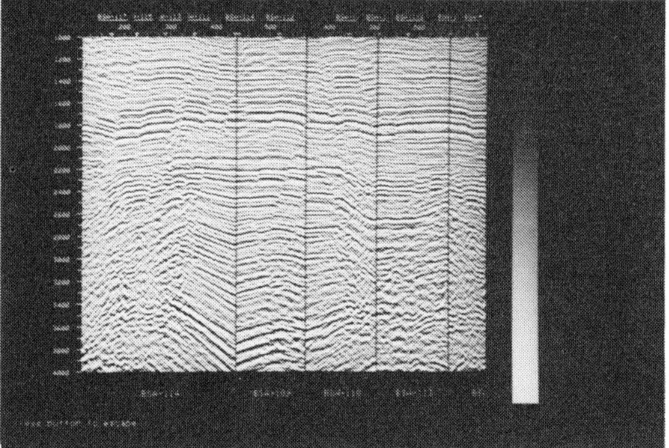

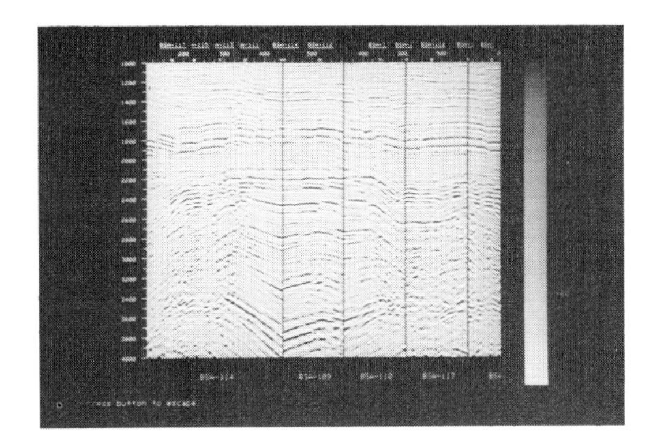

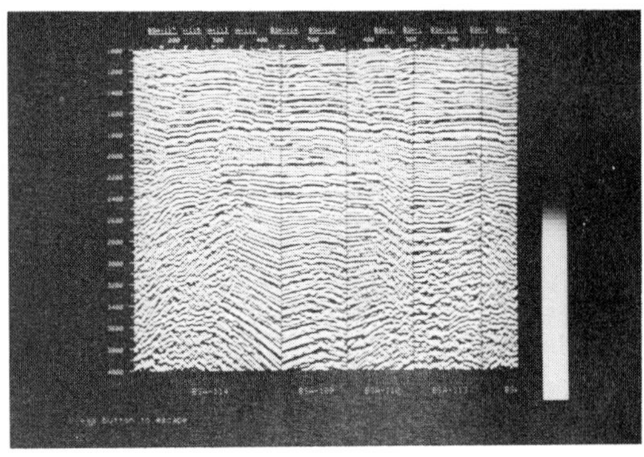

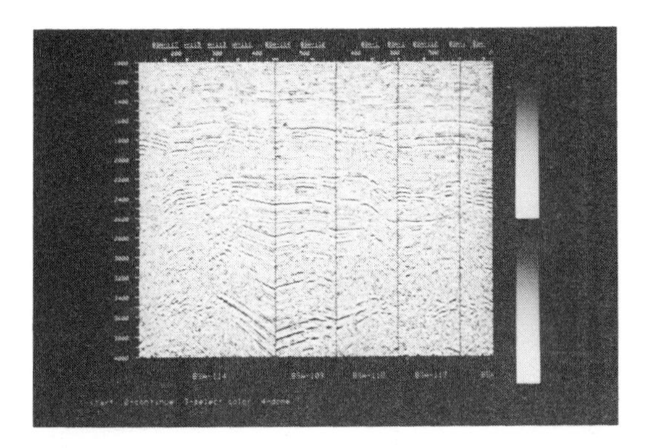

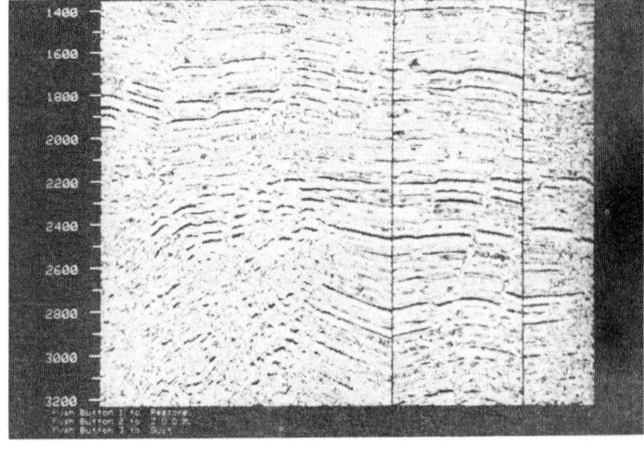
1400
1600
1800
2000
2200
2400
2600
2800
3000
3200

Figure 7. Figure 4 has been zoomed instantly using pixel replication. Sections also can be zoomed by interpolating linearly between traces or samples, or by replicating traces or samples.

Figure 8. This is one page of listing of horizons and faults included in this 2D interpretation project. Each horizon has unique color, as well as unique automatic picking parameters.

Figure 9. First of four-picture sequence showing pixel zooming on sections with some picks posted. Picks on time-series data are stored as single-valued horizons. These files can include faults or seismic horizons.

Figure 10. Second of four-picture sequence showing pixel zooming on sections with some picks posted. Horizons are picked in a point-to-point mode, or automatic mode, and can be edited by switching to delete mode, repicking in point mode, or converging to nearest peak, trough or zero crossing in automatic mode.

Figure 11. Third of four-picture sequence showing pixel zooming on sections with some picks posted. Faults are discernible, but horizons do not stand out on black and white prints.

Figure 12. Fourth of four-picture sequence showing pixel zooming and panning. Faults show blocky steps, which does not happen when zooming is done with linear interpolation.

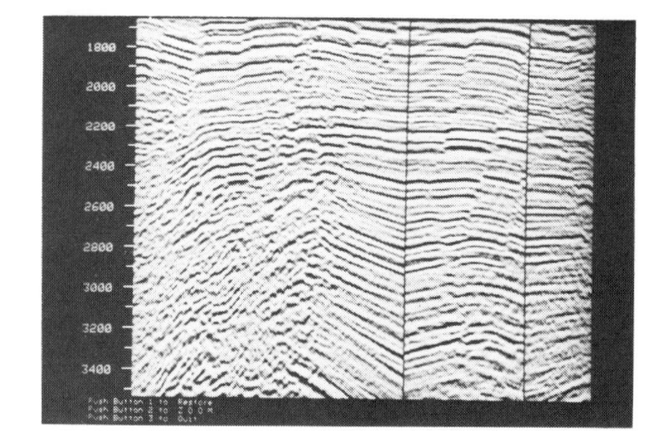

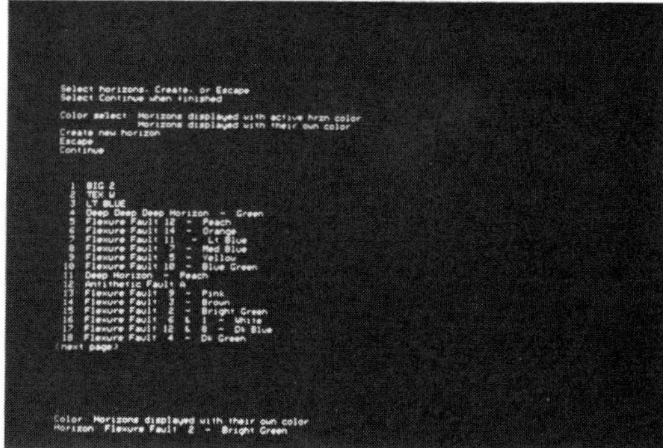

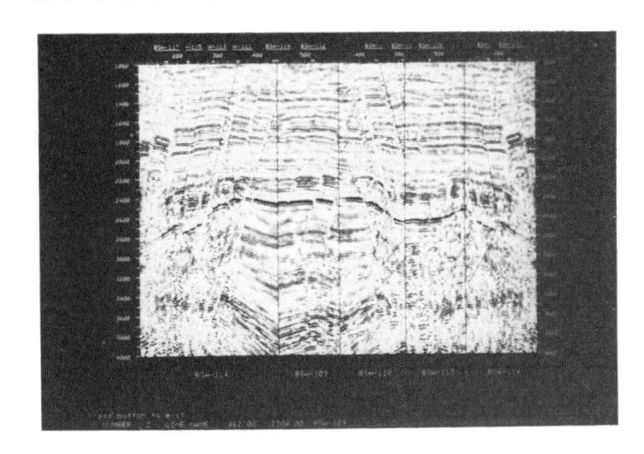

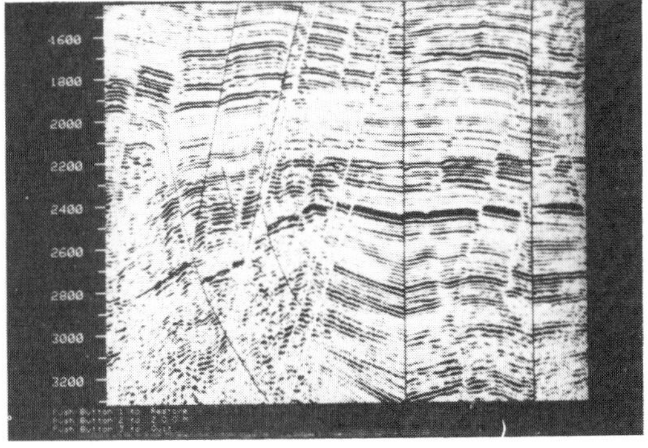
1600
1800
2000
2200
2400
2600
2800
3000
3200

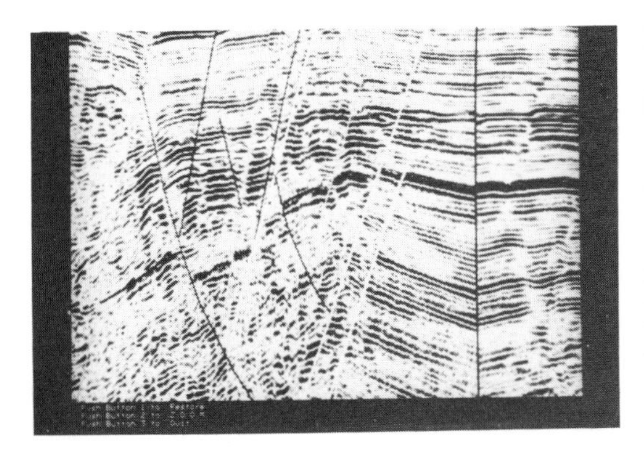

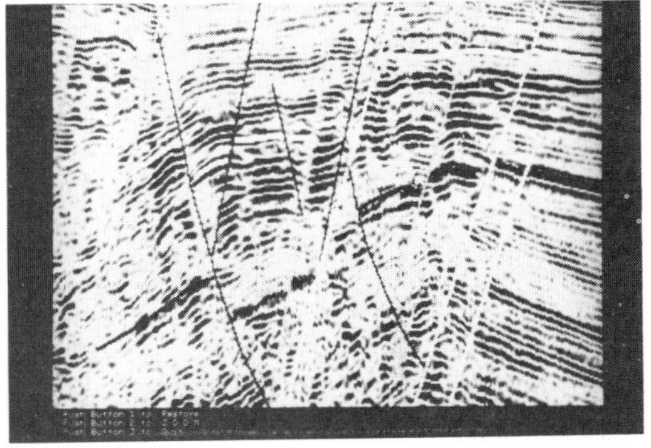

Figure 13. Dip-direction section displayed in wiggle-trace format. Display options include differing amount of overlap of contiguous traces or changing the amount of deflection (in pixels) of highest amplitude wiggles.

Figure 14. Pixel-replication zoom of upper-right corner of wiggle-trace display shown in Figure 13.

Figure 15. Variable-density display of same data illustrated in Figure 14. This display went from black to white to red.

Figure 16. Wiggle traces overlain on variable density, Figure 15. Wiggles are yellow, and do not show up well on black and white print.

Figure 17. Horizons overlain on wiggle-trace display shown in Figure 14.

Figure 18. Window of data at about center of display has been "dragged" to left for across-fault correlation of horizons. This is portion of Figure 13, down and to left of where Figure 14 was located.

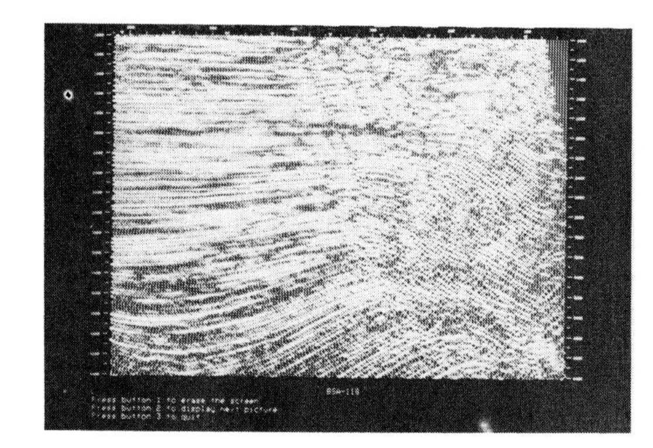

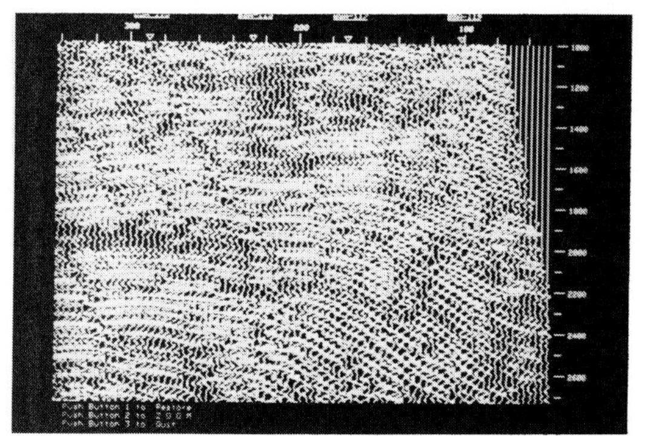

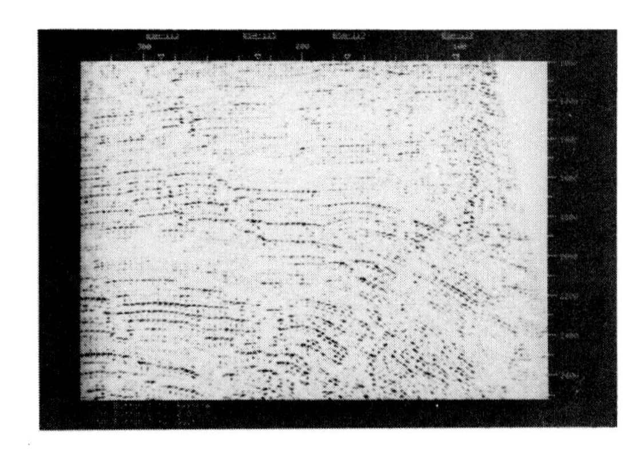

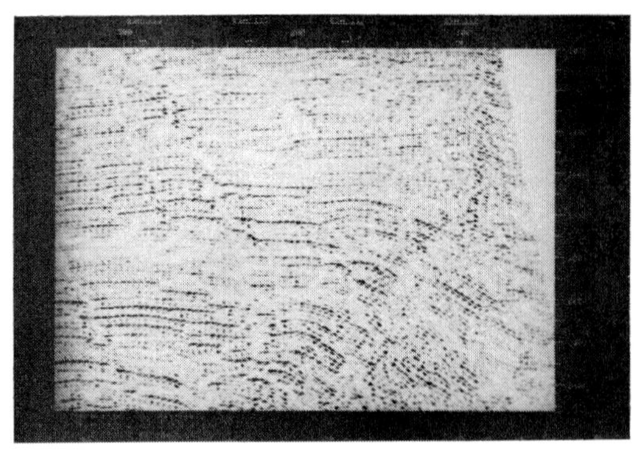

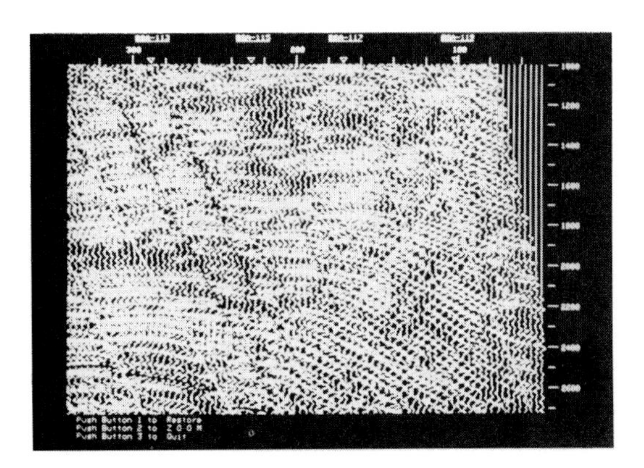
Push Button 1 to Restore
Push Button 2 to ZOOM
Push Button 3 to Quit

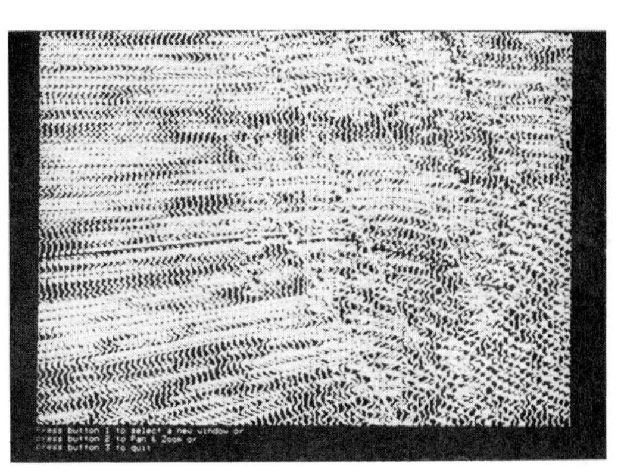
Press button 1 to select a new window or
Press button 2 to Pan & Zoom or
Press button 3 to quit

Figure 19. Instantaneous frequency section of same data shown in Figure 13. Colors used include spectrum from red for low frequency (4-12 Hertz), green for higher frequencies (28-36 Hertz), and blues for highest frequencies.

Figure 20. Figure 19 has been zoomed using pixel replication. There is complete post-stack processing package that can be used to do any of Hilbert transform attribute studies, AGC, filtering, deconvolution, wavelet processing, etc.

Figure 21. Flattening on horizon that is white on Figure 10 (horizon is set at 2600 ms in far left fault block, and between 2200 ms, 400 ms on most of undatumed section). Flattening allows paleogeologic evaluations, in similar manner to hanging well logs on geologic marker.

Figure 22. Flattening on horizon that is black on Figure 10 (horizon is set at 2800 ms in far left fault block, and between 2400 and 2600 ms on most of undatumed section). Appending section displays flattened on different horizons allows interpreter to step through a depositional history.

Figure 23. Flat contour map of upper horizon, same horizon used to flatten on in Figure 21. These maps can be annotated and displayed on CRT (cathode-ray tube) or output to pen plotter at any desired scale.

Figure 24. Flat contour map of lower horizon, same horizon used to flatten on in Figure 22. This was generated with a software package named SURFAS before implementation of fault handling in that package on micro.

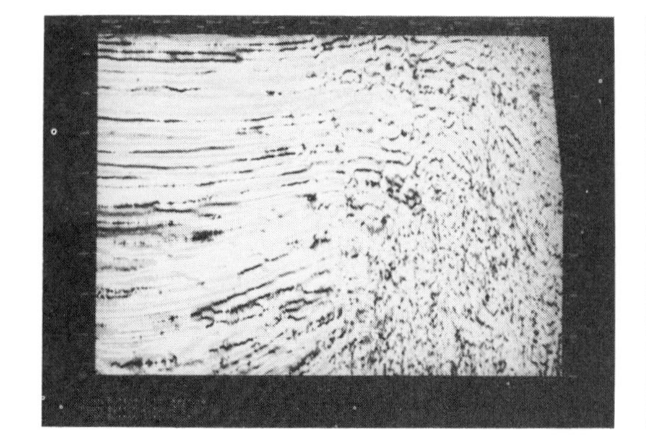

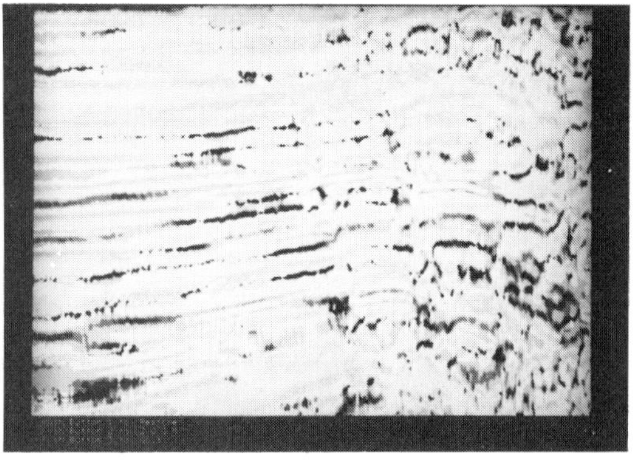

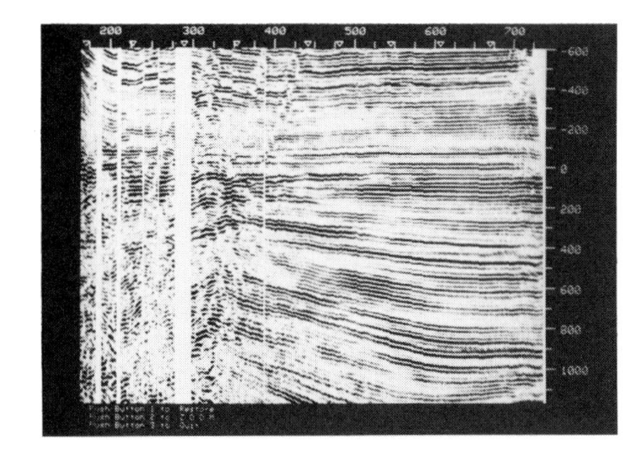
200
300
400
500
600
700
-600
-400
-200
0
200
400
600
800
1000

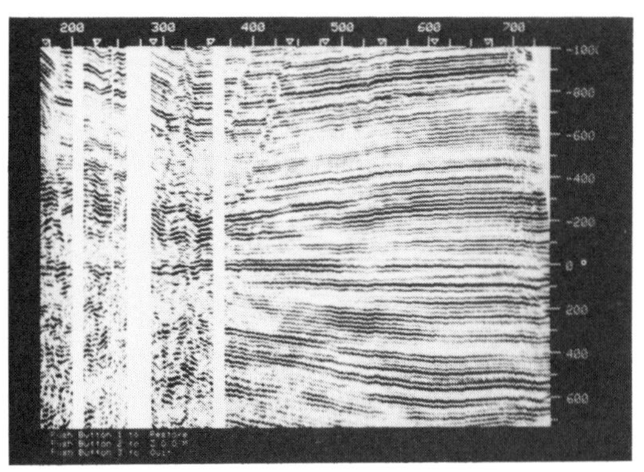
200
300
400
500
600
700
-800
-600
-400
-200
0
200
400
600

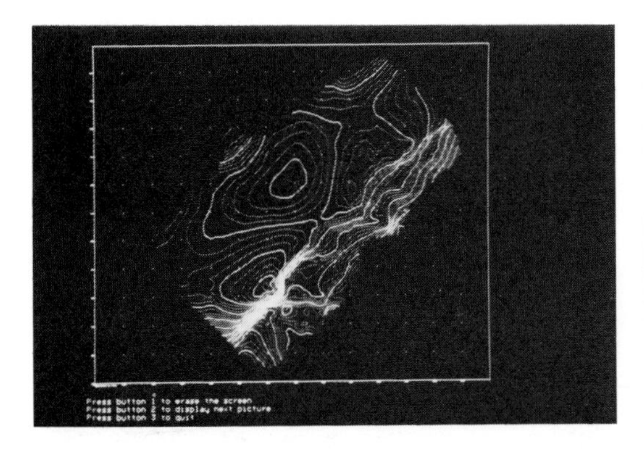
Press button 1 to erase the screen
Press button 2 to display next picture
Press button 3 to quit

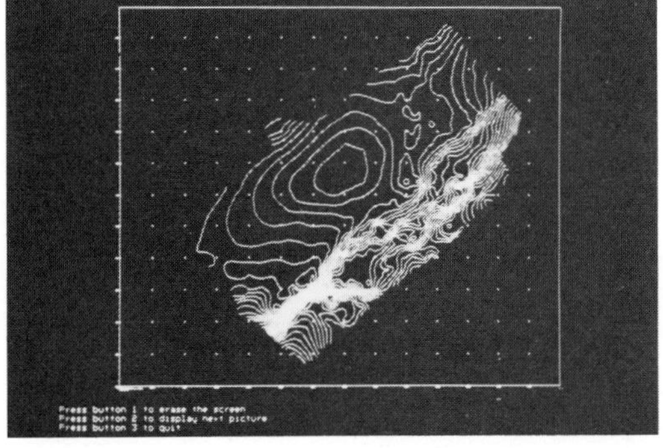
Press button 1 to erase the screen
Press button 2 to display next picture
Press button 3 to quit

Figure 25. Flat contour map of isochron between horizons displayed in map Figures 23 and 24. Extracted amplitudes, residuals or other interactively derived data also can be contoured.

Figure 26. Isometric display of upper horizon (Fig. 23), lower horizon (Fig. 24) and isochron (Fig. 25) on surface of cube; for management or partners presentations.

Figure 27. Pixel-zoomed display of instantaneous phase section from North Sea 3D survey. User can change instantly this display using linear color adjust option to give illusion of phase-shifting data. Figures 27 through 44 are from interpretation of this same survey.

Figure 28. Arbitrary lines can be extracted from data volume to study geologic changes between wells, evaluate true-dip sections, etc. This set of five lines generally extends toward northeast (left to right).

Figure 29. Once area of interest is identified, composite pictures can be made to evaluate several sections simultaneously. Linear trace and sample interpolation has been used to enlarge scale.

Figure 30. Simultaneous display of several time-slice sections is illustrated here. Also, time-slice or time-series displays can be preformatted as animation files and evaluated as movies.

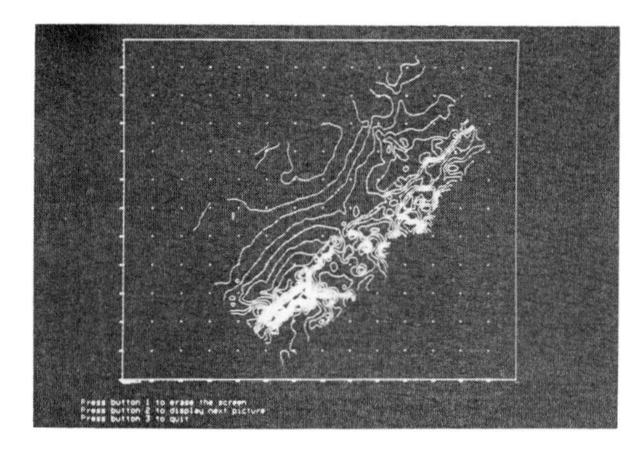

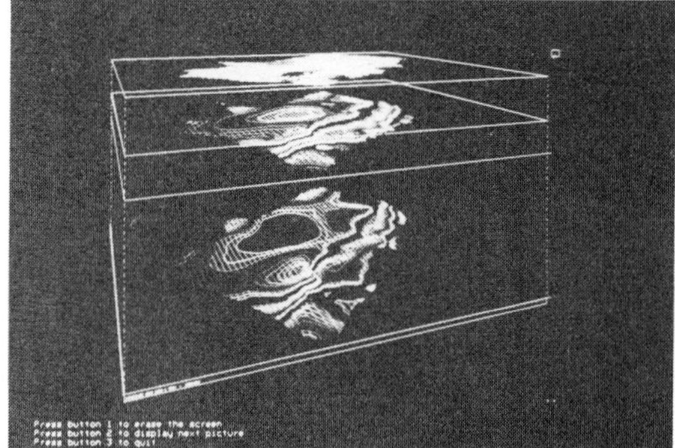

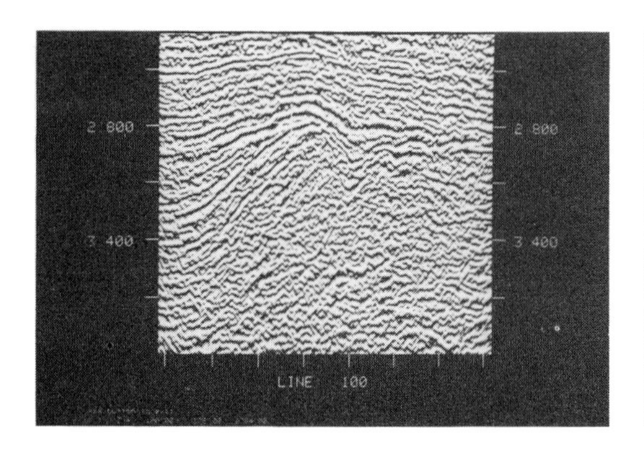
2 800
2 800
3 400
3 400
LINE 100

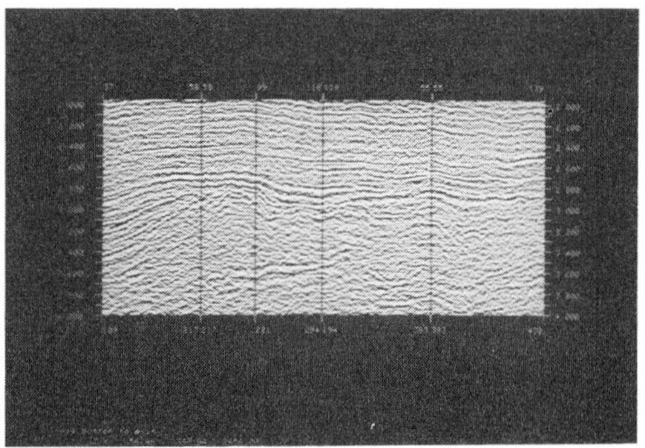

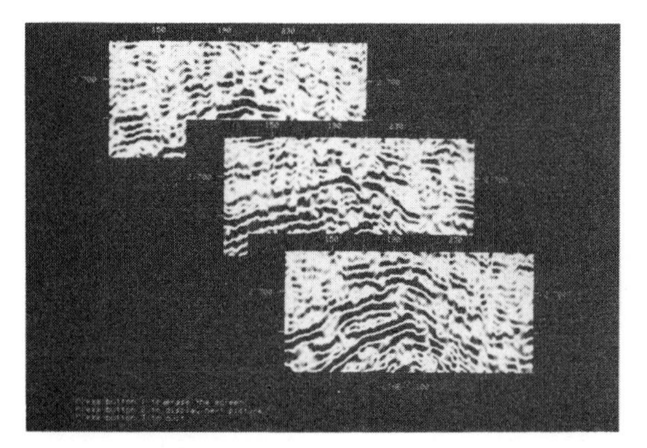

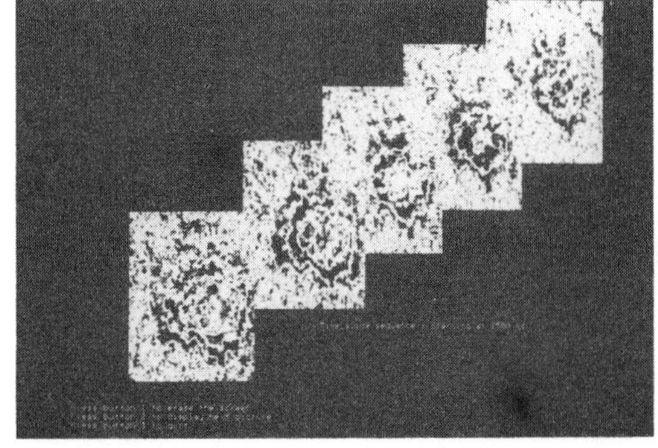

Figure 31. Chair display allows time-slice and time-series data to be picked in concert. This time slice is at 2800 ms and is connected to cross-line sections 50 at bottom and 400 at top. Color table used is designed to enhance structural dip.

Figure 32. This chair display also is at 2800 ms but is connected to cross-line sections 150 and 350. Black-to-white color table has had visual AGC applied using exponential adjust color adjust.

Figure 33. Display in Figure 27 has been stretched by interpolating linearly one trace between each two data traces, then pixel replication zoomed one step. Three horizons are overlain on data. Annotation is red, but this does not show up because it is black-on-black background in this black and white print.

Figure 34. Instantaneous phase section, with same display parameters described for Figure 33, shows subtle structural trends somewhat better.

Figure 35. Data in Figure 33 have been flattened on second horizon. Isochron from first to second horizon is overlain, illustrating horizon recognition on flattened section.

Figure 36. Instantaneous phase data in Figure 34 has been flattened on second horizon. Isochron horizon can be converted back to second time horizon by adding it to upper horizon.

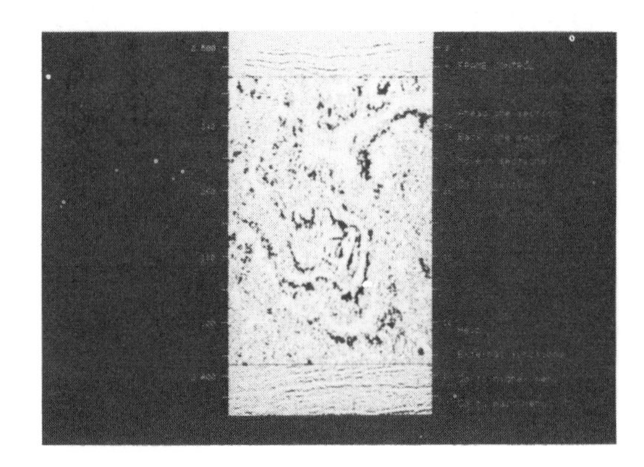

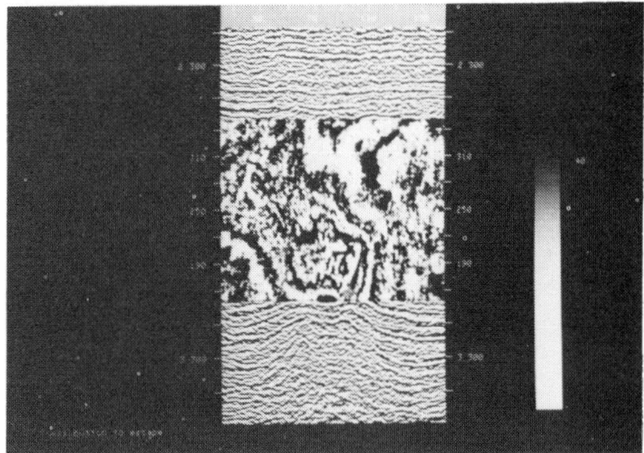

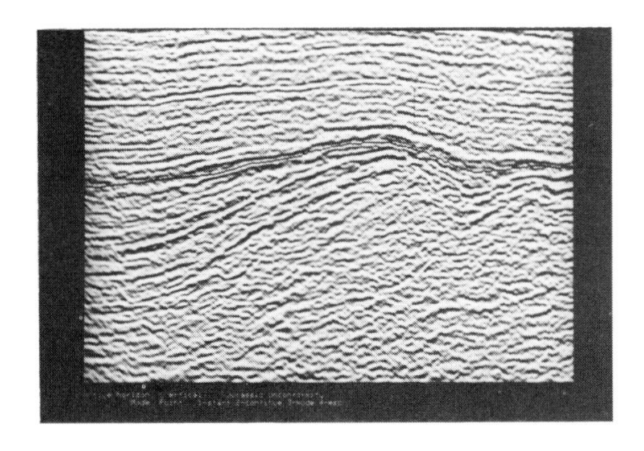

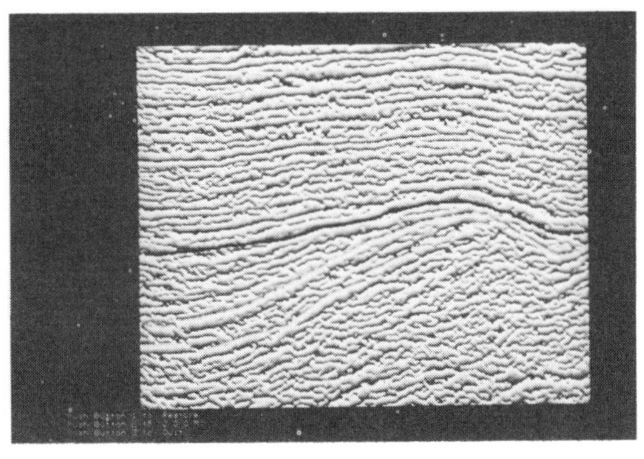

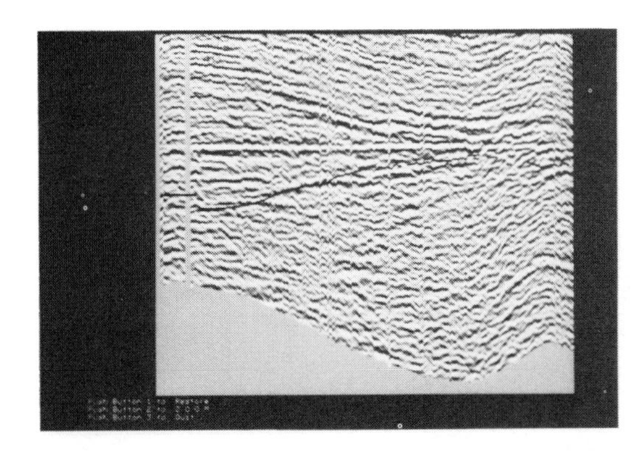

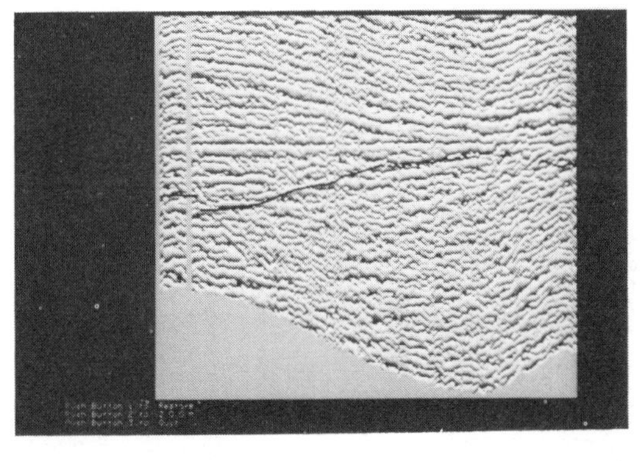

Figure 37. Map view of horizon used for flattening in Figures 35 and 36 shows status of picks. 3D survey already is gridded, and picks can be displayed in map view at any point of interpretation. White is shallowest and black deepest time, with a marker placed at 2848 ms.

Figure 38. Marker in Figure 37 has been moved instantly to 2908 ms. Dynamic changes in pseudocontour allow detailed evaluation of closure extent.

Figure 39. Marker in Figure 38 has been moved instantly to 3000 ms. In addition to marker, 64 separate colors can be assigned to highlight different times or depths.

Figure 40. Color map has been modified to have six 100 ms white-to-black bands, with red marker highlighting the 2848 contour.

Figure 41. Contour maps are better for presentations when colored. This contour map goes from reds to yellows to blues, each color grading bright to dark over a 100 ms window. Horizon displayed here was picked at Hunting Geological and Geophysical in London.

Figure 42. Contour map shown in Figures 42-44 has been colored white-to-black, pixel zoomed, and overlain with contours every 40 ms from same horizon. This concept can be expanded to mix information from different type maps, that is time, isochron, amplitude, residual, etc.

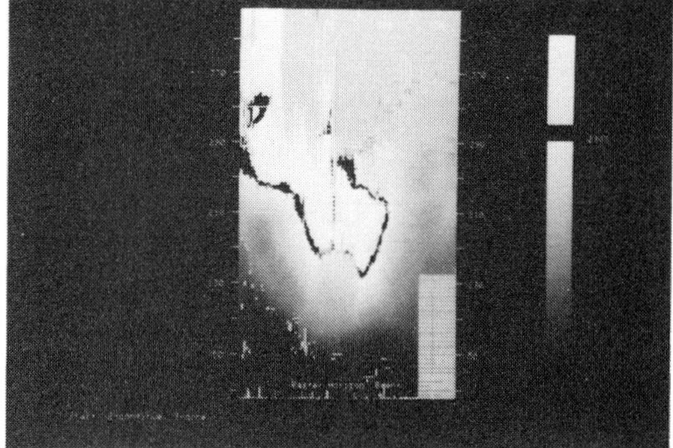

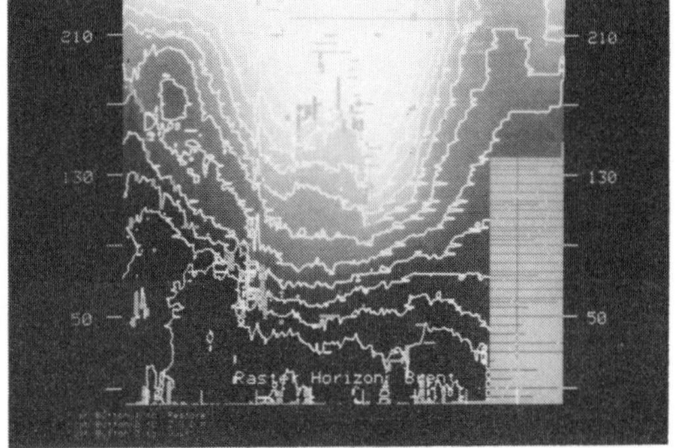
210
210
130
130
50
50

Figure 43. Isochron horizon between top two horizons on Figure 33 has been overlain with time contour map of second horizon. This file can be input to SURFAS for perspective displays.

Figure 44. Map view of extracted amplitudes at second horizon have been smoothed with spatially square operator and overlain with time contours. Variations are shades of red.

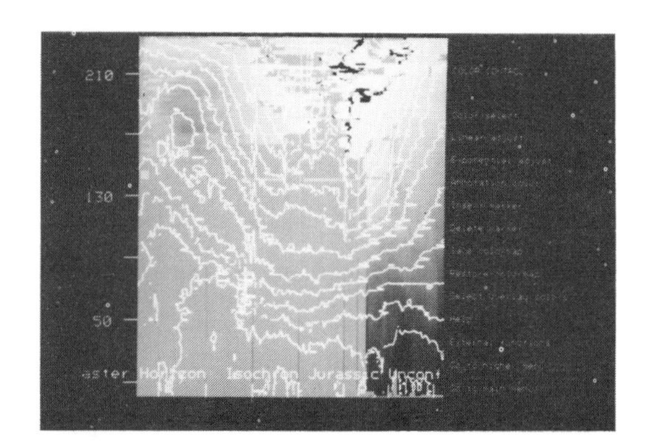

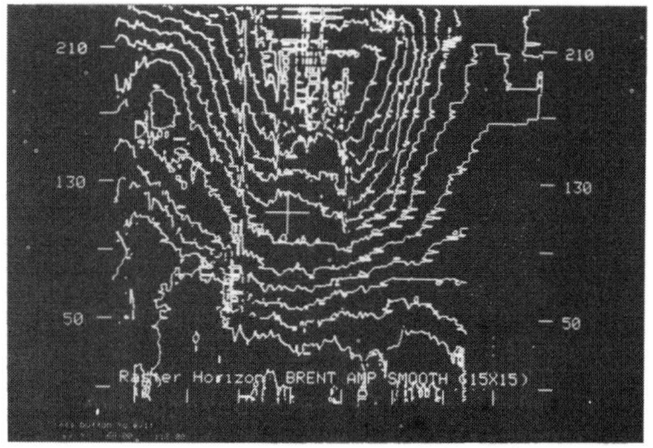

Computer-assisted Drafting of Down-hole Data

J. P. Reed

RockWare, Incorporated

ABSTRACT

Prior to the introduction of microcomputers, graphic drill-hole logs were laboriously plotted by hand. Today, geologists and engineers use portable computers and "off-the-shelf" log-plotting software to draft presentation-quality strip-logs directly at the drill-site. These logs also may be transferred instantaneously (via a phone modem) to a home or client office for subsequent replotting, additional analysis, or transmission to other locations.

Currently available log-plotting software contains the following capabilities:

(1) natural language recognition; lithologic descriptions are translated automatically into the appropriate graphic patterns,
(2) automatic translation of abbreviated descriptions into longhand,
(3) variable graphic formats; all types of existing logs may be duplicated or enhanced to display additional data; and
(4) user-definable patterns such as lithology, alteration, mineralization, fossilization, fracturing, veining, etc.

INTRODUCTION

Historically, log-plotting software has been confined to applications in which the programs were custom designed on a project-by-project basis. These systems usually involved large computers operated by specially trained personnel. Geologists and engineers would present their data to these individuals for entry and processing. These programs involved substantial acquisition, development, and maintenance costs affordable only to large corporations. Consequently, true computing power was controlled by the data-processing staff within these organizations.

Today, geologists and engineers use inexpensive, yet powerful, microcomputers and log-plotting software on a daily basis without the frustration being concerned with a data-processing staff. Unlike most mainframe software, the new generation of log-plotting programs are designed for a high-volume market. These programs are flexible enough to suit almost any type of application (for example, petroleum and mining), yet they are designed for the novice user who has little or no computer experience.

This paper will describe the capabilities of currently available log-plotting software. An emphasis will be placed upon flexibility, specifically in regards to graphic formats, translation logic, and remote operations. In addition, several new approaches concerning data preparation and user-adaptable software will be presented.

TERMINOLOGY

The terms "automated log-plotting" and "log-analysis" are not synonymous. Log-plotting software is limited to performing tasks that otherwise would be done by a draftsman, whereas log-analysis software provides the user with information that is not apparent otherwise from a visual review of the raw data.

An example of using log-analysis software in conjunction with log-plotting software involves geologists who use microcomputers to generate strip-logs of raw values and log-analysis programs to process these data. Once this information has been analyzed, and new data sets (such as d-exponent) have been generated, new types of "analysis logs" are plotted by using the log-plotting software.

ON-SITE LOG-PLOTTING

In the past, the bulk of drilling-related geologic analysis involved drafting. Today, geologists and engineers use "transportable" computers directly at the drill-sites to generate quickly presentation-quality strip logs as the data are being collected. This approach has eliminated the need to plot manually several versions of the same log, such as rough-draft, daily log, and final draft. As a result, geologists and engineers now are able to concentrate their efforts upon observation and analysis.

A typical on-site computer system includes a small microcomputer and a graphics printer. These machines are operated from within a logging trailer, field vehicle, motel, or base camp where the data are entered as they are collected. When required, a graphic strip-log of any portion of the hole then may be plotted by the graphics printer at a rate of five to ten minutes per page.

TRANSFERRING DATA

Geologists and engineers who use microcomputers to plot drill-hole logs directly at the well-site also can transfer data electronically to other computers for subsequent replotting. Transfer can be accomplished via a variety of telecommunications media including land-line phone networks, radio phones, and satellite "up-links." Unlike telecopying, this transfer of quantitative data guarantees an exact duplication at the receiving end. Additionally, the data may be subjected to additional processing, replotting at different scales, or retransmission to other locations without any loss in resolution.

Typical data transmissions require less than ten minutes in contrast to the old practices of sending logs through the mail, using couriers, or phone dictation. For many operators, microcomputer systems have paid for themselves the first time somebody said "Circulate until I've had a chance to look at the logs."

This telecommunications capability requires at least one additional microcomputer at the receiving end. Both the sending and the receiving computer must be equipped with electronic devices termed "modems" (modulator/demodulators), which allow the data to be transferred through conventional communications networks.

The ability to transfer data accurately from one location to another has proven to be indispensable to the well-site service companies who now use phone modems to update their clients several times per day. By plotting these data within the office as the hole is being drilled, the project managers are able to make timely and informed decisions.

NATURAL LANGUAGE DATA ENTRY (NLDE)

In the past, automated log-plotting software required that all qualitative data (for example, lithologic descriptions) be "precoded" prior to entry into the computer system. In order to precode the data, a user would compare the geologic field notes with a code index table. For example, if a geologist wanted to plot a "silicified meta dolostone," he would refer to the code index table. If a match occurred between the observed rock type and a code number, then the user would enter the code number into the computer. If there was no match, then the program would have to be modified by a software engineer in order to recognize the new rock type.

Many individuals within the industry accused this process of being laborious and counterproductive. Others learned the precoding scheme was effective provided that a computer programmer was assigned to the project on a full-time basis in order to update continually the program to recognize new rock types.

An alternative approach that now has gained wide acceptance within the geosciences is a combination of two techniques usually referred to as "keyword recognition" and "pattern association". Keyword recognition involves a process whereby lithologic descriptions are scanned for specific keywords. If a keyword occurs within a description, then the appropriate pattern is plotted.

Table 1. Keyword/pattern-association table.

Keyword	Pattern	Pattern Description
LIMESTONE	1	brick pattern
SANDSTONE	2	offset dots
SHALE	3	horizontal lines

For example, in the simple keyword-index table, as depicted by Table 1, numbers are associated with specific keywords. These numbers refer to patterns described within the right-hand column. Keyword/pattern-association tables define the logic by which log-plotting software systems convert lithologic descriptions into graphic charts.

As an example of how Table 1 might be used, consider the following lithologic description:

"LIMESTONE: arenaceous, gray, very fine grained, fossiliferous."

A typical log-plotting program would scan this description to determine if it contains any of the keywords defined within Table 1. A match would occur for the keyword "LIMESTONE," causing the program to plot pattern 1 (brick symbol) within a designated interval.

This new approach eliminates the precoding requirement. Instead, the geologist or engineer simply enters the lithologic descriptions in a manner identical to conventional field notes. This technique also is referred to as "natural language data entry." It has been determined that the natural language data-entry technique reduces the amount of training that a geologist must undergo to use an automated log-plotting program. This is important especially in a multiple-user environment where new or inexperienced employees must be able to enter quickly the data without specialized training.

ABBREVIATED DESCRIPTION TRANSLATION (ADT)

Most geologists and engineers record their data in an abbreviated format, therefore, it is essential that automated log-plotting software be capable of converting abbreviations into longhand prior to scanning for keywords and plotting the graphic log. This translation process is accomplished by an "abbreviation equivalence table" (Table 2) similar to the keyword/pattern-association table described in the previous section.

Table 2. Abbreviation equivalence table.

Abbreviation	Longhand Equivalent
arg.	argillaceous
f.	fine
fos.	fossiliferous
gr.	grained
gy.	gray
ls.	limestone
v.	very

In the sample abbreviation equivalence table, geologic shorthand terms are associated with "longhand equivalents". The abbreviation-translation portion of the log-plotting software will scan each description for the abbreviations within Table 2 and replace them with the associated longhand word(s).

To illustrate the role of an equivalence table, consider the following abbreviated description:

"Ls., arg., gy., v. f. gr., fos."

The log-plotting program would scan this description and produce the following description:

"Limestone, argillaceous, gray, very fine grained, fossiliferous."

USER-DEFINABLE PATTERNS

Despite many attempts to establish a norm, there are no accepted standards universally, for symbol usage within the geosciences. This problem is well illustrated by a comparison of petroleum, coal, and mining logs. The petroleum geologist might require twenty different patterns for depicting various types of carbonates, the coal geologist might want ten different coal patterns, and the metals geologist needs twelve different rhyolite patterns. Therefore, it is impossible for a programming team to anticipate every possible combination of lithologic, alteration, and mineralization patterns. What is more, the programmer would need to impose a standard upon the user to eliminate conflicting symbol usage.

Thus, of particular importance in evaluating or designing log-plotting software is the capability for creating new patterns. Unless the user is able to add new patterns, the system is worthless. Additionally, the actual process of creating a new pattern must be relatively easy (for example, by interactive graphic-pattern editors).

For example, a mining company in Florida purchased a log-plotting software package to draft phosphate test holes. Although the system initially did not contain the desired phosphate patterns, the built-in "pattern editor" allowed the users to design new, project-specific symbols within one day.

VARIABLE SCALES AND PLOTTING PARAMETERS

The vertical scales at which drill-hole logs are plotted differ considerably. As a result, automated log-plotting software must be adjustable to suit any desired output scale.

For example, geologists in the past usually had trouble correlating drill-core logs with photocopied electric logs because the scales did not match due to scale distortions produced by the photocopying of the electric log. By simply adjusting the scale within the plotting program to the same distorted scale as the geophysical plot, geologists now are able to generate quickly new "working" logs for direct correlation with the photocopied logs.

Variable plotting scales also reduce the amount of redundant work within a mine operation. For example, many mine geologists use the variable-scale capability of log-plotting software to produce two logs for every hole. A log plotted at a small scale (for example, 1:50) is used as a "working copy" for detailed correlation with adjacent test holes and cross-section generation, whereas a large-scale log (for example, 1:600) log is printed for inclusion in reports and archival files. By entering the data only once, and changing the plotting scales, the additional log is plotted with just a few minutes of additional effort.

Some users have discovered new applications for the variable plotting scale capability within automated log-plotting systems. A geologist logging precious-metal holes in the Mojave Desert plotted detailed portions of drill logs at a scale of 1:1 for direct comparison with cored intervals of importance. Copies of these "sublogs" then were pasted inside the core boxes for archival purposes.

In addition to variable vertical scales, it also is necessary for the user to be able to change the horizontal scales at which quantitative data is being plotted. Many coal geologists use this capability to change the horizontal scale at which the BTU values are plotted depending on the overall range within a project area.

Another important consideration involves the general format of the drill-hole log. For example, many coal geologists plot a lithologic column on the left side of a log and display the quantitative data (for example, BTU's, ash content, moisture content) to the right of the lithology column. Other geologists prefer the opposite convention. Figure 1 and Figure 2 illustrate two possible formats which may be generated by a single log-plotting software package. This variable format capability is a fundamental requirement within any log-plotting program in order to prevent immediate obsolescence if a new format is required.

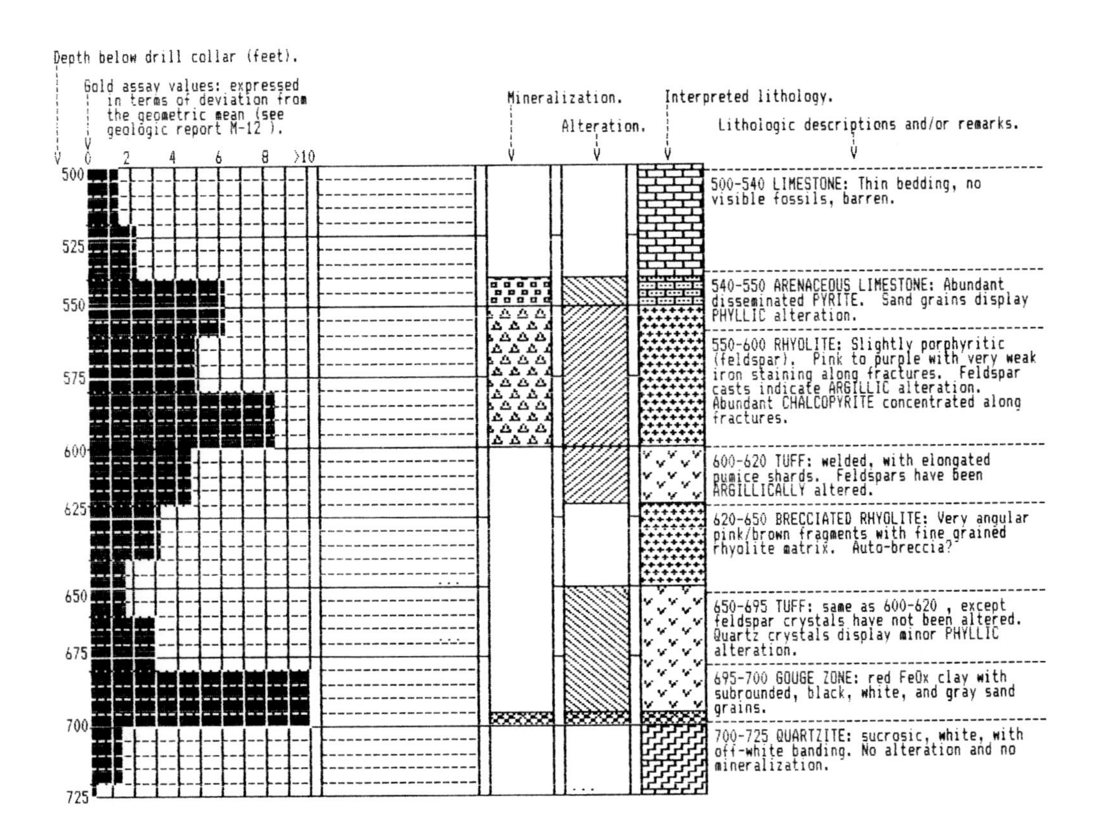

Figure 1. Example of computer-generated mining log.

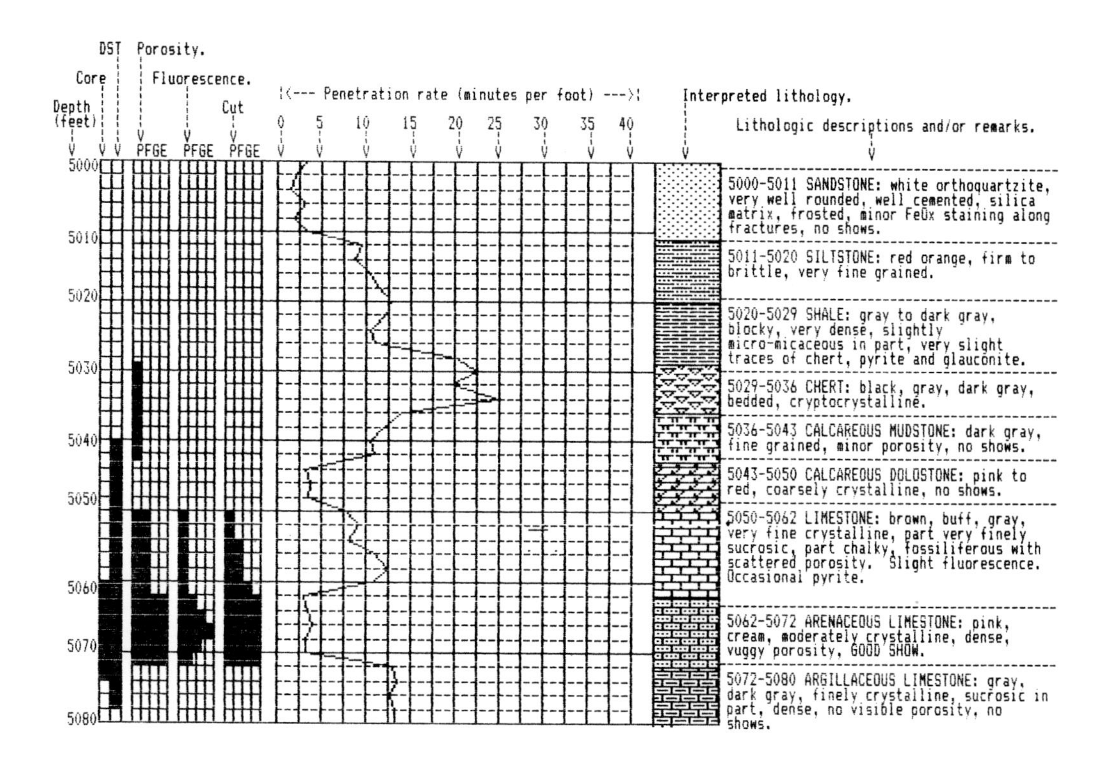

Figure 2. Example of computer-generated petroleum log.

ADAPTIVE SYSTEMS

Adaptative log-plotting software provides the user with a capability to modify the logic by which the program associates patterns with keywords (for example the keyword/pattern-association table). This "self-tailoring" feature may be used to configure the software to suit almost any type of specialized application (for example, coal, petroleum, uranium, metals, industrial minerals) or personal and corporate preferences. In addition, the user is able to convert readily the system to recognize a foreign language.

For example, while drilling about 400 test-holes on a talc exploration project in the Trans-Pecos Region of West Texas, the project geologists encountered a number of rock types (for example, phyllite, metadolostone, and numerous talc types) that had not been included within the initial keyword set. Because the log-plotting software which was used for this project included a keyword/pattern-association editor, modifying the system to recognize the new rock types took less than two minutes per modification.

This adaptability also must extend to the abbreviation equivalence table and the pattern definitions for the same reasons as those outlined for the keyword/pattern-association table. This provides the user with the ability to modify the logic which is used by the log-plotting software.

OUTPUT DEVICES (PRINTERS VERSUS PLOTTERS)

The decision as to what type of output device is to be used for plotting a strip-log usually involves the choice between a dot-matrix printer and a pen-plotter. When comparing these two alternatives, there are four major factors that must be considered: cost, portability, speed, and the quality of graphic output:

COST: In general, a plotter that is capable of plotting continuous geologic logs will cost an order of magnitude more than a dot-matrix printer.

PORTABILITY: Pen-plotters are not transportable readily, being rather bulky and delicate, whereas dot-matrix printers are smaller and better suited for transportation to and from the field area.

SPEED: Pen-plotters require more time to generate graphic products than do dot-matrix printers. Pen-plotters must either move the pen or the paper from one line endpoint to another, whereas dot-matrix printers transfer an image to the printer as a series of horizontal strips. The resultant difference in the time required to print a detailed drill-hole log is considerable. Logs that require hours on a printer may require days on a plotter.

QUALITY: The quality of graphic output from a pen-plotter is superior to that of a dot-matrix printer. This difference is due to the fact that pen-plotters are capable of drawing relatively straight diagonal lines whereas dot-matrix printers construct diagonal lines by offsetting a series of small dots (hence the "stair-step" appearance). In relation to log-plotting, many users consider the quality of high-resolution, dot-matrix graphic products (see Figs. 1 and 2) to be acceptable in light of the other advantages of cost, portability and speed.

COMPATIBILITY WITH OTHER SOFTWARE

Log-plotting software must be capable of reading data from an ASCII (American Standard Code for Information Interchange) text file. Some users store the data for more than one drill hole by using "off-the-shelf" database programs.

Most of these programs allow the data for an individual drillhole to be copied into an ASCII text file. This ASCII compatibility minimizes the problems associated with managing multiple drill-hole files.

For example, a user in Nevada who had been measuring geologic sections at outcrops and entering the data into spreadsheet data files used the "print-file" command from within the spreadsheet program to create ASCII files for each section. These data then were read by a log-plotting program and automatically plotted in a strip-log format.

The capability to read ASCII text files also provides a way by which microcomputer logging programs can access data within minicomputer databases. For example, many corporations and government offices are using minicomputers to store large drill-hole databases. In order to plot a selected log, the data are transferred to a microcomputer (via a phone modem) where it is plotted by inexpensive microcomputer software.

The ability of the log-plotting software to read ASCII files also provides the user with an of upgrading to new log-plotting software without reentering the data. This capability especially is important when large amounts of data are to be archived. For example, many state surveys and large corporations have discovered that it is more efficient to store drill-hole data in computer files (as opposed to steel cabinets). By adhering to an ASCII format, these files may be retrieved and plotted by programs developed in the future.

COSTS AND PORTABILITY

A typical computer system for generating geologic logs directly within the field would involve the hardware and software listed in Table 3.

Table 3. Hardware/software costs.

Item	Costs
Portable computer	2,100
Dot-matrix printer	400
Power filter	100
Word-processing software ...	50
Log-plotting software	1,000
Totals:	$3,650(US$)

These costs assume that the log-plotting software is to be licensed from a software vendor. If the software is to be developed "in-house," a time allotment of 2,500 to 3,500 hours for development and "debugging" will be required. For example, RockWare's "LOGGER" program has involved more than three years of development. In short, if you plan on writing a log-plotting package, be sure to allot a generous amount of time, money, and patience.

SUMMARY AND CONCLUSIONS

The new generation of microcomputers and user-adaptable software has provided the petroleum and mining industries with a cost-effective method for automatically plotting drill-hole logs in a fraction of the time normally required for manual drafting. These systems now are small enough to allow geologists to plot logs directly at the drill site. Data from the drill-site may be transferred instantaneously to a home or client office via conventional phone lines for subsequent replotting.

EXPLOR and PROSPECTOR — Expert Systems for Oil and Gas and Mineral Exploration

R. M. Maslyn

Consulting Geologist

ABSTRACT

Artificial intelligence programs are said to exhibit "convergent thinking." That is, many inputs (factors) are examined to produce a single output (decision). Expert systems are programs that incorporate the knowledge of experts on a specific subject and then act as consultants on that subject. They are certainly the most exciting and immediately practical of the various artificial intelligence developments.

Expert systems, including EXPLOR and PROSPECTOR, contain a series of "rules" encoded as "if-then" or "true-false" type statements that follow logical paths to arrive at decisions. The input data determines the directions the program branches along each node of the decision tree.

PROSPECTOR, a mineral-exploration program, incorporates the knowledge of a number of experts on 26 types of mineral deposits. The user inputs field observations in word form such as "dolomite", "barite", etc. and the computer then prompts the user for additional information before classifying the type of deposit being examined or outputting what types of mineral deposits might occur in the geologic terrain described by the user. At any point, the user may ask for an explanation of why PROSPECTOR is asking a particular question, or for a more detailed description of the type of deposit that may occur in the area being examined. PROSPECTOR has been used successfully to locate a hidden but suspected molybdenum deposit.

EXPLOR is an expert system for oil and gas exploration which outputs oil and gas lead areas for possible prospect generation by an exploration geologist. The geologist is provided with high-graded leads resulting in greater output and efficiency.

EXPLOR outputs five types of lead areas: anticlinal closures, anticlinal noses, stratigraphic pinchouts, possible by-passed wells, and productive water saturation closures. These leads may be used for prospect generation or initial lease evaluations.

EXPLOR differs from PROSPECTOR and some other expert systems in three major respects. First, EXPLOR uses an internal feedback system requiring no additional information once the data has been input. Secondly, because rock formation information usually is available only from well logs (discrete samples of the area's geology), EXPLOR approximates data trends using polynomial trend surfaces to provide values throughout the area being evaluated. Thirdly, because many of the data values used in exploration are numerical rather than verbal, the program is written in BASIC and compiled for increased speed.

Although EXPLOR was first tested on sandstones in the D-J Basin, the exploration parameters can be adjusted to allow it to evaluate channel sandstones anywhere in the world.

AI AND EXPERT SYSTEMS

Computer programs using artificial intelligence (AI) are characterized by "convergent thinking." That is many inputs are taken together to produce a single output. This is analogous to the way a human brain would examine several factors (inputs) in arriving at a decision (output).

Perhaps the most exciting and certainly the most practical of the recent AI developments are "expert systems." These are programs which after incorporating the knowledge of experts in a specific field act as consultants in that field. Expert-system programs utilize "rules" for specifying which direction the program branches along decision trees. Each rule is a "yes-no" or "true-false" type of question. The input from the user determines both the direction the program branches and the next question the computer asks the user in order to reach a conclusion. Use of expert systems is becoming more widespread ranging into fields as diverse as medicine, production engineering, and geology (Ennis, 1983). Most such programs are written in a word-oriented computer language such as LISP (LIST Processor), although LISP is not well-suited for mathematical computations.

CREATION AND EVOLUTION OF AN EXPERT SYSTEM

The first step in the process of constructing an expert system involves the construction of a "knowledge base" which contains the modified knowledge of experts on a specific subject. This knowledge is not only a description of the factors necessary for a given conclusion, such as whether a given mineral deposit or an oil accumulation is likely to be present, but also the intuitive process used to reach that conclusion. Many times the expert is not conscious of the factors and processes involved in reaching a conclusion until forced to list them for input into an expert system (S. A. Adams, pers. comm., 1984).

Inasmuch as expert systems are examples of artificial intelligence, their development mimics the process of learning as well. Once the expert system is constructed, it is tested on known situations to determine how accurately it describes actual situations, and if there are additional factors involved in locating a mineral or oil and gas accumulation that are not included already in the expert system. This becomes an iteractive process as the expert system grows in knowledge. For more detail on the process of constructing an expert system the reader is referred to McCammon (1983).

USES OF AN EXPERT SYSTEM

Expert systems have two direct uses. The major one is as an expert consultant on a specific subject; secondarily, these programs are teaching devices. An indirect benefit of formulating the questions the program asks and testing the program with known situations is the examination of why those factors are necessary or important for a given type of mineral deposit to be present. If the program results do not match the actual situation then factors other than those identified already must be acting. Their identification may lead to a better identification of the processes that form a mineral deposit or an oil accumulation.

Therefore, in some situations expert systems may not provide any new insights or information to individual users who are themselves experts in a field, the expert systems being computer-based may evaluate large areas quickly. This advantage is utilized by the EXPLOR program in increasing the area that an exploration geologist can examine quickly.

PROSPECTOR EXPERT SYSTEM FOR MINERAL EXPLORATION

PROSPECTOR, a mineral-exploration program, is one of the best known of the expert systems (Dunda, 1980, 1981; Rebon, 1981). This program incorporated the knowledge of a number of experts on 26 separate types of mineral deposits. The program prompts the user for information on host-rock type, alteration, mineral assemblages, diagnostic features, etc. The user inputs his geologic field observations in simplified word form such as "dolomite," or "barite," etc. The computer, using full sentences, then asks for more specific information relating to the previous geologic information. In this way the program checks for each of several geologic factors associated with a given type of deposit. The questions the computer asks draw upon the "knowledge base" of geologic information supplied to it by experts on each type of deposit. At any time during the questioning the user can request an explanation of why PROSPECTOR is asking a particular question or obtain additional information about a particular conclusion.

After searching through the decision trees, three forms of output are available. First, PROSPECTOR can classify the type of deposit described, possibly suggesting a search for additional minerals or rock alterations typical of this type of deposit. Second, the program can list the types of deposits that are most likely to occur in the geologic environment described. Thirdly, the program can be used to organize a drilling program to define a suspected mineral deposit. PROSPECTOR has been used successfully to lay out a test drilling program which resulted in the discovery of a commercial molybdenum deposit (Campbell and others, 1982).

Many tests have been run of the PROSPECTOR program on known areas of mineralization. Campbell and others (1982) attempted a "blind" test of the program on an area that had data available but was not defined fully. Their test involved the Mt. Tolman area in eastern Washington. Molybdenum mineralization had been noted in the area and a molybdenum porphory deposit was suspected. At that time PROSPECTOR included information on two types of porphyry molybdenum deposits, the "hood zone" type and the "vertical cylinder" type.

Geochemical anomalies (Fig. 1) and hydrothermal alteration information (Fig. 2) were input to PROSPECTOR. Drilling subsequent to PROSPECTOR's evaluation confirmed that "The area rated favorably by PROSPECTOR completely overlies the deposit as now known...The weakest part of PROSPECTOR's evaluation was its failure to recognize the full extent of the deposit," (Campbell and others, 1982). Figure 3 is a map of area of the Mt. Tolman molybdenum deposit.

An additional iteration, or reworking of the questions and parameters used in the PROSPECTOR program to arrive at its results probably would have resulted in better definition of the Mt. Tolman ore deposit. New data, questions, or cut-off parameters could have been added to the program and PROSPECTOR in a sense would "learn" from the experience.

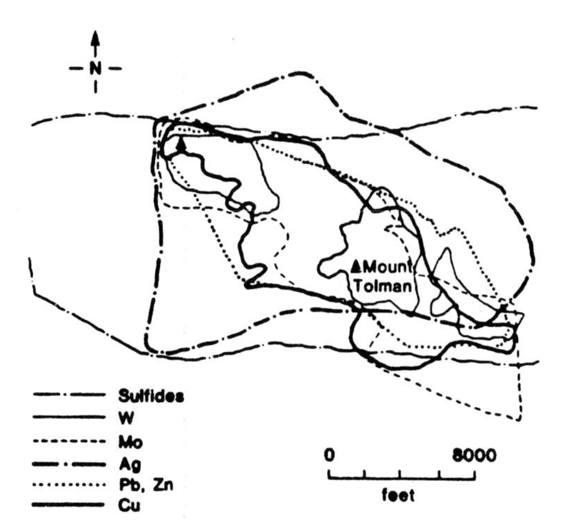

Figure 1. Geochemical anomalies around Mt. Tolman, Washington area (from Campbell and other, 1982).

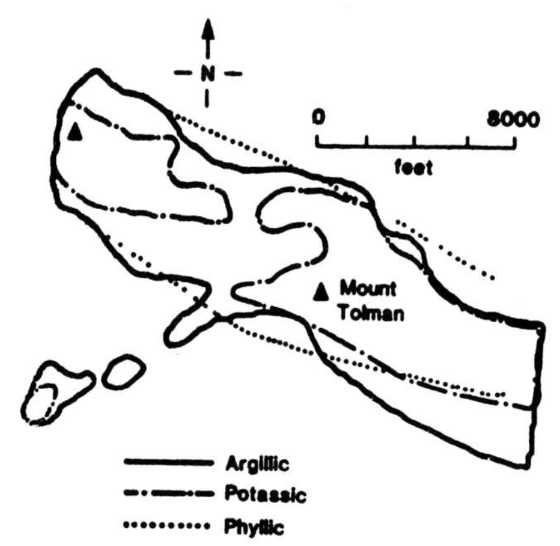

Figure 2. Hydrothermal alteration map of Mt. Tolman area (from Campbell and others, 1982).

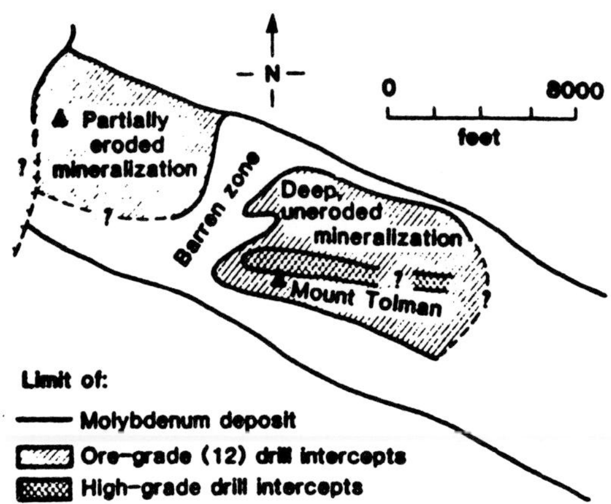

Figure 3. Map of Mt. Tolman molybdenum deposit after drilling to determine its extent (from Campbell and others, 1982).

EXPLOR EXPERT SYSTEM FOR OIL AND GAS EXPLORATION

EXPLOR is an expert-system program for reconnaissance oil and gas exploration. The program selects lead areas which then are evaluated by an exploration geologist. EXPLOR eliminates much of the drudgery and improves the efficiency of an exploration geologist. EXPLOR differs from PROSPECTOR in three major ways. First, it uses an internal feedback system which requires no additional information from the user after the data has been input. Second, because rock-formation data are available only from electric logs which are samples of the area's geology, EXPLOR contours the data, thus providing values for every 40 acre tract (16 points per square mile). Third, because much of the data used in oil exploration is numerical (such as structural elevations), the program is written entirely in BASIC for ease in mathematical operations.

At the current time, EXPLOR is configured for channel sandstones such as the "D" and "J" sands of the Denver-Julesburg Basin (Fig. 4). These are lower Cretaceous deltaic sands overlain and underlain by marine shales. Both of these sands show complex lateral stratigraphic variations that are beyond the capabilities of the program to model. Therefore, the sands are treated as discrete units which thin and thicken laterally. Any stratigraphic variations are left for the exploration geologist to detail once a lead area has been output by the program. Both stratigraphic and structural trap lead areas are identified by the program.

As EXPLOR only operates on one township at a time data are entered via a spread-sheet program containing one township of data per disk file. The data-file column headings are the well status, well spot location, section, structure, net sand, water saturation, and oil and gas shows. Oil and gas shows from cores or tests in the D and J sands are noted by numeric codes which indicate the type of shows. The data then are saved as a disk file to be used by other modules of the program.

As many of the wells in the basin were drilled before the widespread use of modern porosity log-tools, the water saturation is calculated approximately from the static SP and mud-filtrate-resistivity values. These calculations are made by the spreadsheet program prior to running EXPLOR. Unfortunately, many of the Rmf values must be calculated from the mud-resistivity values on the log headings. This introduces a potential source of error, as does the use of one formation water-resistivity value for each sand within a township. For these reasons, the calculated water-saturation values must be considered as approximate.

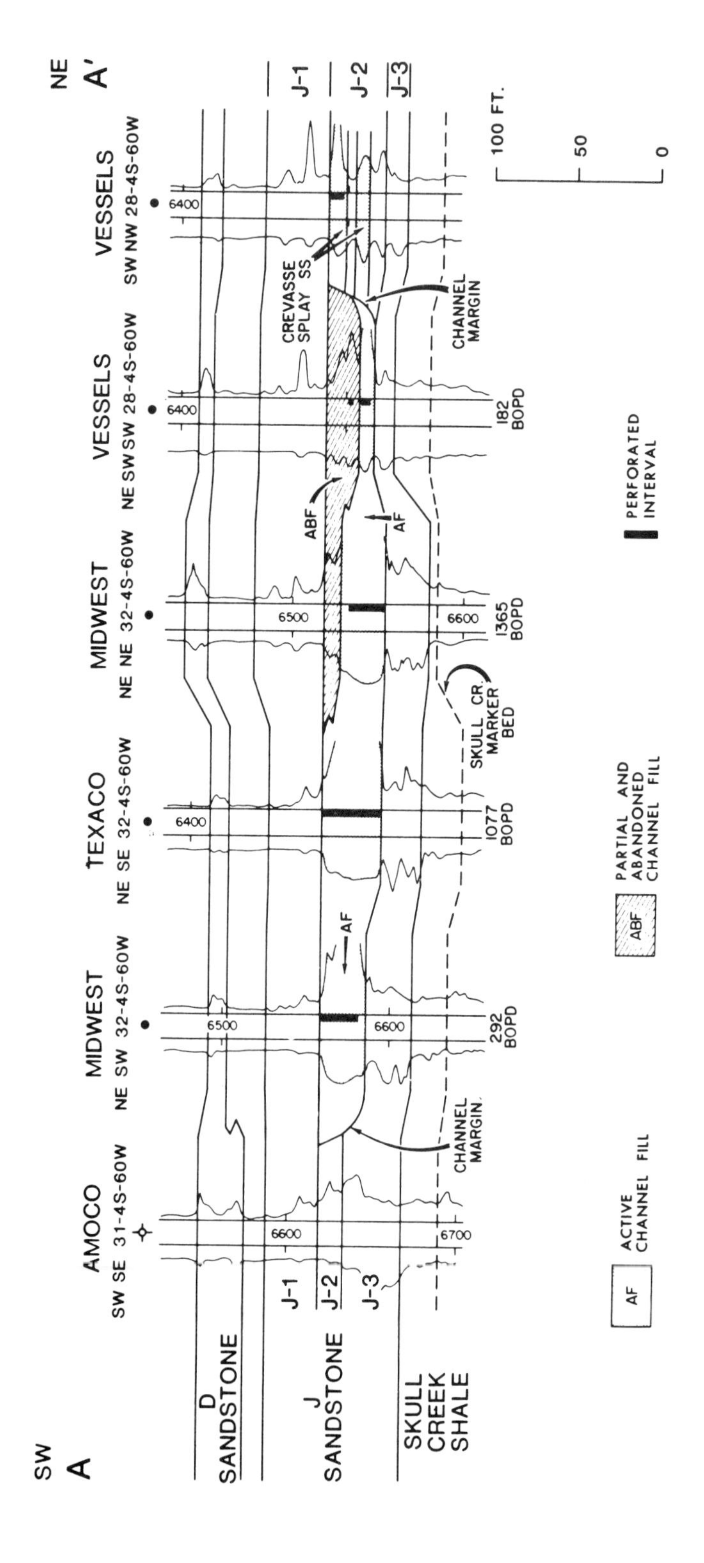

Figure 4. Electric-log cross section of "D" and "J" sands in Peoria Field, Denver Basin (from Land and Weimer, 1978).

The data then are contoured and output on a series of maps. Figure 5 is a map of the D-sand structure for the test township. The gently overall westward structural dip is apparent from the map. Figure 6 is a map of the net D-sand thickness in the same township showing the lateral variability of the sand. Figure 7 is a map of the calculated water-saturation values using the method described here. Some of the wells with high water-saturation values and indicated as producers are producing from the J-sand. A few values show the variability of values derived from this method as compared with actual production.

TOWNSHIP MAP OF D STRUCTURAL VALUES IN TEST AREA

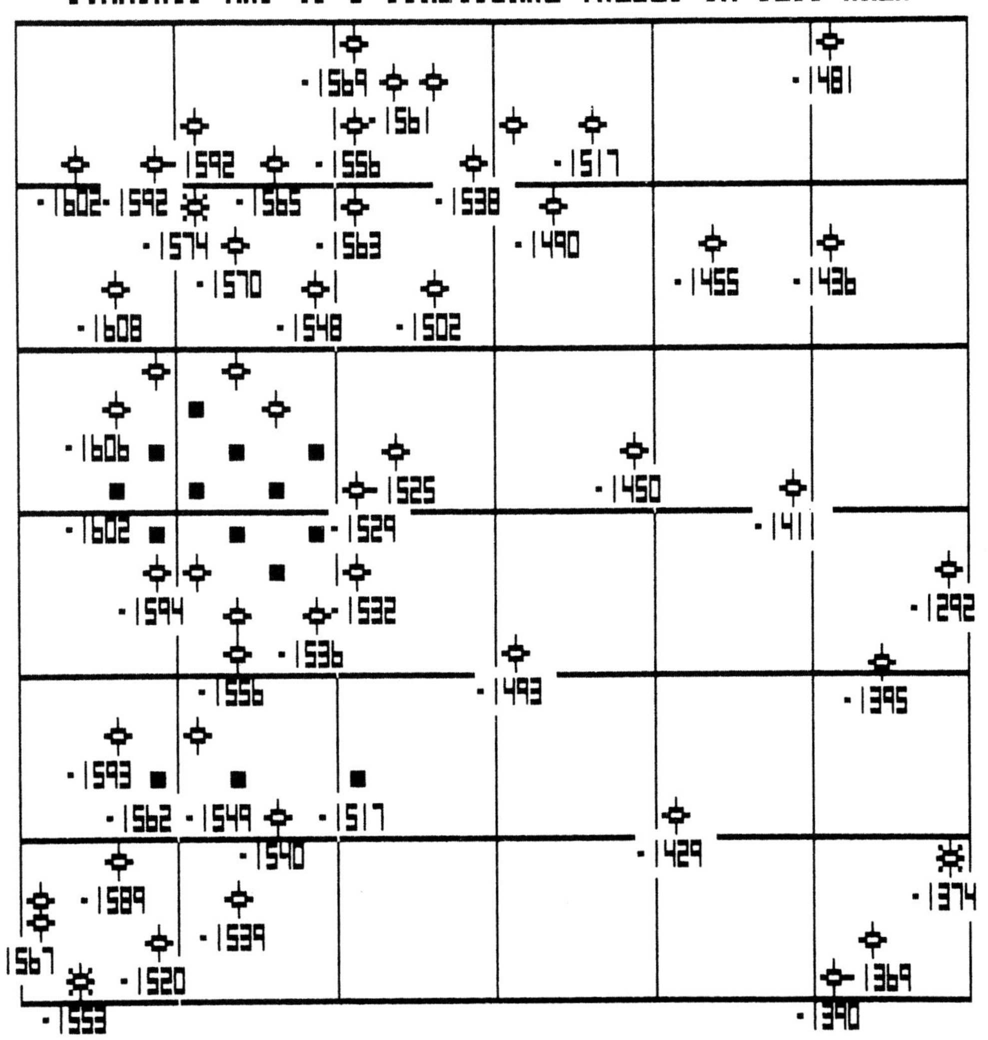

Figure 5. "D" sand structure in test township area.

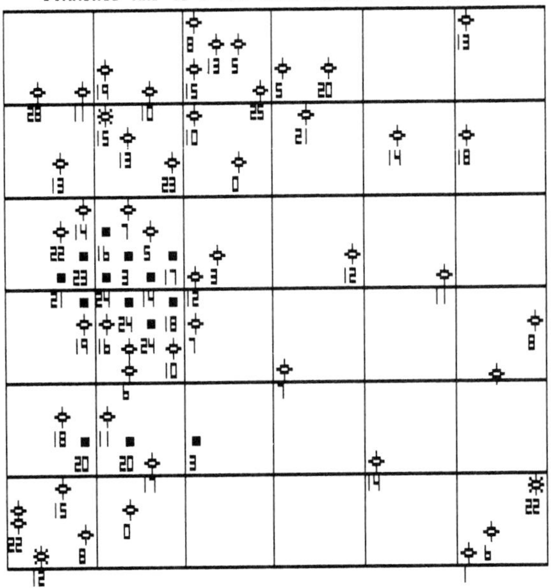

Figure 6. Net "D" sand thickness in test township.

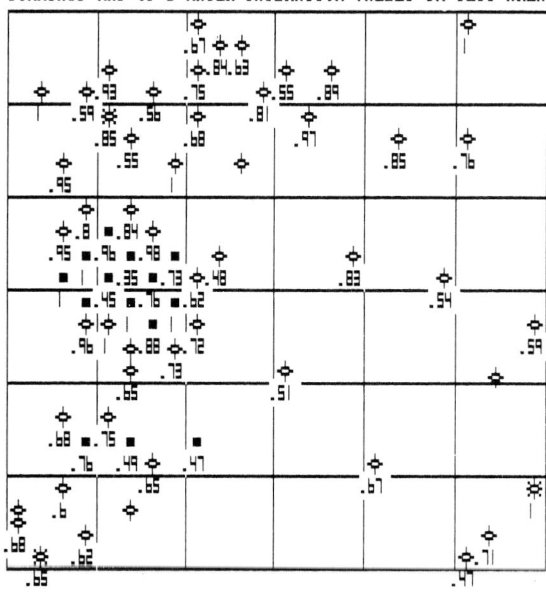

Figure 7. Calculated "D" sand water-saturation values in test township. Calculations made using method described in text.

Most contouring algorithms involve gridding routines which draw polygons connecting the data points and then interpolate contour values between the points. This method although effectively honoring the data, has the disadvantage that the highest and lowest data points on the posted base map become the highest and lowest values for the contouring. This is contrary to oil and gas exploration where structural trends are projected from the data points to locate undrilled structurally favorable areas such as anticlines and structural noses. In order to avoid this difficulty, EXPLOR contours the data points using polynomial trend surfaces. These are mathematical approximations or "best-fits" to the data points. Because a trend surface is the best fit to all existing data points, it can interpolate structural dips or other formation values between the existing data points and locate new areas of structural closure. This technique has been successful in several areas. It is limited, however, to areas without major faulting or structural complexity. This is true for much of the D-J Basin. Figure 8 is an example of a sixth-degree trend-surface map produced from the D-sand structural data. Figure 9 is a fifth-order trend-surface map of the net sand values, Figure 10, a fifth-order trend-surface of the water saturation values.

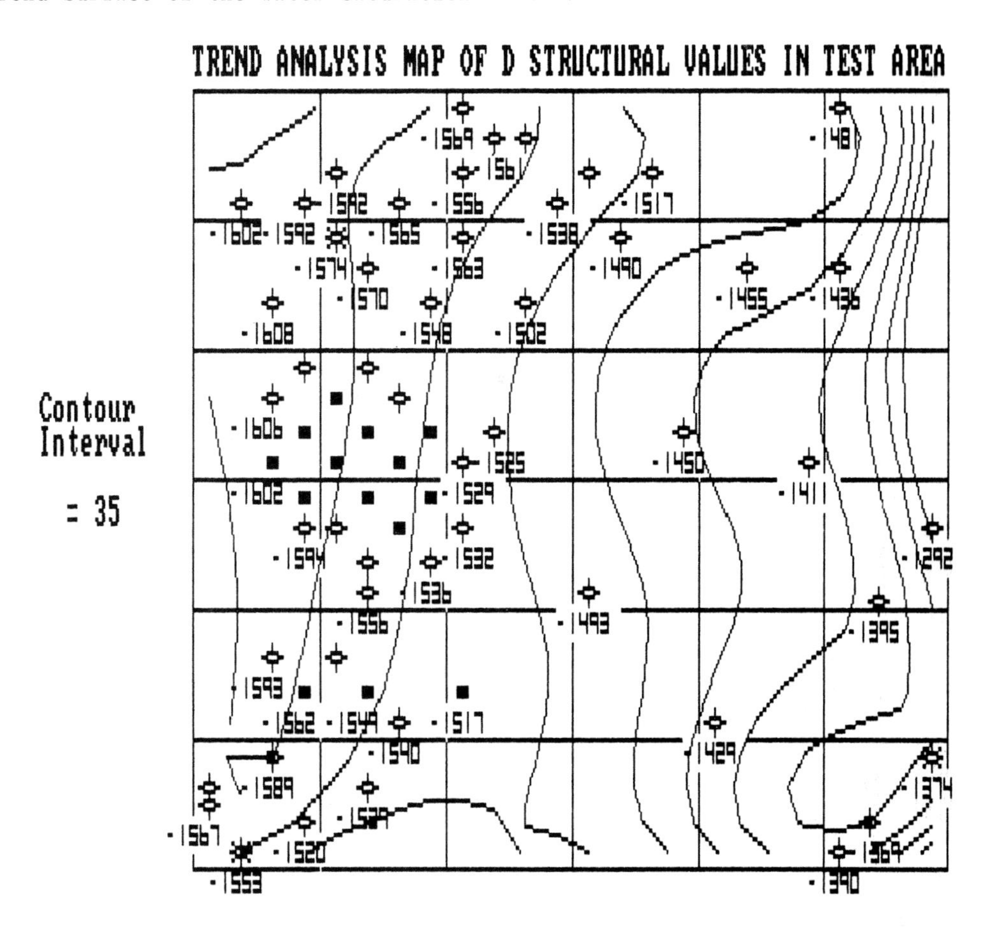

Figure 8. Sixth-degree polynomial trend-surface map of "D" sand structure in test township.

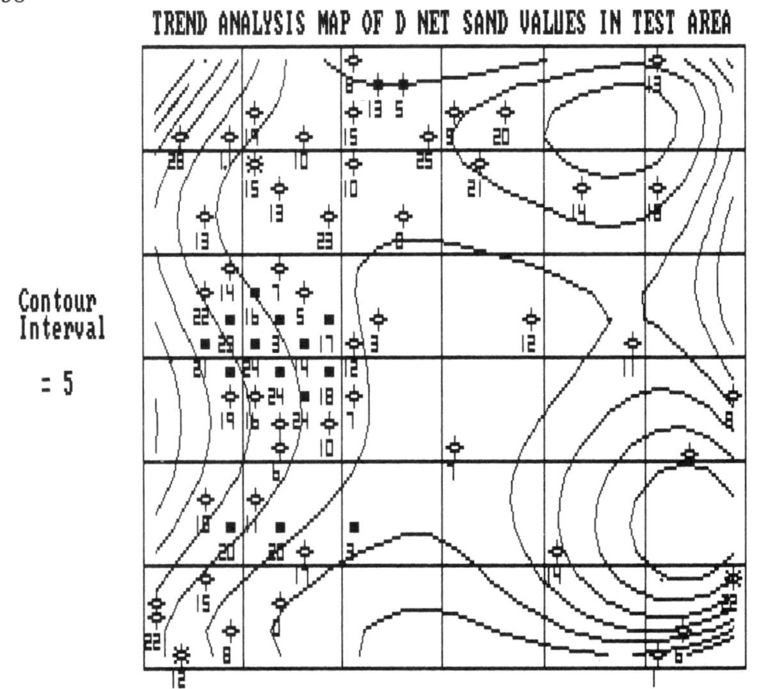

Figure 9. Fifth-order polynomial trend-surface map of net "B" sand values in test township.

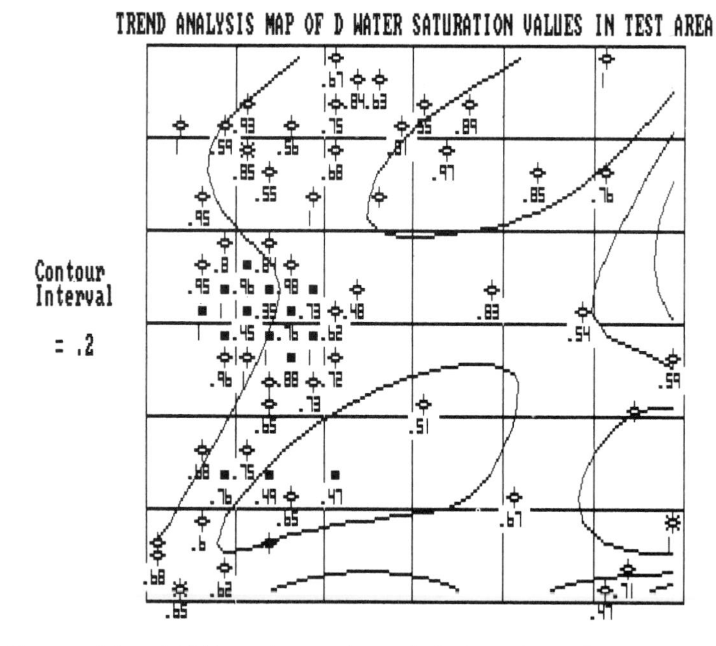

Figure 10. Fifth-order polynomial trend-surface map of water-saturation values in test township.

The program evaluates every 40 acre tract in the township to check if it meets preprogrammed criteria. This is the heart of the EXPLOR program. Within this final program module are the "if-then" rules that comprise the decision tree of the expert system. Among the factors directly evaluated are the following:

(1) Is the location drilled or not drilled?
(2) Is there sufficient net sand to contain an economic oil field?
(3) Is there 3 or 4-way structural closure?
(4) Is there a sand pinchout updip?
(5) Are there drillstem test, core, or log oil and gas shows?

Each of these factors is used to select routes along the branches of the decision tree. If a path is followed to a successful conclusion then that area is output from the program as a lead area. In this way the expert system is following the same lines of reasoning as any geologist familiar with the D-J Basin might follow to develop a prospect. The program has three advantages. The first one is the contouring by trend surfaces. Because the resultant map is a mathematical best-fit to the data, subtle trends in rock-formation structure may be apparent. The final map also is not biased by any preconceived interpretations of structural grain or pattern. The second is the water-saturation calculations. These values, which can be derived on a hand calculator, are derived quickly by a computer and are an additional tool which can be used to evaluate an area. The third advantage is the speed of a compute. Once the data are input, EXPLOR can contour, evaluate, and output quickly a township area.

Those areas which qualify are output as favorable geologically lead areas for further evaluation by an exploration geologist. Four types of lead areas can be output. These are:

(1) Structural closures,
(2) Stratigraphic traps,
(3) Possible bypassed wells,
(4) Closed areas where the water-saturation values indicate productivity.

Figure 11 is the lead-areas output map for the D-sand in the test township. This is accompanied by Figure 12, the description of each lead area includes the type of lead, the 40-acre location description, net-sand thickness, and DST, core, or log shows present in the nearest well with an oil or gas show.

SUMMARY

Expert systems are computer programs that incorporate the knowledge of experts on a specific subject and then act as consultants on that subject. After the initial data entry, expert systems use either external or internal-feedback to answer a series of "if-then" questions and reach a decision. PROSPECTOR is an expert system for minerals exploration which uses external feedback to prompt the user for additional information and then reach a conclusion. The EXPLOR expert system uses internal feedback to evaluate every 40-acre tract in a township for oil and gas leads. Four types of lead areas are output. These are structural closures, stratigraphic traps, possible bypassed wells, and productive water-saturation closures. Each of these leads then is evaluated by an exploration geologist.

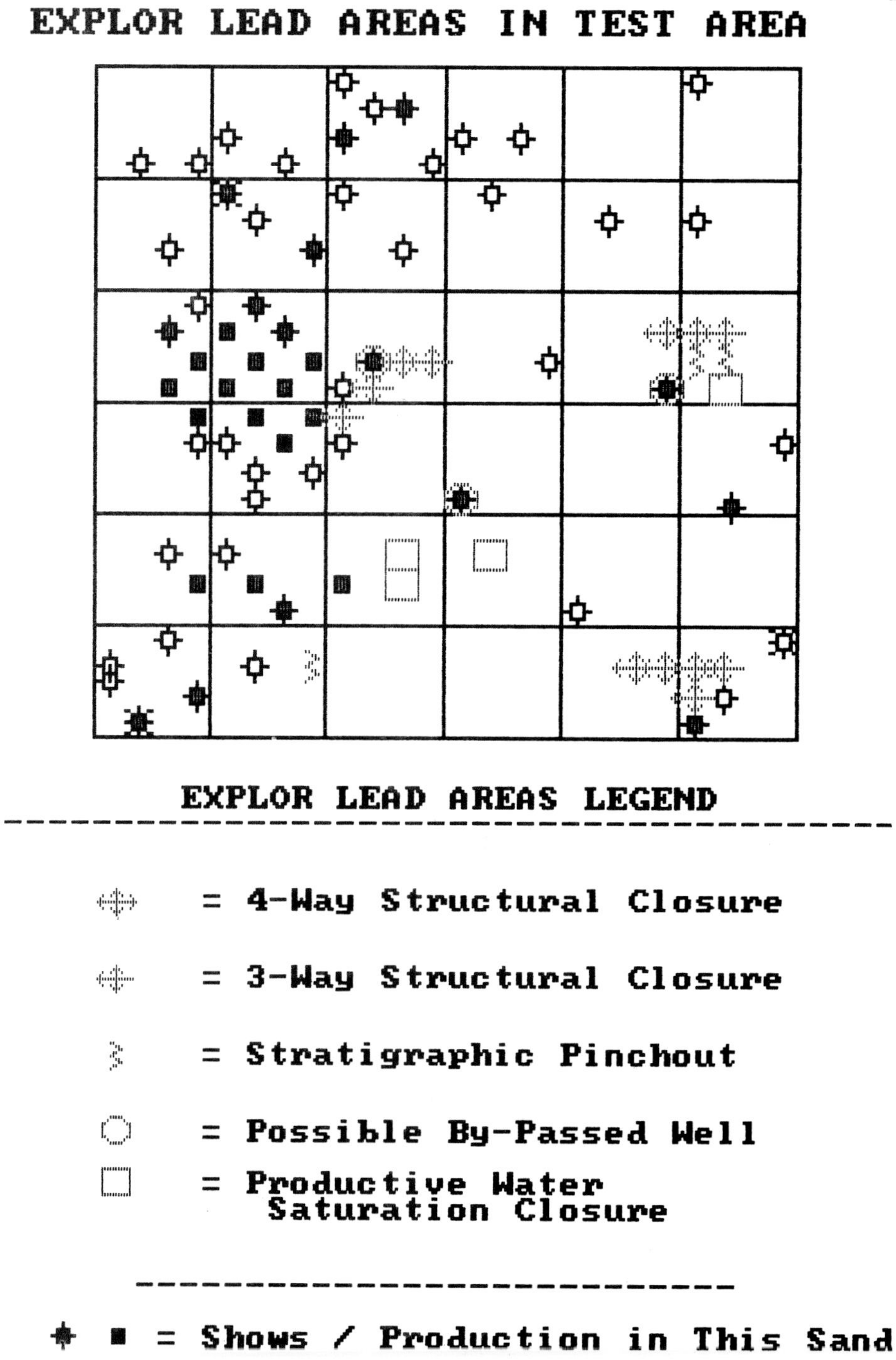

EXPLOR LEAD AREAS IN TEST AREA

EXPLOR LEAD AREAS LEGEND

= 4-Way Structural Closure

= 3-Way Structural Closure

= Stratigraphic Pinchout

= Possible By-Passed Well

= Productive Water Saturation Closure

= Shows / Production in This Sand

Figure 11. EXPLOR "D" sand lead-areas output map for test township.

```
                        D LEAD AREAS IN TEST AREA
--------------------------------------------------------------------------------

     THE FOLLOWING  22  LEAD AREAS WERE DERIVED USING THESE PARAMETERS

                     CHANNEL WIDTH =  1  MILES
          MINIMUM SAND REQUIRED FOR AN ECONOMIC WELL =  3  FT
                SAND PINCHOUT THICKNESS =  2  FEET (MAXIMUM)
                    WATER SATURATION CUTOFF =  55  %


        THERE IS A STRUCTURAL LEAD AREA WITH 3-WAY CLOSURE
     LOCATED IN THE SE / NE OF SECTION  14 OF TEST AREA
     THERE ARE  10  FEET OF NET SAND AT THIS LOCATION.
     DOWNDIP LOG SHOW IN D/A WELL IN SESE SECTION 14 OF TEST AREA


        THERE IS A STRUCTURAL LEAD AREA WITH 3-WAY CLOSURE
     LOCATED IN THE SW / NW OF SECTION  13 OF TEST AREA
     THERE ARE  9  FEET OF NET SAND AT THIS LOCATION.
     DOWNDIP LOG SHOW IN D/A WELL IN SESE SECTION 14 OF TEST AREA


        THERE IS A STRUCTURAL LEAD AREA WITH 3-WAY CLOSURE
     LOCATED IN THE SE / NW OF SECTION  13 OF TEST AREA
     THERE ARE  7  FEET OF NET SAND AT THIS LOCATION.
     DOWNDIP LOG SHOW IN D/A WELL IN SESE SECTION 14 OF TEST AREA


        THERE IS A POSSIBLE BY-PASSED WELL - WATER SAT. VALUE =  .48
     LOCATED IN THE NE / SW OF SECTION  16 OF TEST AREA
     THERE ARE  9  FEET OF NET SAND AT THIS LOCATION.


        THERE IS A STRUCTURAL LEAD AREA WITH 3-WAY CLOSURE
     LOCATED IN THE NW / SE OF SECTION  16 OF TEST AREA
     THERE ARE  8  FEET OF NET SAND AT THIS LOCATION.
DOWNDIP OIL AND GAS SHOW IN D/A WELL IN NESW SECTION 16 OF TEST AREA


        THERE IS A STRUCTURAL LEAD AREA WITH 3-WAY CLOSURE
     LOCATED IN THE NE / SE OF SECTION  16 OF TEST AREA
     THERE ARE  7  FEET OF NET SAND AT THIS LOCATION.
DOWNDIP OIL AND GAS SHOW IN D/A WELL IN NESW SECTION 16 OF TEST AREA


        THERE IS A STRATIGRAPHIC LEAD WITH DOWNDIP SHOW
     LOCATED IN THE NW / SW OF SECTION  13 OF TEST AREA
     THERE ARE  10  FEET OF NET SAND AT THIS LOCATION.
     DOWNDIP LOG SHOW IN D/A WELL IN SESE SECTION 14 OF TEST AREA
```

Figure 12. Printed descriptions of each of lead areas on lead-areas output map. (Continued)

```
     THERE IS A STRATIGRAPHIC LEAD WITH DOWNDIP SHOW
LOCATED IN THE NE / SW OF SECTION  13 OF TEST AREA
THERE ARE  8  FEET OF NET SAND AT THIS LOCATION.
DOWNDIP LOG SHOW IN D/A WELL IN SESE SECTION 14 OF TEST AREA


     THERE IS A STRUCTURAL LEAD AREA WITH 3-WAY CLOSURE
LOCATED IN THE SE / SW OF SECTION  16 OF TEST AREA
THERE ARE  9  FEET OF NET SAND AT THIS LOCATION.
DOWNDIP OIL AND GAS SHOW IN D/A WELL IN NESW SECTION 16 OF TEST AREA


     THERE IS A POSSIBLE BY-PASSED WELL - WATER SAT. VALUE =  .54
LOCATED IN THE SE / SE OF SECTION  14 OF TEST AREA
THERE ARE  11  FEET OF NET SAND AT THIS LOCATION.


     THERE IS A PRODUCTIVE AREA ON WATER SATURATION CONTOUR MAP
LOCATED IN THE SE / SW OF SECTION  13 OF TEST AREA
THERE ARE  9  FEET OF NET SAND AT THIS LOCATION.
DOWNDIP LOG SHOW IN D/A WELL IN SESE SECTION 14 OF TEST AREA


     THERE IS A STRUCTURAL LEAD AREA WITH 3-WAY CLOSURE
LOCATED IN THE NW / NW OF SECTION  21 OF TEST AREA
THERE ARE  11  FEET OF NET SAND AT THIS LOCATION.
DOWNDIP OIL WELL SHOW IN OIL WELL IN SWSE SECTION 17 OF TEST AREA


     THERE IS A POSSIBLE BY-PASSED WELL - WATER SAT. VALUE =  .51
LOCATED IN THE SW / SW OF SECTION  22 OF TEST AREA
THERE ARE  7  FEET OF NET SAND AT THIS LOCATION.


     THERE IS A PRODUCTIVE AREA ON WATER SATURATION CONTOUR MAP
LOCATED IN THE SW / NE OF SECTION  28 OF TEST AREA
THERE ARE  7  FEET OF NET SAND AT THIS LOCATION.
DOWNDIP OIL WELL SHOW IN OIL WELL IN NWSW SECTION 28 OF TEST AREA


     THERE IS A PRODUCTIVE AREA ON WATER SATURATION CONTOUR MAP
LOCATED IN THE SE / NW OF SECTION  27 OF TEST AREA
THERE ARE  8  FEET OF NET SAND AT THIS LOCATION.
DOWNDIP LOG SHOW IN D/A WELL IN SWSW SECTION 22 OF TEST AREA


     THERE IS A PRODUCTIVE AREA ON WATER SATURATION CONTOUR MAP
LOCATED IN THE NW / SE OF SECTION  28 OF TEST AREA
THERE ARE  6  FEET OF NET SAND AT THIS LOCATION.
DOWNDIP OIL WELL SHOW IN OIL WELL IN NWSW SECTION 28 OF TEST AREA


     THERE IS A STRATIGRAPHIC LEAD WITH DOWNDIP SHOW
LOCATED IN THE SE / NE OF SECTION  32 OF TEST AREA
THERE ARE  3  FEET OF NET SAND AT THIS LOCATION.
DOWNDIP OIL AND WATER SHOW IN D/A WELL IN SWSE SECTION 29 OF TEST AREA
```

(Figure 12 continued.)

```
   THERE IS A STRUCTURAL LEAD AREA WITH 3-WAY CLOSURE
LOCATED IN THE SW / NE OF SECTION  35 OF TEST AREA
THERE ARE  15  FEET OF NET SAND AT THIS LOCATION.
DOWNDIP LOG SHOW IN D/A WELL IN SWSW SECTION 36 OF TEST AREA


   THERE IS A STRUCTURAL LEAD AREA WITH 3-WAY CLOSURE
LOCATED IN THE SE / NE OF SECTION  35 OF TEST AREA
THERE ARE  17  FEET OF NET SAND AT THIS LOCATION.
DOWNDIP LOG SHOW IN D/A WELL IN SWSW SECTION 36 OF TEST AREA


   THERE IS A STRUCTURAL LEAD AREA WITH 4-WAY CLOSURE
LOCATED IN THE SW / NW OF SECTION  36 OF TEST AREA
THERE ARE  19  FEET OF NET SAND AT THIS LOCATION.
DOWNDIP LOG SHOW IN D/A WELL IN SWSW SECTION 36 OF TEST AREA


   THERE IS A STRUCTURAL LEAD AREA WITH 3-WAY CLOSURE
LOCATED IN THE SE / NW OF SECTION  36 OF TEST AREA
THERE ARE  19  FEET OF NET SAND AT THIS LOCATION.
DOWNDIP LOG SHOW IN D/A WELL IN SWSW SECTION 36 OF TEST AREA


   THERE IS A STRUCTURAL LEAD AREA WITH 3-WAY CLOSURE
LOCATED IN THE NW / SW OF SECTION  36 OF TEST AREA
THERE ARE  10  FEET OF NET SAND AT THIS LOCATION.
DOWNDIP LOG SHOW IN D/A WELL IN SWSW SECTION 36 OF TEST AREA
```

(Figure 12 continued.)

REFERENCES

Campbell, A.N., Hollister, V.F., Duda, R.O., and Hart, P.E., 1982, Recognition of a hidden mineral deposit by an artificial intelligence program: Science, v. 217, no. 4563, p. 927-929.

Duda, R.O., 1980, The PROSPECTOR system for mineral exploration: Final Report, SRI Project 8172, Artifical Intelligence Center, SRI International, Menlo Park, California, 120 p.

Duda, R.O., 1981, Operation manual for the PROSPECTOR consultant system: Final Report, SRI Project 8172, Artificial Intelligence Center, SRI International, Menlo Park, California, 75 p.

Ennis, S.P., 1983, Expert systems - an emerging computer technology: Oil and Gas Jour., v. 81, no. 30, p. 184-188.

Land, C.B., and Weimer, R.J., 1978, Peoria Field, Colorado - J sandstone distributary channel reservoir, in Energy resources of the Denver Basin: Rocky Mtn. Assoc. Geol. Guidebook, p. 81-104.

McCammon, R.B., 1983, Operation manual for the PROSPECTOR Consultant System: U.S. Geol. Survey Open File Rept. 83-804, 29 p.

Reboh, R., 1981, Knowledge engineering techniques and tools in the PROSPECTOR environment: Final Report, SRI Projects 5821, 6415, and 8172, Artificial Intelligence Center, SRI International, Menlo Park, California, 149 p.

A Hybrid Microcomputer System for Geological Investigations

F. W. Jennings, J. M. Botbol and G. I. Evenden

U.S. Geological Survey

ABSTRACT

The Hybrid S-100 is a medium-sized microcomputer utilizing 64K bytes of memory and an IEEE-696(S-100) bus. It can be configured and used for a variety of computing applications in geologic research.

Five applications of the Hybrid S-100 are described: laboratory automation, process control, data inventory, field-data reduction, and word processing.

Experience has shown that the Hybrid S100 is a flexible device useful for many applications. A list of system components, vendors, and photographs of a transportable Hybrid S-100 are presented.

INTRODUCTION

Geologic investigations require computing support in three fundamental domains: the field, laboratory, and office. Typically, a geologic project uses computers in at least two of these domains. Although the role of the microcomputer may differ with each application, recent advances in microcomputers have made them a valuable tool in each of these domains.

Microcomputers are appealing because of their lower cost, portability, and ease of use compared with larger mainframe systems. Consequently, increasing numbers of microcomputers are finding their way into Earth-science applications.

This report describes a microcomputer system equipped with generally accepted standard accessories including floppy-disk drives, magnetic-tape drive, printer, and terminal. This system, termed the "Hybrid S-100", has been applied to geologic projects in the field, laboratory, and office.

The Hybrid S-100 has replaced an obsolete minicomputer at the U.S. Geological Survey, Woods Hole, Massachussetts Laboratory, as the offline controller of a high-resolution rotating-drum microdensitometer used in analysis.

Because the physical location of the Branch of Atlantic Marine Geology's Data Library makes direct-line access to the Branch multiuser system impractical, a Hybrid S-100 is used there for word processing and data storage and retrieval.

In the sediment analysis laboratory, the Hybrid S-100 controls the flow of data from the analytical devices to the appropriate database, which is maintained in the main Branch multiuser system.

A Hybrid S-100 is being used aboard ship to display data as they are recovered from submersible ocean-bottom data-capture devices, permitting modification and refinement of subsequent device deployments.

The Hybrid S-100 is used on various Branch projects to prepare manuscripts and encode programs.

A wide range of S-100 compatible components is available, and can be integrated as needed into the system. Once in use, the S-100 is a reliable, simple, and dependable system.

To assist those who might decide to pursue a similar endeavor, basic system features and brief descriptions of some typical applications are presented. A list of vendors and component parts is provided in the Appendix A.

OVERVIEW

For applications discussed in this report, the Hybrid S-100 is a single user system that employs an 8-bit microprocessor. The nucleus of this system is a Z-80 Central Processing Unit (CPU) complimented with 64K bytes of memory and various peripheral devices. Internally, an IEEE-696 (S-100) bus is used to interconnect the controllers of the peripheral components with the CPU and power supply. Major peripheral components, on separate circuit boards that plug into the S-100 bus, include a 9-track tape drive and an analog-to-digital (A/D) converter. Peripherals connected directly to the Z-80 computer board include two 8-inch floppy-disk drives, a printer, and a cathode-ray tube (CRT) terminal. When portability is required, the Hybrid S-100 chassis, together with power supply, and floppy-disk drives and 9-track tape drive, is housed in a shock-mounted rack in a hard fibreboard case fitted with four casters. Not included in the case are the terminal and printer, which are transported separately.

The S-100 bus was selected for two principal reasons: first, the ability to expand, change, and otherwise provide a dynamic system, and second, the wide selection (hundreds of boards) of modules available for the S-100 bus. The bus itself is merely a standardized medium by which component parts can communicate. The bus provides for power distribution and consists of a series of integrated connector sockets for insertion of circuit boards. These circuit boards are the functional parts of the system and they can perform a wide variety of tasks, such as containing the CPU, memory, and peripheral interfacing. Indeed, the S-100 can consist of only one such board (which contains a microprocessor, memory, and some peripheral ports), or it can consist of more than 20 circuit boards, each capable of performing tasks that rival the capabilities of minicomputers.

The principal point to be made here is that a bus-structured system such as the S-100 can be tailored to the needs of the application without overkilling the job with unnecessary equipment or sacrificing real needs because of the lack of expansion capability.

The major disadvantage of bus systems such as the S-100 is the burden placed on the end user to have the available expertise in both hardware and software with which to assemble and maintain effectively a viable system. A good working knowledge of digital electronics (and occasionally, analog) is required not only to resolve occasional bus-interface problems, but also to interface the bus-circuit boards with external devices. In addition, all S-100 boards are not created equal, and the user must be careful to ensure that components will interface properly on the bus before acquisition. Even though the Institute of Electrical and Electronic Engineering (IEEE) standards have done much to alleviate S-100 interface problems, some exist yet. After successful electrical interfacing, the specialized software needed to control the components on the bus may be complex. It may require a considerable knowledge, of not only the operating system and assembly language programming, but also of the peculiarities of the physical devices. Although some manufacturers provide software to be employed with their S-100 circuit boards, this cannot be taken for granted nor can the user be confident that the software supplied is appropriate to his task.

DESCRIPTION OF COMPONENTS

It is accepted generally that the best computer hardware is useless without appropriate software to drive the system and serve the user. Thus, it is wise to utilize software that is as close to accepted standards as possible. Therefore, we selected the 8-bit CP/M* operating system with C as the principal programming language, the WORDSTAR word-processing program, dBASE II database-management system, and a number of in-house and proprietary utility programs. The tape drives are driven through a series of in-house, C-language programs required by specific applications. At present, the equipment discussed in the following paragraphs is entirely operational.

Central Processing Unit

The CPU is a Teletek Systemaster Z-80 (8-bit) microcomputer. The microcomputer and 64K bytes of memory occupy a single board which was designed for an S-100 bus. This board is a complete stand-alone computing system; it has two RS-232 serial ports and two parallel ports. It contains circuitry to drive either 8-inch or 5-inch floppy disk drives, or a mixture of both. This board (with software, printer and terminal) can function as a stand-alone word processor or even as a programming-development tool (again, with appropriate software).

Disk Drives

Two Qume DT-8 floppy-disk drives were installed in the Hybrid S-100. Eight-inch floppy-disk drives were selected because of the broad selection of standard software available on 8-inch single-sided, single-density, floppy disks. These disks which are format standardized to a greater extent than are other floppy-disk systems, can be used as a primary data and program dissemination (interchange) medium. The Qume drives are capable of reading both the standard single-sided disks and double-sided, double-density disks.

Tape Drive

The Cipher F880X 9-track tape drive was selected as the most desirable tape drive for this system because it generates tapes that can be read by any other computer (including large mainframe CPU's) using American National Standard Institute (ANSI) compatibility. It is a compact, horizontal, front-loading, self-threading device that can be rack-mounted and included in the transportable fiberboard case discussed later.

The tape controller is an Alloy ITS-100 (S-100 board) that is contained completely on one printed circuit board. The convenience of this interface greatly influenced the selection of this tape drive.

Chassis

The chassis of the Hybrid computer is an Integrand 800RV/7, complete with power supply and S-100 bus. It provides for our current and anticipated needs, in this application of up to seven circuit boards. If required, other chassis configurations can be selected with as many 20 positions.

*) Any use of tradenames in this publication is for descriptive purposes only and does not constitute endorsement by the U. S. Geological Survey.

Terminal

The choice of terminal is the least critical decision made for system components. For simplicity and price, we selected a TeleVideo 910 terminal (CRT) with optional speeds up to 19.2M baud. This nonprinting terminal can be configured for many programming applications (for example, special key commands for word processing), and past experience has shown it to be durable and reliable.

Printer

An Epson RX-80 printer, equipped with a 2K buffer and an RS-232 serial interface, was added to the system. This provides dot-matrix impact output on 9 1/2 inch perforated paper.

Rack and Case

To contain the basic components of the system we designed a case that contained a foam-insulated, industry-standard 19-inch rack. The case has removable front and back panels, and is fitted with four casters for mobility. The case was built by ATS Cases, Inc.

APPLICATIONS

The U.S. Geological Survey's Branch of Atlantic Marine Geology requires a full range of scientific computing for the broad spectrum of marine geologic projects that are generally in process. Most of the computing is carried out on in-house multiuser systems. There, however, are a few applications that are particularly well suited to the Hybrid S-100. These include laboratory automation, process control, data inventory, field-data reduction, and word processing.

Laboratory Automation

The sediment-analysis laboratory performs grain-size analyses on samples collected from the many regions surveyed by the Branch. Both a multichannel particle counter and a rapid sediment analyzer are used to determine the desired range of grain sizes in each sample. The particle counter outputs digital data directly into the Hybrid S-100. The rapid sediment analyzer generates voltages that are fed directly to an A/D converter S-100 board mounted in the Hybrid system. The digital data then are reduced and formatted for direct remote entry in the larger multiuser computer system that hosts the sediment-analysis database. The operator executes a telecommunications program, which allows him to link with the database host. He then transmits the data to the host machine, where the data are merged with an existing database and further processing is performed.

The unique aspects of this application are: (1) the incorporation of an S-100 compatible analog-to-digital converter as a peripheral device; (2) that data are reduced and reformatted; and (3) that by using a telecommunications program, the S-100 computer is used as a terminal for remote data entry to a host computer. BASIC and C were the principal programming languages used for generating the necessary applications programs.

Process Control

One of the devices used in the analysis of image data is a high-resolution, rotating-drum microdensitometer. In the past, the device was driven by a minicomputer for which neither spares nor support is available. Rather than drive the device from a valuable port in our large system, we decided to drive it with the Hybrid S-100. In this way the microdensitometer is a stand-alone device that will not interfere with other Branch processing and unnecessarily occupy an I/O port on a busy system. Z-80 Assembler and C are the principal programming languages used.

Data Inventory

The Branch data library is situated such that a direct line to the principal Branch computer system is impractical. Further, only a few programs are needed to support the data-library activity. Because of this, and given the highly specific nature of the application, the S-100 was selected as the principal support computer for data-inventory processing within the Branch. This application further reduces the workload on the main system and provides dedicated support for the library, utilizing principally dBASE II and WORDSTAR.

Field-Data Reduction

Most Branch data are collected at sea. Some of the data-capture devices, mounted in watertight recoverable containers, are designed to sit on the ocean bottom and record data on internal magnetic-tape cartridges. After recovery, some mechanism is required to play-back immediately and partially reduce the collected data so that the scientist-in-charge can determine parameters for subsequent deployments.

The Hybrid S-100, interfaced to a peripheral drum plotter and cartridge reader and utilizing an Alloy Engineering Co. S-100 interface card, was selected to perform the task of shipboard play-back and reduction. The system has been tested successfully in the laboratory, and at-sea testing will be undertaken in the near future. This application was programmed primarily in the C language with some Z-80 assembler language.

Word Processing

This application has become popular with the programming and scientific staff for composing and editing letters, reports, and draft scientific manuscripts. It is used also to examine acquired data and to generate source code during application-program generation. Although word-processing programs are available widely for all classes of large and small computers, the personal aspects of the microcomputer seem to have universal appeal. Thus, when data are not being processed, the S-100 may be used to prepare a manuscript or encode a program. Word processing usually is done with WORDSTAR.

CONCLUSIONS

The Hybrid S-100 is a valuable single-user computing supplement to the multiuser computing systems that handle the bulk of in-house computing in a geologic research organization.

The Hybrid S-100 requires the availability of a person knowledgeable in digital electronics, particularly when the intended use involves a variety of S-100-compatible peripherals. Portability is one of the key factors in the utility of the Hybrid S-100, as is its flexibility of configuration, which makes it viable for many applications.

APPENDIX A

HYBRID S-100 MICROCOMPUTER: LIST OF COMPONENTS.

Component	Supplier
CPU	Teletek 4600 Pell Drive Sacramento, CA 95838 (916) 920-4600
Chassis	Integrand 8620 Roosevelt Ave. Visalia, CA 93291 (209) 651-1203
Disk Drive	Qume Corporation 2350g Qume Drive San Jose, CA 95131 (408) 942-4000
Tape Drive	Cipher Data Products, Inc. 10225 Willow Creek Road San Diego, CA 92138 (619) 578-9100
Tape Interface	Alloy Engineering, Inc. 100 Pennsylvania Ave. Framingham, MA 01701 (617) 875-6100
Terminal	TeleVideo Systems, Inc. 1170 Morse Ave. Sunnyvale, CA 94086 (408) 745-7760
Printer	Epson America, Inc. 3415 Kashiwa Street Torrance, CA 90505 (213) 539-9140
Rack and Case	ATS Cases, Inc. 25 Washington Street Natick, MA 01760 (617) 653-6724
Modem	Hayes Microcomputer Products, Inc. 5923 Peachtree Industrial Blvd. Norcross, GA 30092 (404) 441-1617
dBASEII	Aston-Tate 10150 West Jefferson Blvd. Culver City, CA 90230 (800) 437-4329
WORDSTAR	Micropro International Corp. 33 San Pablo Ave. San Rafael, CA 94903 (415) 499-1200

APPENDIX B

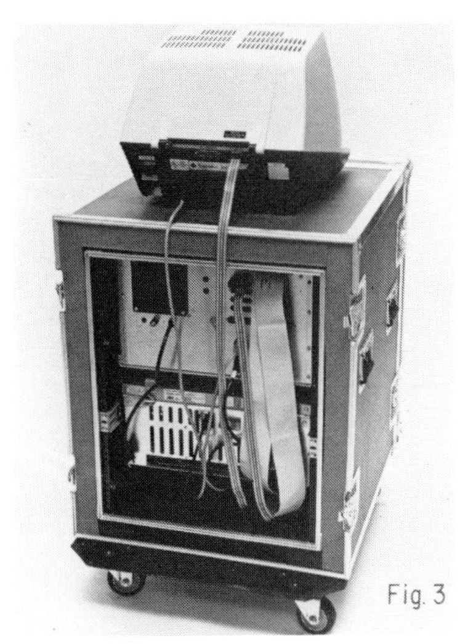

Figure 1. Completely enclosed Hybrid S-100 prepared for shipment. Terminal and printer are shipped in separate case.

Figure 2. Complete Hybrid S-100 ready for operation. Note foam padding around equipment rack.

Figure 3. Hybrid S-100, rear view, showing cables and ports.

Figure 4. Hybris S-100 CPU chassis disk drives are integral to chassis.

Figure 5. Hybrid S-100, inner view, showing (A) power supply, (B) disk drives, and (C) slots for peripherals.

Figure 6. Hybrid S-100 showing tape drive open and in service position. Reel of tape is shown in loaded position. Tape is loaded through access door on front panel. Threading is automatic.

Management of Earth-science Databases and a Small Matter of Data Quality

J. D. Bliss

U.S. Geological Survey

ABSTRACT

An Earth-science database should only be established if: (1) it has a clearly defined mission, (2) a principal user(s) is known and available for consultations, and (3) adequate and appropriate data are known to be available. The database structure is best when it follows the conceptual needs to be made with clear identification of relevant data, presence of a specific user-recall sequence, condensation of redundancy, and grouping data elements in time and scale. Both the principal user and the database manager are accountable for a successful database. Maintaining data quality requires careful evaluation of data sources, minimal error introduction during data entry, and uniform data formatting. Data must conform to physical and chemical laws when applicable or be evaluated with quality indexes (QI) when appropriate. Data elements with finite responses may be controlled tightly by using either open or closed tables of values. Users need to be warned if data are not subject to quality control or evaluation. Documentation needs to be present (1) to record decisions of data managers regarding how data are excluded and handled, and (2) to set standards and rules for both data contributors and users to ensure the database is used correctly.

INTRODUCTION

Managers of Earth-science databases are faced with a host of problems, beginning with creation of a database and continuing through its operation. Management needs to assure the presènce of reliable and useful data in Earth-sciences databases. For that reason, the problems of maintaining data quality are of primary concern here. Databases developed from discrete data will be addressed; this excludes data and data problems relating to high frequency or analog observations and related data-reduction schemes. However, many of the conclusions made are valid for other types of databases. No attempt is made to discuss particular databases or data-management languages.

DEFINITIONS OF TERMS USED

Earth-science database--a collection of interrelated, geologically based data housed at a central location. Example: collection of analytical results from rock samples.

Entity--basic unit for which data will be stored. Example: a specific rock sample.

Record--a group of attributes usually for an entity.

Data element--smallest unit within the database which has meaning to the user. Example: sodium content in ppm of a specific rock sample.

Attribute--property or characteristic of a data element to be stored with each entity. Example: sodium content (data element) in ppm is given as 15.3 (attribute) for a given rock sample (entity).

DATABASE CONSTRUCTION

The effectiveness of database management depends, to some extent, on database design. The design of the database and the structure of the records that make it depend on the database-management system used. However, data independence should be strived for because it is desirable. Much can be, and should be, considered independent of the database-management language and is the primary topic of the following discussion. The process which creates databases actually starts at the time the researcher begins to impose simplifications on a complex world--one which has an unknown natural structure. This conceptual structure is a useful approximation of part of the natural structure. The design and data content of the approximate model can be translated into a physical structure. This, in turn, is the blueprint used in the database design (Longe and others, 1978).

Data elements should group as they are in the conceptual model in time and scale; modifications are made only to accommodate a clearly defined anticipated recall sequence. Database designers always should consider creating several databases to correspond to the several levels of data detail. Files containing related material should share one or more common data elements that contain uniformly defined attributes. This will allow retrievals to bracket several databases. Database families of this type can reduce redundancy even though retrieval mechanics will be somewhat more complex where several databases are needed to describe completely the phenomenon.

Ackoff (1967) suggests that condensation will help to make databases more useful, and this is an important consideration during design. Condensation can be effective inasmuch as most information contains redundancy. Organizing the proposed data elements into several different schemes can help one recognize these redundancies. Care must be taken that condensation does not lead to confusion in meaning: not only may some redundancy be necessary but meaning also may lie in the way data elements are associated (Longe and others, 1978).

Harris, Winczewski, and Umphrey (1982) suggest that database design for spatial data needs to be keyed to the collection site. The site should have two identifiers, one recognized by the reporter, and the second assigned by the database manager. The second identifier will be common for all records describing material or observations made repeatedly at the same geographic coordinates. Under their scheme, the site always could be identified uniquely, particularly if addressed in several records by several reporters, each perhaps with their own site identifiers. They also show that database-management operations should be divided clearly into (1) data requirements and (2) application requirements. The former should be generalized and the second specific. Assisting users in applications programming could become dangerous to the long-term health of the system if it becomes the database staff's primary function. With good management, software for a limited number of repeated applications can and should be provided.

Database managers should be wary of the suggestion by Harris, Winczewski, and Umphrey (1982) of allowing users to enter descriptions, remarks, or comments with little or no restriction. Uncontrolled data entry in general is a potentially dangerous practice and should be controlled carefully to ensure against abuse. A list of Harris and others' principles is given in Appendix A.

One aspect of spatially located databases that usually is not identified is the uncertainty of the exact location. Uncertainty includes both the degree of precision and inaccuracies of the measurement. Longe and others (1978) propose that uncertainty be reported with the geographic coordinates used. They also give a table relating a minimum radius of uncertainty to the map scale (Table 1). Because the map scale falls by a factor of five (say from 1:5,000 to 1:25,000) the uncertainty radius on the ground increases by a factor of one-fifth. Therefore, in databases that give map scales, one can obtain some measure of inherent uncertainty in the location. To my knowledge, most databases do not report uncertainty; the databases operated by the Geological Survey of Austria routinely do identify uncertainty in location, and it is a substantial concern to them (Dr. W. Schnabel, oral communication, 1981).

Table 1. Approximate uncertainty in location as function of map scale as given in Longe and others (1978).

Map Scale		Minimum Radius of Uncertainty on the Ground	
1 inch to the following:	Ratio	Feet	Meters
1000 feet	1 : 12,000	82	25
1320 feet	1 : 15,840	82	25
2640 feet	1 : 31,680	82	25
0.789 mile	1 : 50,000	82	25
1 mile	1 : 63,360	105	32
2 miles	1 : 126,720	207	63
3.95 miles	1 : 250,000	410	125
7.89 miles	1 : 500,000	820	250
15.78 miles	1 : 1,000,000	1640	500

Database design can be complex; this is partly due to the complexity of the phenomena being described. There also is a tendency to computerize more facts than needed. Scientific researchers perceive their critical deficiency to be a "lack of relevant information" (Ackoff, 1967). Although this cannot be discounted as an important problem for database managers, Ackoff (1967) also observes that an "over abundance of irrelevant information" is even a greater problem. Those who design the data structure may lack the knowledge needed to sort out the relevant from the irrelevant. Researchers who consult those in database design tend to believe that any data associated with a phenomenon are valuable, including some data that rarely are available or that may never be used. Unfortunately, the irrelevance of data cannot be anticipated specifically. This determination might be made only, for example, by examining the history of a database. The inability to define irrelevant data also is due to poor understanding of the phenomenon under inspection; a situation which may occur in research. Although a database so constructed possibly would offer opportunities to explore the data directly, that usually does not happen. A database designed with data elements for all the variables will require additional work both to acquire and to enter the large number of items, and overall quality and quality assurance efforts will suffer.

Data extracted from the scientific literature should never be so extensive as to make one believe that the database actually will replace the published source. This is a serious problem in several public databases and represents a special situation of the previous discussion. Of course, some published sources are of limited availability and the process of including extensive comments may capture important points. Nevertheless, the databases should meet well-defined standards and contain a relevant subset of the published source. Each record should contain a data element requiring the clear identification of its source in the literature.

DATABASE USERS--THE GOOD, THE BAD, AND THE UNINFORMED

A valid database can be established only after the future users are defined clearly. Databases that are not used, or are underused because they are irrelevant to the decision process of the users, may reflect too little consultation with users during design and construction, or perhaps the phenomenon is not yet understood sufficiently to allow for meaningful systematization of related data. (The latter situation is difficult to recognize considering the complexity of most processes in which researchers are interested.) There should be a known _principal_ user who has specific decision processes for which the database is to be used (Ackoff, 1967).

Those in database design need to press the user systematically and persistently for guidance during the design stage, before expensive mistakes are made. Obviously, deficiencies in a database will become apparent in the best of situations, but this can be minimized by insisting that a detailed document be prepared and circulated among the principal users. The users should be among those accountable for the successful implementation of the database. In the beginning, a pilot system should be built so that the users can experience its operation and then recommend corrections.

Although there is a need for flexibility in database operation, database administrators can be far more helpful if they understand what questions and decision paths the researcher will be using. When database operations become too removed from the primary users and their needs, incompatibility will occur between the two. Some managers are attempting routinely to meet several radically different needs in a single database. This is possible only if one set of needs is a subset of a broader set routinely supported in database operations. Multiple use of databases in areas of incomplete overlap will make the system seem "useless" to some users, particularly those interested in subordinate database. It can be assumed, unfortunately, that novice users will attempt to ask questions that the database was never designed to answer. Adequate documentation will not preclude this type of abuse.

The user is one of the three primary groups involved with database operations, which also include the data contributors and the database administrator and staff. If the data user and contributor are identical, substantial advantages are added. One disadvantage is that a user-contributor tends to have a narrow view point, one considerably less than the scope of the database, and will submit data that fall only within that viewpoint and disregard other readily available data. The user as a user rather than a contributor comes first, and it is the function of management to see that all procedures are evaluated as to their impact on the user.

DATA QUALITY

One of the problems faced in database operations is the extreme inconsistency in the quality of source-document data. Selecting between good and bad data may not be possible or even recognized if database administrators simply do not have the expertise to make the selection. Furthermore, each database contains errors introduced by the process of data entry. Although specific procedures of record checks can be devised that reduce the number of introduced errors, the problem of bad data in the source literature is more difficult to access. Data tables in publications can contain more inconsistencies and mistakes than would be suggested by the quality of the text. Usually researchers take greater care in preparing the text than they do in compiling tables of the supporting data. Stockmayer (1978) notes that the uncritical acceptance of bad scientific information can lead to social penalties. Although bad data in databases for the geological sciences may or may not have similar effect, the presence of several sets of numbers or observations that are incompatible will frustrate the user. Stockmayer (1978) suggests that the cost of searching and evaluating critical computerized data are "less than 0.2 percent of the original research and about one-quarter of the cost of initial publication." The potential benefit to database users of prior evaluation should justify easily the expenditure of this amount.

Unfortunately, many databases lack either quality control or evaluation, although some methods have been developed which attempt to do so. Milne and others (1982) identify the need to alert the user to the potential deficiencies in a given database. They also make use of internal checks for quality control when possible. Alternatively, they evaluate data by computing what is termed a quality index (QI) factor for data types where quality control is not possible. The degree of detail or resolution are factors that can be expressed by QI. New records describing features present already in the database are compared by the QI factors and only the one with the highest values is kept. Some databases containing water chemistry have begun to use percent ionic imbalance for evaluation purposes. Other operations attempt to exclude what is perceived to be poor-quality data. Further problems in assuring quality arise because some data in the database are anecdotal or are derived from unique, nonrecurring events. In addition, a great amount of geologic data also is interpretive--it will become dated. About all a manager can do in this situation is the ensure that the various types of data fall within broad but reasonable boundaries, conform to the literature from which they were extracted, and are formatted systematically. The user should be warned explicitly about which databases contain data that can neither be subject to quality control nor be evaluated. Surprisingly, the lack of uniform data formatting is a problem in some public databases.

One strategy that can improve greatly data quality is what Longe and others (1978) term a "table of value" or what some systems term a "dictionary." The table of value gives a list of all permissible attributes to a specific data element. This ensures uniformity and makes retrievals keyed to that data element more reliable. Tables of value can be closed where they contain a fixed number of valid responses or open where additions to the responses listed can be made. The penalty for using this tool is loss of flexibility and occasional arbitrary classifications (Longe and others, 1978).

An important yet neglected task is to record decisions made by the database manager in handling problems which may be unimaginable when the database was established. During a database life time, several managers will make decisions on how data are to be included, excluded, or handled. Undocumented decisions will result in confusion and frustration, not to mention wasted resources and actions, in managers who follow. Additionally, those who produce geologic data are not necessarily the primary users. It is, therefore, critical that both contributors and users understand the rules of the game and what database elements represent. The need for a standards and criteria document for each database or system seems self-evident yet such documentation is seldom prepared.

OPERATIONS

The operation of databases is characterized by inertia. Flaws in structure and other shortcomings generally increase in visibility. Modification becomes increasingly difficult as database size increases. One solution proposed to handle some problems that are unsolvable due to database size is the distributive database, in which parts are located at several centers. Although accessibility and utilization would improve surely, the uniformity needed to meet the collective goals of the total database will be difficult to ensure particularly during a long period of time. Local scientific viewpoints gradually would cause the various parts to diverge. This has been a problem even under a single management addressing a similar set-up inhouse. It is neither reasonable for users to expect nor cost-effective to support high rates of turn-around for users of large databases. In fact, data subsets or indices either on or offline may assist answering many user questions.

INSTITUTIONAL SETTING

An increasingly serious problem is the growing competition between funding needs of database operations and research needs of the organization in which they are housed. Loughridge (1981) notes that institutional data management continues to be the best strategy for public databases but increasingly is difficult to support. In part, this is due to institutions which do not recognize that databases are an institutional resource that require continuous support. It should not be forgotten that data are as valuable a product of research as publications. Databases, therefore, are a national resource which institutions hold in trust; they are "perishable if not properly preserved and protected" (Loughridge, 1981). It is clear that redundant collections of data housed in several government agencies leads to expenditures of funds which could be better spent on data analysis if the data were housed centrally. The ideal world would be one where there would be a single data center supported by several user institutions. This is unlikely to happen for some types of databases, because agreement on what constitutes acceptable data is variable and controversial.

SUMMARY

Earth-science databases will be most successful if they are designed with the user's needs in mind. Database design should conform as closely as possible to the physical structure which has been translated from the conceptual model which a user has devised to approximate a part of the real world. Several databases may be required so that data will be grouped at approximately the same level of detail and in the same temporal context--as close as possible to the conceptual model. Only a specific recall sequence should modify this format. The final design blueprint will be improved greatly by systematic documentation of users' requirements and the establishment of a pilot database. Once design is completed, the management of the database environment is of primary concern. Four clearly recognized areas of function are involved: data acquisition and entry (encoding); quality assurance; operations involving data extraction, updates, and backups; and user assistance. The successful orchestration of these functions should meet users' needs (or the primary mission of the databases) as well as protect and support the contained databases.

Although none of the functional areas are unimportant, the absence of quality assurance efforts can be detrimental. Attributes need to be tested to determine if they fall within realistic limits. Physical and chemical laws should be obeyed. Tables of values giving all permissible attributes for specific data elements should be established when appropriate. Alternatively, if such tests are not possible, some attempt needs to be made to evaluate the data. For example, records with different resolution or detail may be assigned different values (quality index). If no quality assurance is present, or not possible, as in the situation of interpretive data, the database user needs to be so informed. Documenting operation decisions can provide useful insight to users and future database managers.

REFERENCES

Ackoff, R.L., 1967, Management misinformation systems: Management Science, v. 14, no. 4, p. 147-156.

Harris, K.L., Winczewski, L.M., and Umphrey, H.R., 1982, Computer management of geologic and petroleum data at the North Dakota Geological Survey: North Dakota Geological Survey Report Invest. No. 74, 34 p.

Longe, R.V., Burk, C.F., Dugas, J., Jr., Ewing, K.A., Fergusion, S.A., Gunn, K.L., Jackson, E.V., Kelly, A.M., Oliver, A.D., Sutterlin, P.G., and Williams, G.D., 1978, Computer-based files on mineral deposits: guidelines and recommended standards for data content: Geol. Survey Canada Paper 78-26, 72 p.

Loughridge, M.S., 1981, Afterthoughts, in Frontiers in data storage, retrieval and display: Proceedings of a marine geology and geophysics data workshop: National Geophysical and Solar-Terrestrial Data Center, Boulder, Colorado, p. 145-146.

Milne, G.W.A., Fisk, C.L., Heller, S.R., and Potenzone, R., Jr., 1982, Environmental used of the NIH-EPA chemical information system: Science, v. 215, no. 4531, p. 371-375.

Stockmayer, W.H., 1978, Data evaluation: a critical activity: Science, v. 201, no. 4356, p. 577.

OTHER REFERENCES

Doszkocs, T.E., Rapp, B.A., and Schoolman, H.M., 1980, Automated information retrieval in science and technology: Science, v. 208, no. 4439, p. 25-30.

Hittelman, A.M., ed., 1981, Frontiers in data storage, retrieval and display: Proceedings of a marine geology and geophysics data workshop, National Geophysical and Solar-Terrestrial Data Center, Boulder, Colorado, 151 p.

Lide, D.R., Jr., 1981, Critical data for critical needs: Science, v. 212, no. 4501, p. 1343-1349.

Roederer, J.G., 1981, Considerations in the development of a national geophysical data policy: EOS, v. 62, no. 27, p. 569-570.

Wade, N., 1981, Computer data banks: the delights and dangers: Science, v. 214, no. 4522, p. 772-773.

APPENDIX A

The following list gives all of the GEOSTOR principles as presented in Harris and others (1982, p. 2):

(1) GEOSTOR principles apply to the design of systems for the management of map-oriented data.

(2) The needs of, or the demands on, the user, as data collector or data analyst must be considered above that of the machine.

(3) Each item expected to be retrieved on a regular basis should be defined as a data element, a "pigeonhole" to be filled with a value. Generalized supplementary data elements should be made available for unexpected items.

(4) The user should be able to enter description, remarks, or comments with little or no restriction.

(5) The data elements, supplementary elements, and comments should be grouped into a data structure exploiting the natural association among the variables or the anticipated recall demand on them.

(6) A separate, custom-designed data structure should be prepared for each natural association of variables.

(7) All data structures for a collecting site should be keyed to that site and its location data elements.

(8) Each collecting site should have two identifiers. The first should be unambiguous, preferably assigned by the system. The second should be one assigned by and easily recognized by the user.

(9) The quality of maps and cross sections produced from map-oriented data depends heavily on the quality of the location values used in all phases of processing, from data collection, through coordinate conversion, to final map product.

(10) Except for rare situations, justified by high demand, the processing of the retrieved values should be outside the retrieval system.

Micro-GRASP, a Microcomputer Data System

R. W. Bowen

U.S. Geological Survey

ABSTRACT

Micro-Grasp, which was developed on a 64K INTERTEC* SuperBrain microcomputer, is a scaled-down version of the U.S. Geological Survey's Geologic Retrieval And Synopsis Program (GRASP). The GRASP system is used extensively on a variety of small- to large-scale host computers to access many diverse types of data. Micro-Grasp retains the major features of GRASP and provides access to private (or nonshared) databases of small to medium sizes (less than 300K bytes). Micro-Grasp is a command-oriented system which operates on ASCII data files of fixed-field character and numeric data. The files are in a matrix form where each row is a record and each column is a field. Micro-Grasp offline-data-input facilities can be used to reduce the maintenance cost of host-resident GRASP databases.

INTRODUCTION

During the last few years, a software system termed GRASP has been used within the U. S. Geological Survey for accessing data files. The GRASP system, and the files it accesses, are stored on various timeshare host computers which support a broad spectrum of users. As workloads increase, the fixed resources of the host computer must be allocated to users in units of decreasing size. Available disk-storage space decreases and response times become longer. Thus, the host computer seems smaller to individual users.

Recent advances in microprocessor technology have resulted in the development and availability of microprocessor-based terminals. Many of these devices can be used as terminals to a host computer as well as small independent computers. These small computers are termed micros. Currently, the most popular programming language for micros is BASIC.

With the development of appropriate software, written in BASIC, much of the work done on timeshare host computers can be processed on micros. One such micro is the SuperBrain. Micro-Grasp was developed on the 64K INTERTEC SuperBrain to provide an alternative mechanism for accessing and maintaining data files.

*) Any use of tradenames in this publication is for descriptive purposes only and does not constitute endorsement by the U. S. Geological Survey.

AVAILABILITY

Micro-Grasp has been implemented on IBM PC-compatible microcomputers, the Tektronix 4050 series microcomputers, the INTERTEC SuperBrain and various other CP/M-based systems. The IBM PC-compatible and the SuperBrain versions are available through the following address:

U. S. Geological Survey
Information Systems Division
Office of Customer Services, Mailstop 802
12201 Sunrise Valley Drive
Reston, VA 22092
tele: (703) 860-6585

SYSTEM OVERVIEW

The user controls the execution of Micro-Grasp by entering a series of commands. A total of 13 commands can be used. These commands allow the user to enter new data records, establish selection criteria, create histograms, create subfiles, calculate summary statistics, display data values, perform arithmetic operations, calculate a table of correlation coefficients, review the status of the system, and terminate the session. The commands are summarized as follows.

The FILE command is used to change the current data-file name thus establishing a new data definition (dd). This dd may be in the header section of an existing file or it may be entered by the user. The INPUT command is used to create a new data file or to add records to an existing data file. Each value is entered when the system prompts with the field name and size. The DEFINE command is used to create new numeric-type variables by entering a definition in the form 'name=expression.' The expression may contain field names, constants, arithmetic operators, and intrinsic functions (abs, sqrt, log, exp, mod, int, sin, cos). The CONDITIONS and LOGIC commands are used to select specific records which satisfy some selection criteria. The LIST command is used to display specified values from a set of selected records. The HISTOGRAM command is used to generate a histogram for a selected item. The histogram may be displayed on the screen or directed to the printer. The CORRELATION command calculates a table giving all pairwise correlation coefficients for as many as seven numeric items. The STATISTICS command is used to determine the number of values, range, sum, mean, variance, and standard deviation of specified variables (fields) in selected records. The CONSTRUCT command is used to create subfiles containing selected fields from selected records. The REVIEW command is used to provide the user with information on the current status of the session. The NAMES command is used to output the current dd. The dd may be written as a file header, directed to the printer, or displayed on the screen. The QUIT command is used to end a session.

Data Files

Each Micro-Grasp data file starts with a data definition (dd). The dd gives a description of the fields in the data records. The data records immediately follow the dd in a data file. Data values may be numeric, character strings, or blank and must be in a fixed-field format.

The data records can be thought of as a large table of rows and columns. Each row is a record and each column is a field. Hence records are divided into a fixed number of fields. The dd (which defines the data records) is formatted as follows:

The first record contains two values (N & L) separated by at least one blank character. N is the number of fields in a data record and may have a maximum value of 42. L is the total number of characters in a data record and may have a maximum value of 255. This first record is followed by N field-description records. Each field-description record has a unique field name of up to 5 characters, a type code (n for numeric and c for character), a beginning-field-position, an ending-field-position, a minimum possible value, and a maximum possible value. If the ending-field-position is not given, it defaults to the position immediately preceeding the start of the next field (or to the value of L for the final field). The minimum and maximum possible values apply only to numeric fields (type code n). If they are not specified, no range checking will be done when using the INPUT command. If they are specified, the ending-field-position must be given. If only the minimum possible value is specified, no check for maximum value is performed.

The following is an example of a Micro-Grasp file:

```
data       --> 5  47
definition     date     n 1     7    800101  891231
               site     c 8     12
               user     c 13    29
               proj     c 30    39
               file     c 40    47
data       --> 800415 R    NWright       ORA         rn01.g
records        800415 R    PSchruben     ORA         smastr
               800415 R    WGrasp        WHGrasp     mas_100
               800415 R    WGrasp        WHGrasp     mas_250
               800416 D    GBrethauer    Geomath_d   yucpz
               800416 R    JBliss        Nrap        ncrib
               800416 R    JBliss        Nrap        nref
               800416 R    MShrestha     Nrap        mas2
               800416 R    MShrestha     Nrap        ncrib
               800416 R    NWright       WLDMIN_R    coxcu
               800417 D    NBridges      Phos        phsbib2
               800417 R    ESivritepe    ORA         info
               800417 R    ESivritepe    ORA         seam
                                      .
                                      .
```

Command Descriptions

The FILE command is used to establish or change the selected data file. A data file must be selected before using any of the data-specific commands (LIST, NAMES, INPUT, CONDITIONS, CONSTRUCT, CORRELATION, HISTOGRAM, STATISTICS, and DEFINE). The system expects a data-file name to be entered when it displays the prompt "ENTER DATA FILE NAME:." The data file must contain at least a data definition (dd). If no data-file name is entered (that is, the user hits only the CR key), the system assumes that a new data file is to be created by entering the dd and data values on the keyboard. The format of the dd is described in the Data Files section. Execution of the FILE command will nullify existing conditions, logic, and any variables which have been created using the DEFINE command.

The INPUT command is used to input data in a 'prompted' mode creating new records. These records may be added to the currently selected data file or written to a new file. The data definition is displayed and the user is given the opportunity to enter an output file name. If the file name entered is the same as the current data-file name, records will be added to the current data file. If the file name is different, records will be written to that file and ANY EXISTING DATA ON THAT FILE WILL BE OVERWRITTEN!

For each field, the name of the field is printed along with an indication of the maximum field size in which the user can enter the value. Missing values are indicated by entering a CR without a value. If a data value is entered which contains more characters than the specified field size, the bell sounds and an error message is displayed. The user then may reenter the value. If minimum and (optionally) maximum values have been specified in the dd, a range check is made on the value entered. The value for a field may be repeated from the preceeding record by entering the caret symbol (∧) or a period (.). The user may backup to a previous field by entering a minus sign (-). After values for all fields (some possibly missing) have been entered, the user is asked if he wants to reenter the values for that record, continue to the next record, or quit the input process and enter a new command.

The DEFINE command is used to define new variables. The user is prompted with a numeric label preceding each definition. Definitions must be entered on a single line as a variable name, followed by an equal sign (=), followed by an arithmetic expression. As many as 10 variables may be defined. The variable name must be new. Existing names cannot be redefined. A question mark (?) may be entered to list existing names. An expression may consist of numeric constants, field names, the arithmetic operators exponential (∧), multiply (*), divide (/), add (+), or subtract (-), and the functions absolute value (abs), square root (sqrt), base e logarithm (log) , power of e (exp), modulo 10 (mod), integer truncation (int), sine (sin), or cosine (cos). Trigonometric function arguments must be in degrees. The total number of names, constants, operators, and functions in one expression may not exceed 20. Parentheses may be used to group the expression or alter the normal order of evaluation. Expressions within parentheses and function arguments are evaluated first, followed by exponentiation, followed by multiplication and division, followed by addition and subtraction. Within the same level, operations are performed left to right. Field names containing special characters, such as (,),.,+,-,*,/, and , may not be used in definitions.

After all definitions have been entered, a CR (with no definition) will end the process of defining new variables. Definitions stay in effect until a new dd is selected (via the FILE command). For example, 'day=10*mod(date/10)+mod(date)' is a definition which defines 'day' as the last 2 digits of the field 'date.'

The CONDITIONS command is used to enter (or add to) a set of 'conditions.' Each condition is labeled with a letter which is later used in the LOGIC command to refer to the condition. Conditions are in the form 'name relation value.' The name must be a valid field name. The relation must be eq (equal to), lt (less than), gt (greater than), le (less than or equal to), ge (greater than or equal to), ne (not equal to), or cs (contains string). The value may be blank (indicating missing data), a number, or a character string. If conditions have been entered previously, the system allows the user to add to the existing set or enter a new set. When all desired conditions have been entered the RETURN key (CR) is used to exit this command. A question mark (?) may be entered to get a list of valid field names. 'date gt 800416' is an example of a condition.

The LOGIC command is used to complete the selection criteria by connecting the conditions which have been entered with logical operators. The valid logical operators are * (and) , + (or) , and - (not). The conditions to be connected are referenced by the letter which was assigned when the conditions were entered.

The logic is entered in the form of an expression containing letters standing for conditions and the logical operators *, +, and -. The - operator will be applied before the * operator which will be applied before the + operator. The order of evaluation can be overridden by the use of parentheses. A 'null' expression (CR only) is entered to indicate that no selection criteria are to be applied. The expression also may be a letter with no operators indicating a single condition as the selection criterion. '-(a+b)*c' and '.not.(a.or.b).and.c' are examples of the same logical expression.

The LIST command provides a method of displaying the values of specified fields for each record which satisfies the selection criteria (that is, conditions and logic). The display format is user-selected as column type or row type. In column-type format, the values for all selected fields are displayed in a single row (wrap-around will occur if the line contains more than 80 characters). In row-type format, the value for each selected field is displayed on a separate line (blank values are not displayed) and a line of *'s is displayed after each record has been processed.

The LIST command will ask the user to enter the pause interval (entering CR suppresses the generation of pauses), the listing format (c or r), the output device, and the names of fields to be displayed. The fields to be displayed are indicated by entering a list of field names. The list of field names most recently entered (by the LIST, CONSTRUCT, CORRELATION, or STATISTICS command) may be selected or a different list may be entered. Lists are entered by typing the name of each desired field (followed by CR) after the system prints a number. The list is ended by entering CR without a field name. If all fields are to be listed, the user may simply enter '.all' when the first number is printed. A question mark (?) may be entered to get a list of valid field names.

During the listing process, lines are displayed until the 'pause interval' is reached, at which time a bell sounds. At this point, the user may continue the listing by entering CR or terminate the listing by entering a nonblank character (followed by CR) when the bell sounds.

The STATISTICS command is used to display summary statistics for each of a number of specified fields. The statistics are calculated using only those nonblank values for specified fields in records which satisfy the current selection criteria. The summary statistics are number-of-values, minimum, maximum, sum, mean, variance, and standard deviation. All summary statistics are computed for numeric-type fields. Only the number-of-values is computed for character-type fields. The list of field names is entered as described in the LIST command.

The HISTOGRAM command is used to construct a histogram for a selected field. The user is asked to enter the field name, minimum and maximum values, and the number of classes desired. The user then is asked if equal-size class intervals are desired. If not, the upper limit for each class, except the last, must be entered. Next the user is asked if the mean of some field should be calculated for each of the intervals. If so, the name of the field is requested. Finally, the user is asked if printer output is desired.

The histogram then is calculated and displayed. The frequency scale is selected automatically to fill the screen horizontally. If means of some other field were requested, their values would be displayed on the right of the histogram.

The CORRELATION command is used to calculate all pairwise correlation coefficients for a list of as many as seven numeric fields. The list of field names is entered as described in the LIST command. Output is in the form of a table (or matrix). The pairwise correlation coefficients are given in the lower triangular portion of the table. The corresponding number of pair values used in the calculation is given in the upper triangular portion of the table. Output may be printed as well as displayed on the screen.

The CONSTRUCT command is used to create Micro-Grasp-compatible subfiles containing selected fields from selected records. The selection can be restricted further by specifying a set of link-key values. These key values are read from a user-specified file and compared with the values from a user-specified key field of the current data file. The constructed output file contains only those records which meet the selection criteria and whose key-field value matches a link-key value. The data fields output are specified by entering a list of field names.

The REVIEW command is a way to determine the selected data file name, the set of conditions and logic currently in effect, the most recent list of field names that was entered, and the names of any defined variables.

The NAMES command is used to output the dd to the screen, printer, or to a selected file. If any defined variables exist, their names will be displayed on the screen.

The QUIT command is used to exit the system and return to operating system.

Individual records of any Micro-Grasp data file may be referenced by number using the intrinsic field name 're.' Each data file contains this hidden field even though it does not occur in the header or data sections of the file. The field name re may be used in any context which is valid for field names.

Lists of field names are required for the LIST, CONSTRUCT, STATISTICS, and CORRELATION commands. These lists are entered ordinarily by the user; however, they may be read from a file. To read a list of field names from a file, the user should enter '.read' followed by the name of the file containing the list when the first list item is requested. For instance, '1. .read myfile' would tell the system to get the list of field names from the file named myfile. Each record of myfile should contain a single field name starting in column one.

Command Summary

The commands which may be issued (and their function) are as follows:

COND- ENTER CONDITIONS ('FIELD REL VALUE') WHICH WILL
BE USED IN FORMING SELECTION CRITERIA.

LOGI- COMPLETE FORMATION OF SELECTION CRITERIA BY
ENTERING A LOGICAL EXPRESSION.

LIST- DISPLAY SELECTED FIELD VALUES.

FILE- SELECT OR CHANGE THE DATABASE TO BE USED.

NAME- DISPLAY FIELD NAMES, THEIR TYPES AND EXTENTS.
ALSO USED TO SAVE THE CURRENT DATA DEFINITION.

CONS- CONSTRUCT A SUBFILE OF SELECTED FIELDS.

STAT- COMPUTE SUMMARY STATISTICS ON SELECTED FIELDS.

HIST- CREATE A HISTOGRAM OF A SELECTED FIELD.

CORR- CALCULATE A TABLE OF CORRELATION COEFFICIENTS.

INPU- ENTER DATA IN PROMPTED MODE ACCORDING TO THE
CURRENT DATA DEFINITION.

REVI- DISPLAY THE CURRENT STATUS OF THE SESSION.

DEFI- DEFINE NEW FIELDS AS FUNCTIONS OF ORIGINAL FIELDS.

QUIT- TERMINATE THE SESSION.

EXAMPLE RUN

The following example illustrates how the various commands can be combined. Throughout the example lines marked with * contain user-entered data (which is in lowercase).

```
WELCOME TO THE USGS MICRO-GRASP DATA SYSTEM.
*ENTER DATA FILE NAME: geochem

*ENTER COMMAND: list
*ENTER PAUSE INTERVAL: 27
*ENTER LISTING TYPE (C OR R): r
*ENTER 1 (FOR PRINTER), 2 (FOR SCREEN), OR 3 (FOR DISK): 1
ENTER THE LIST OF ITEM NAMES:
* 1. .all
id     =d102134
name   =barker f
rlse   =    7212
date   =  680424
proj   =  955270
job    =969
fdnum  =kgr-2
lat    =  39.250
long   = 105.627
local  =08093
fmtn   =unnamed
sio2   =  70.480
al2o3  =  14.780
fe2o3  = 0.33
feo    = 3.03
mgo    = 0.49
cao    = 0.98
na2o   = 3.23
k2o    = 5.62
tio2   = 0.25
p2o5   = 0.1
mno    = 0.04
co2    = 0.1
cl     = 0.01
fpct   = 0.06
*******************************
id     =d102135
*a


*ENTER COMMAND: define
* 1. femag=(feo+0.9*fe2o3)/(mgo+feo+.9*fe2o3)
* 2.


*ENTER COMMAND: statistics
*DO YOU WANT PRINTER OUTPUT? no
*DO YOU WANT TO ENTER A NEW LIST? yes
ENTER THE LIST OF ITEM NAMES:
* 1. feo
* 2. fe2o3
* 3. mgo
* 4. femag
* 5.
```

```
feo      HAS 112 VALUES.
  MIN=0.01    MAX=18.53 MEAN=3.4665
  SUM=388.25 VARIANCE=24.3087 STD DEVIATION=4.9304

fe2o3    HAS 112 VALUES.
  MIN=0.01    MAX=14.2  MEAN=5.3228
  SUM=596.15 VARIANCE=25.2769 STD DEVIATION=5.0276

mgo      HAS 103 VALUES.
  MIN=0.03    MAX=29.2  MEAN=4.9976
  SUM=514.75 VARIANCE=57.1283 STD DEVIATION=7.5583

femag   HAS 103 VALUES.
  MIN=0.0382046486775    MAX=0.991511035654    MEAN=0.7110
  SUM=73.2332338086 VARIANCE=0.0652 STD DEVIATION=0.2553

*ENTER COMMAND: correlation
*DO YOU WANT PRINTER OUTPUT? no
*DO YOU WANT TO ENTER A NEW LIST? no

         feo      fe2o3    mgo      femag
feo      1        107      99       99
fe2o3  -.327584  1        99       99
mgo     .567383 -.043203  1        99
femag  -.320312  .420809 -.705041  1

*ENTER COMMAND: define
* 1 . femag=(feo+0.9*fe2o3)/(mgo+feo+.9*fe2o3)
* 2 .

*ENTER COMMAND: conditions

*A. fmtn ne
*B. fdnum cs ppg
*C. fdnum cs -
*D. femag lt 0.201
*E. femag gt .965
*F. name eq braddock
*G.

*ENTER COMMAND: logic
*ENTER LOGIC: -c*a*(d+e)+b+f
*ENTER COMMAND: list
*ENTER PAUSE INTERVAL:
*ENTER LISTING TYPE (c OR r): c
*ENTER 1 (FOR PRINTER), 2 (FOR SCREEN), OR 3 (FOR DISK): 2
*DO YOU WANT TO ENTER A NEW LIST? y
ENTER THE LIST OF ITEM NAMES:
* 1 . re
* 2 . name
* 3 . fdnum
* 4 . fmtn
* 5 . feo
* 6 . fe2o3
* 7 . mgo
* 8 . femag
* 9 .
```

```
SELECTION CRITERIA FOR THIS OUTPUT IS AS FOLLOWS:
LOGIC: -c*a*(d+e)+b+f
CONDITIONS:
A. fmtn  ne
B. fdnum cs ppg
C. fdnum cs -
D. femag lt  .201
E. femag gt  .965
F. name  eq braddock


re  name      fdnum     fmtn      feo    fe2o3 mgo    femag
 18   braddock 13-3-101              15.4    2.2    5.1 .773132
 19   braddock 13-3-102              10.8    1.3    7.0 .630996
 20   braddock 13-3-104               3.4   0.16   0.96 .786856
 21   braddock 13-3-104               1.6   0.34   0.48 .798827
 22   braddock 13-4-101              12.3    1.5    6.1 .691139
 23   braddock 13-4-102               1.4    1.4   0.16 .943263
 25   sharp wi s71-ppg-  pikes pe     1.3    1.3   0.06 .976285
 26   sharp wi s71-ppg-  pikes pe     9.8    1.3   0.49 .957243
 27   sharp wi s71-ppg-  pikes pe     2.4   0.46   0.12 .9591
 41   snyder g 1028      precambr    0.74   0.75   0.03 .979239
 61   snyder g 1297      precambr    1.58   1.18   0.05 .981427
 74   miesch a 36        sawatch     0.16   0.14    7.2 .0382047
 76   miesch a 46        sawatch     0.16   12.2    0.2 .982363
 77   miesch a 47        sawatch     0.16   12.8    0.1 .991511
 79   miesch a 51        sawatch     0.12   12.2    0.1 .991071
 83   miesch a 65        sawatch     0.12   12.2    0.4 .965218
 87   miesch a 74        sawatch      0.2   0.13    5.5 .0544955
 88   miesch a 77        sawatch     0.16   11.5    0.1 .990575
 89   miesch a 79        sawatch     0.16   12.5    0.4 .96613
 90   miesch a 80        sawatch     0.16    11.    0.3 .971043
 95   miesch a 101       sawatch     0.62   0.01   12.4 .0482769
 99   miesch a 149       sawatch     0.28   0.53    3.1 .196267
 102  miesch a 154       sawatch     0.24   12.7    0.1 .991504

*ENTER COMMAND: logic
*ENTER LOGIC:
*ENTER COMMAND: construct
*ENTER OUTPUT FILE NAME: test.out
*DO YOU WANT TO ENTER A NEW LIST? yes
ENTER THE LIST OF ITEM NAMES:
* 1 . .read test.lst
IS CONSTRUCTION TO BE BASED ON MATCHING VALUES BETWEEN
* A LINK FIELD AND A FILE OF LINK KEYS? yes
*ENTER ITEM NAME: id
*ENTER NAME OF FILE CONTAINING LINK KEYS: test.key

*ENTER COMMAND: file
*ENTER DATA FILE NAME: test.out

*ENTER COMMAND: stat
*DO YOU WANT PRINTER OUTPUT? no
ENTER THE LIST OF ITEM NAMES:
* 1 . .all

name  HAS  23 VALUES.

fdnum HAS  23 VALUES.

fmtn  HAS  17 VALUES.
```

```
feo    HAS  23 VALUES.
  MIN= .12    MAX= 15.4    MEAN= 2.75043
  SUM= 63.26 VARIANCE= 20.6571 STD DEVIATION= 4.54501

fe2o3 HAS  23 VALUES.
  MIN= .01    MAX= 12.8    MEAN= 4.77391
  SUM= 109.8 VARIANCE= 30.6238 STD DEVIATION= 5.53387

mgo    HAS  23 VALUES.
  MIN= .03    MAX= 12.4    MEAN= 2.19348
  SUM= 50.45 VARIANCE= 11.3159 STD DEVIATION= 3.36391

lat    HAS  23 VALUES.
  MIN= 38.633    MAX= 40.987   MEAN= 39.4317
  SUM= 906.928 VARIANCE= .772011 STD DEVIATION= .878642

long   HAS  23 VALUES.
  MIN= 104.877    MAX= 106.777   MEAN= 105.457
  SUM= 2425.51 VARIANCE= .495828 STD DEVIATION= .70415


*ENTER COMMAND: file
*ENTER DATA FILE NAME: michop

*ENTER COMMAND: names
*ENTER 1 (FOR PRINTER), 2 (FOR SCREEN), OR 3 (FOR DISK): 2


 21 216
id        n 1    4
cnty      c 5    32
lat       n 33   41
long      n 42   50
pool      c 51   81
yrdcp     n 82   86
yrabp     n 87   92
npool     n 93   94
era       c 95   106
systm     c 107  122
sries     c 123  138
avtck     n 139  142
prod      n 143  146
prova     n 147  151
crucm     n 152  160
apigr     n 161  169
sulfr     n 170  178
elev      n 179  183
dscwl     c 184  191
depth     n 192  197
frmtn     c 198  216

*ENTER COMMAND: statistics
*DO YOU WANT PRINTER OUTPUT? no
ENTER THE LIST OF ITEM NAMES:
*1. depth
*2.

depth   HAS  96 VALUES
  MIN= 1130   MAX= 10142  MEAN= 4099.64
  SUM= 393565  VARIANCE= 2.74601E+06 STD DEVIATION= 1657.11
```

```
*ENTER COMMAND: histogram
*ENTER ITEM NAME: depth
*ENTER MIN VALUE: 1000
*ENTER MAX VALUE: 11000
*ENTER NUMBER OF CLASSES: 5
*DO YOU WANT EQUAL SIZE CLASS INTERVALS? y
*DO YOU WANT MEANS OF ANOTHER ITEM OVER THESE INTERVALS? y
*ENTER ITEM NAME: avtck
*DO YOU WANT PRINTER OUTPUT? n
                                frequency                          mean of
depth                   10         20         30         40         50 avtck
 1000         +---------+---------+---------+---------+---------+
               XXXXXXXXXXXXXXXXXXXXXXXX                            3.7083
 3000         +
               XXXXXXXXXXXXXXXXXXXXXXXXXXXXXXXXXXXXXXXX            12.975
 5000         +
               XXXXXXXXXXXXXXXXXXXXXXXXXXXXX                       62.379
 7000         +
               X                                                   158
 9000         +
               X                                                   20
 11000        +


*ENTER COMMAND: histogram
*ENTER ITEM NAME: depth
*ENTER MIN VALUE: 1000
*ENTER MAX VALUE: 7000
*ENTER NUMBER OF CLASSES: 6
*DO YOU WANT EQUAL SIZE CLASS INTERVALS? n
ENTER UPPER LIMIT FOR CLASS  1
*? 1500
ENTER UPPER LIMIT FOR CLASS  2
*? 2000
ENTER UPPER LIMIT FOR CLASS  3
*? 3000
ENTER UPPER LIMIT FOR CLASS  4
*? 5000
ENTER UPPER LIMIT FOR CLASS  5
*? 6000
*DO YOU WANT MEANS OF ANOTHER ITEM OVER THESE INTERVALS? n
*DO YOU WANT PRINTER OUTPUT? n

                          frequency
depth              10         20         30         40         50
 1000      +---------+---------+---------+---------+---------+
            XXXXX
 1500      +
            XXXXXX
 2000      +
            XXXXXXXXXXXXX
 3000      +
            XXXXXXXXXXXXXXXXXXXXXXXXXXXXXXXXXXXXXXXX
 5000      +
            XXXXXXXXXXXXXXXXX
 6000      +
            XXXXXXXXXXXX
 7000      +
```

```
*ENTER COMMAND: conditions

*A. pool eq berea
*B. depth gt 7000
*C. npool ne
*D.

*ENTER COMMAND: logic
*ENTER LOGIC: a+b+c


*ENTER COMMAND: list
*ENTER PAUSE INTERVAL: 15
*ENTER LISTING TYPE (C OR R): c
*ENTER 1 (FOR PRINTER), 2 (FOR SCREEN), OR 3 (FOR DISK): 2
*DO YOU WANT TO ENTER A NEW LIST? y
ENTER THE LIST OF ITEM NAMES:
* 1. depth
* 2. npool
* 3. pool
* 4.

SELECTION CRITERIA FOR THIS OUTPUT IS AS FOLLOWS:
LOGIC: a+b+c
CONDITIONS:
A. pool    eq berea
B. depth   gt  7000
C. npool   ne


depth   npool   pool
  1860          berea
  1895          berea
  7077          niagaran reef
  5240    3     richfield
          5     richfield
  3062    2     traverse
  5027    3     richfield
 10142          richfield
          4     richfield
  5139    3     richfield
  3837          berea
*ENTER COMMAND: file
*ENTER DATA FILE NAME: sites
```

```
*ENTER COMMAND: input
THE FOLLOWING DATA DEFINITION WILL BE USED FOR DATA INPUT:
6 160
org      c 1   34
systm    c 35  60
cntct    c 61  77
addr     c 78  116
city     c 117 148
cntry    c 149 160
*ENTER OUTPUT FILE NAME: temp
*DO YOU WANT TEMP TO CONTAIN THE RECORDS IN SITES (y/n)? n
AFTER THE LAST FIELD FOR EACH RECORD ENTER 1 TO REENTER,
 2 TO CONTINUE OR 3 TO END INPUT.
A DATA VALUE OF '-' SKIPS BACK A FIELD.
A DATA VALUE OF '.' COPIES FIELD VALUE FROM LAST RECORD.

*org    = Hungarian Geophysical Institute
*systm  = RIAD-35 under IBM's VOS
*cntct  = Dr. L. Zalahi-Sebess
TOO MANY CHARACTERS ENTERED FOR contact REENTER VALUE.
*cntct  = Dr. Zalahi-Sebess
*addr   = Columbus u. 17-23
*city   = Budapest XIV
*cntry  = Hungary
*REENTER, CONTINUE, OR QUIT (1, 2, or 3)? 2

*org    = Environmental Systems Research
*systm =  PRIME 400
*cntct  = Mr. J. Dangermond
*addr   = 380 New York Street
*city   = Redlands, CA 92373
*cntry  = USA
*REENTER, CONTINUE, OR QUIT (1, 2, or 3)? 3

*ENTER COMMAND: list
*ENTER PAUSE INTERVAL:
*ENTER LISTING TYPE (C OR R): r
*ENTER 1 (FOR PRINTER), 2 (FOR SCREEN), OR 3 (FOR DISK): 2
ENTER LIST OF ITEM NAMES:
* 1. .all
org    =Hungarian Geophysical Institute
systm  =RIAD-35 under IBM's VOS
cntct  =Dr. Zalahi-Sebess
addr   =Columbus u. 17-23
city   =Budapest XIV
cntry  =Hungary
*****************************
org    =Environmental Systems Research
systm  =PRIME 400
cntct  =Mr. J. Dangermond
addr   =380 New York Street
city   =Redlands, CA 92373
cntry  =USA
*****************************


*ENTER COMMAND: quit
```

An Interactive Program for Creating and Manipulating Data Files Using an Apple Microcomputer System

J. C. Butler

University of Houston

ABSTRACT

The proliferation of relatively inexpensive microcomputer systems in industry and academia has resulted in considerable experimentation. If the individual has previous computing experience (or as a novice gains experience in programming), such experimentation eventually leads to questioning the practicality of using a micro for building and manipulating files and allowing a mainframe computer to do the actual processing. Most geoscientists amass a considerable volume of data early in their careers, and many believe that there would be an advantage in being able to play an active role in data manipulation (sorting, rearranging, adding files, adding new variables, adding new samples, etc.) either as an end in itself or as a prerequisite to some formal analysis.

The commercial database-management systems that the author has experimented with (such as PFS:File and PFS:Report, published by Software Publ. Co.) work well but the structure of the files is fixed, and in many such systems it is not possible to use these files as input to other programs. Thus, many geoscientists currently are planning, or already are implementing a home-made file-management program.

In this paper a number of considerations facing the do-it-yourselfer are discussed--file structure (random-access versus sequential), the organization of pertinent information within the file, the question of compatibility with existing file structures (such as DIF and DAISY), editing the file, flexibility of searching, adding or subtracting information, selecting subsets, and sorting. Of prime concern is how to take full advantage of the amount of memory available. the author's system was written in Applesoft BASIC for an Apple IIe with 64K. With the disk-operating system (DOS) loaded on the language card some 48+K are available for actual operation. Using the machine-language program CHAIN (available on the SYSTEM MASTER disk that comes with the Apple system) to link relatively short programs, a system has been developed that allows a matrix with up to 8,000 elements (the product of the number of rows and columns) to be worked within an interactive mode.

The program that is described grew during a series of fits and spurts and no claim is made to its uniqueness. What is important is that the investigator is comfortable with the design of a system and it meets the individual needs of the investigator. In general, such a system will require some degree of design by the individual and, for most commercial systems, tinkering is prohibited.

INTRODUCTION

During the past three years the number of microcomputer systems in the Department of Geosciences (14 full-time faculty members) at the University of Houston has increased from zero to seven. Although all seven systems now are dedicated to various tasks (data gathering, word processing, general department record-keeping and student/faculty problem solving), we went through a stage characterized by considerable experimentation. The first systems were not much more than novelties and usually served as a way of releasing tensions through a game of skill or chance or the calculation of a biorhythm prior to a meeting with the Dean. Gradually, those individuals with previous programming experience started translating old programs into BASIC and using the micros as a part of their class exercises or research efforts. Within a short period of time, we began to experiment with various commercial database-management systems. PFS:File, Report and Graph work well and are in use as aids in keeping track of alumni and short-course participants. However, these packages, as well as others with which we have experimented, lack the flexibility that most individuals desire. In general, the files, once created, are not compatible with formats required by other programs that one may wish to use as part of a data-processing system. Most geoscientists amass a great volume of data early in their careers and desire an easy to understand, easy to modify, and easy to install system that facilitates creating files and manipulation of the data either as an end in itself (such as sorting) or prior to some formal analysis (such as calculating binary or unary statistics). Given our experiences and those of colleagues both in academia and industry, it is not surprising that many individuals currently are implementing their own home-made file-management systems. A number of listing of such programs have been published--such as GEOFILE (Burwell and Topley, 1982)--and many of these were examined to see if they would serve as the basis for a management system for general use in our department. It soon became apparent that the desire to personalize the system made it more efficient to begin from scratch.

Many of the ideas and concepts discussed here were incorporated into FILEMASTER--an interactive file-management system written in Applesoft BASIC for an Apple IIe (the software is available from the author or COGS at cost). The purpose of this paper is to consider some of the options and strategies open to the designer of a file-management system and not to present new or unique approaches as there are not any.

DATA FORMAT

Kalish and Mayer (1981) present a cogent assessment of the need for a standard data format to facilitate data exchange between systems and users. Even short-time users are most likely familiar with the frustration that is experienced when it is realized that a carefully constructed data file for a program X is not compatible with programs Y and Z and, in fact, the file may be unique to program X.

Experience to date suggests that a file structure that looks similar to a more familiar form of the data (such as a table in the text) may be preferred to a soundly designed yet somewhat cumbersome format. In developing GEOFILE, for example, Burwell and Topley (1982) used a matrix structured such that the rows contain information pertaining to the measured properties (variables) whereas each column represents an object--a structure that is the reverse of what the author prefers. Kalish and Mayer (1981) argue that DIF, a format for data interchange (developed by Software Arts, Inc.), would be an ideal universal format, and note that it has been adopted by a number of the commercially available software packages (such as VISI-CALC-written by Software Arts, Inc.-and DB MASTER-from Stoneware Microcomputer Products). However, the DIF structure seems to be complicated unduely (see example in Kalish and Mayer,

1981). In spite of various pleas for a universal format, one does not exist at this time and, given the independent nature of most investigators, it is unlikely that an agreement can be made on a truely universal system.

Kalish and Mayer (1981) note that the user has three options when faced with data in an incompatible format: (1) retype the data; (2) write a program to reformat the data; or (3) modify the programs to accept data in different formats. They conclude that all three processes are inefficient and tedious. Given a belief in personalized design and the lack of a universal format, option (2) seems to be a realistic alternative. It may be inefficient but it is not tedious. Utility programs are available that increase the speed of loading and saving of files (such as DIVERSI-DOS by Bill Basham). Even a relatively long file (some 8000 elements in the matrix) can be reformatted in a few minutes.

FILEMASTER creates files in which the information is stored in sequence as follows:

(1) NN--the number of elements in the matrix;
(2) ND--the number of objects (rows) in the matrix;
(3) NV--the number of variables (columns) in the matrix;
(4) LA$(NV)--a vector containing NV column labels;
(5) TT$(ND)--a vector containing ND object labels;
(6) X(ND,NV)--the raw-data matrix: NV columns and ND rows.

Most of the data sets processed by the author contain ratio or interval data, but it is possible to mix in nominal or ordinal data by defining a suitable coding strategy.

STATISTICS WITH DAISY (written by Kevin C. Killion and available through Rainbow Computing Inc.) is a versatile general statistics program with high-resolution graphics output. This system uses its own file-structure system:

(1) ND (rows);
(2) NV (columns);
(3) LA$ (column labels);
(4) X (raw data);

and a simple program can be written to <u>translate</u> FILEMASTER files into DAISY files (or vice versa). Many of the <u>commercially</u> available systems describe the file structure in detail sufficient to allow writing a translator program (for example APPLE PLOT--available through Apple Computer, Inc.) but some go to considerable length to prevent examination of the organization of information within their files.

RANDOM-ACCESS VERSUS SEQUENTIAL FILES

Many microcomputer systems support both random-access and sequential data files. In general, sequential files require that the entire file be loaded into memory in order to use or modify any part of the file whereas any location can be accessed in a random-access file without having to read the entire file. Most of our applications require the presence of the entire file so a sequential file is satisfactory.

MAXIMIZING AVAILABLE MEMORY

Programs such a FILEMASTER tend to grow as <u>topsy</u> because one modification may lead to another. The last large version occupied some 70 sectors, required some 14K to store and allowed working with matricies with less than 1,800 elements. It became obvious that further expansion was possible only by an even further reduction in the maximum allowable size of the raw-data matrix.

Also, even with the DIVERSI-DOS utility, loading of the program took almost 30 seconds.

A reasonable solution was to break up the large program into several smaller subprograms and to use the machine-language utility program CHAIN (available on the Apple System Disk) to load and execute each subprogram without clearing the portion of memory in which the current values of the variables are stored. The current FILEMASTER (seven subprograms) now can work with matricies with up to 8,000 elements.

BUILDING THE FILE

Experience gained through instructing noncomputer users was incorporated into the file-building subprogram. Making the input sequence correspond to typing a data table seems to make file development a nonformidable task. Column labels (names of the variables) are input first with verification immediately following. The matrix is filled in a row-by-row fashion. Object designators (titles, sample number, etc.) are input followed by a prompt for the value of each variable in the proper sequence. File building is terminated by pushing the return key when asked for the title for the next object.

GENERAL CONSIDERATIONS

If various options are to be included in a file-management system some thought needs to be given as to the manner in which the user is informed of the options that are available. A simple menu with clear and concise terminology seems to be the most effective way to communicate options. After completing one option, control should return to the section in which the menu is displayed. If possible, all of the pertinent information should appear on the screen--manditory frequent reference to a manual should be avoided. Usually, some length of time will elapse between issuing an instruction and return of control to the keyboard. Rather than leave the screen blank during this time, making the user wonder if the system has failed, it is advisable to provide a flashing statement such as WORKING on the screen.

Efficient handling of errors that arise during execution is an essential requirement. Experience suggests that many of the typical errors occur during the reading of a file or during the transfer of a file from memory to the disk. The Apple ONERR GOTO function should be an integral part of a program such as FILEMASTER so as to trap errors before execution is halted. In most situations valuable time spent in building a file can be saved if care is given to the structuring of the error-handling routine(s).

Total time could be reduced significantly by executing a machine language or compiled version of FILEMASTER. The compiler program available to the author (EINSTEIN from the Einstein Corporation), however, will not allow variables to occur in the arguments of DIMension statements. Thus, a considerable part of the flexibility of the current version would be lost. If a user knows that a given size of the raw-data matrix will not be exceeded, then it would make sense to fix the dimensions and execute the compiled version. In general, an increase in total processing time by a factor of three to five can be realized.

EDITING

An efficient editing system is a must in any file-management package. The user should have the ability to change column labels in addition to editing the entire file or individual rows in the matrix. Our preference to use prompts consisting of the sample descriptor and appropriate column labels so that it is clear which element in the matrix is being examined. The current value is displayed and a return indicates acceptance. Typing the corrected value and pushing the return-key changes the stored value.

MODIFYING EXISTING FILES

Rarely does a file remain intact during the course of an investigation and a file-management system must allow for updating and rearrangement. Options should be available for adding new samples and new variables to an existing file.

Options should be included to allow creating a file with the same samples but a subset of the variables (a ternary subset, for example) or a file with all of the variables but a subset of the samples.

Modification options encourage experimentation with data sets and have proven to be useful as a part of formal courses as well as in independent research.

SELECTING SUBSETS

A subprogram that allows selection of samples that meet certain user-defined criteria has proven to be a most useful addition to the file-management system. For example, one could select all of those rock samples in which MgO is greater than 5.5%, FeO is greater than 2.00% and less than 4.00%, and TYPE (a coded indicator of tectonic environment of emplacement) is equal to 2. Again, such an option has encouraged users to experiment with their data sets and in the process they seem to gain insight into the structure of their data.

After selection of a subset the current version of FILEMASTER replaces the original file with the subset. If the user wishes to reclaim the original larger set the version stored on disk must be read in again.

TRANSFORMATION

Many familiar data sets consist of raw data combined with new variables that are simple functions of the raw data. In a set of major-element chemistry of igneous rocks, for example, users might want to include the magnesium number, the sum of the alkali oxides, or the ratio of ferric to ferrous iron. A FILEMASTER subprogram (TRANS) allows the user to create linear sums, ratios, special variables (such as the Mg number), and percentages. Also, the user may elect to transform the data according to one of a number of predefined options--row normalization, column normalization, proportional to the range, chi-squared, etc. Such a subprogram should be structured so that new transformations and options can be added with a minimum of difficulty. This can be supported by using a Transformation Menu with a numeric code and the ON J GOTO command.

COMMUNICATION WITH PROCESSING PROGRAMS

The current version of FILEMASTER includes subprograms that computes summary unary and binary statistics and prepares binary and ternary scatter diagrams. If the user's microsystem has a GRAPPLER print interface card (from Orange Micro Systems) any high-resolution display on the screen can be dumped to the printer with one or two commands. There is no limit as to the number and variety of subprograms that can be added to the basic file-management system as long as the system makes use of a utility such as CHAIN that does not change the current value of all variables when a new subprogram is executed.

OUTPUT OPTIONS

Saving the current file to the disk must be one of the options on the general menu as one may wish to same one or all of the modified versions of the raw-data file. As noted previously, errors may occur during the writing the file to the disk and an error trapping routine should be active during execution of the SAVE option.

FILEMASTER allows the user to sort the data according to the value of a given variable (and carry the other variables) as well as alphabetizing the string row descriptors. The rearrangement option is included in the general printing subroutine as most users think about the process at the time that hardcopy of the data set is requested. Assuming that the typical data set contains more than a few rows and columns, the sorting routine should not be of the bubble-sort variety. A quick sort routine is part of the print subprogram in FILEMASTER.

As the Apple IIe (and II and II+ as well) lack a PRINT USING routine one was modified to allow for printing neat output as well as for allowing variables stored in memory in exponential form to be printed in decimal form. A user-defined function that rounds to the desired number of places to the right of the decimal point should be part of the PRINT USING routine.

SUMMARY AND CONCLUSIONS

If the past three years serve as a way of forecasting the future, there should be a marked increase in the number of microcomputers in colleges and universities during the next three years. At the University of Houston the administration plans to have a large number of workstations in place within three years. Each workstation will consist of a microcomputer that can serve as a stand-alone system or as a smart terminal for communicating with one of the larger mainframe computers on campus. We believe that efficiency will be maximized if the building of files, their maintenance and modification are tasks that are undertaken on the microcomputer.

In making a decision to purchase one of a large number of commercial file-management systems or in deciding to develop ones own personalized system there are a number of questions that should be answered. For example, what is the structure of the file required and produced by the system? Can the file be translated easily into another format? Can a commercial system be modified so as to develop a more personalized system? How large a file will the system handle? How much time is required to undertake routine tasks (such as sorting)? How easy is it to edit the file? Can new information be added to the file? Can files with similar structure be added together?

The same set of questions also must be answered before one decides to develop a file-management system from scratch. For some routine tasks (such as a system to store addresses, sort, obtain subsets and print mailing labels) there are a number of commercial systems available that will get the process going for a relatively small investment of both money and time. Unless the user gains a great amount of satisfaction from programming, the decision to develop a personalized file-management system should be made only after a thorough analysis of need and up-to-date knowledge of existing systems.

REFERENCES

Burwell, A.D.M., and Topley, C.G., 1982, GEOFILE: an interactive program in BASIC for creating and editing files: Computers & Geosciences, v. 8, no. 3-4, p. 323-334.

Kalish, C.E., and Mayer, M.F., 1981, DIF: a format for data exchange between applications programs: Byte, v. 6 , no. 11, p. 174-184.

Program ROCALC: Norms and Other Geochemical Calculations for Igneous Petrology

J. C. Stormer, Jr., M. H. Dehn, W. P. Leeman and D. J. Matty

Rice University

ABSTRACT

ROCALC is a program which transforms rock analyses in oxide weight percent to molecular and ionic equivalents, and calculates CIPW and other "norms" along with various ratios, ternary projections, and other chemical indices. It is written in BASIC for the APPLE II microcomputer (an IBM PC version is available). ROCALC illustrates the particular advantages of the microcomputer environment for these types of calculations, especially optimization of interaction with the user for flexibility of input and output options.

ROCALC is part of a library of programs under development for the analysis of chemical data from igneous rocks. Other programs in the library create and edit files of analytical data, create output files for plotting variation diagrams, calculate densities and viscosities of silicate liquids, and solve linear least-squares mixing problems (fractionation and assimilation modeling).

INTRODUCTION

Microcomputers are suited ideally to the types of calculations that are used routinely for the transformation and analysis of chemical data from igneous rocks and minerals. The CIPW norm calculation, in which the analysis, given in terms of oxide percentages, is transformed into percentages of idealized mineral components is one example. The calculation of the structural formula from a mineral analysis is another. These calculations require a relatively lengthy series of steps with several branches, but each step is simple mathematically and the calculations can be performed on a series of relatively small data arrays. These data can be entered from the keyboard or files on diskette as required. A minimal memory is required for active data, and is accomodated easily by a low-cost microcomputer.

In the microcomputer environment there is no real benefit in optimizing the code for processing speed or storage efficiency for these calculations. No charge is incurred for CPU and I/O time. The physical limitations of the user limit the speed of the input and comprehension of the output. The greatest advantage to be gained is, therefore, in increasing convenience for the user and reducing errors in data entry.

The petrology package we have been developing consists of a number of programs written in Applesoft BASIC. These programs also make extensive use of a package of machine language subroutines ("Amperware", copyright 1982 by

Scientific Software Products, Inc., 5726 Professional Circle, Suite 105, Indianapolis IN 46241). These provide a number of powerful extensions to Applesoft BASIC. Our package is designed to run on an APPLE II+ or IIe with 64K or more of memory, at least one disk drive, DOS 3.3, and a printer. (We have modified ROCALC to work on IBM PC-compatible machines.)

ROCALC CALCULATIONS

Most of the mathematical calculations used in ROCALC should be familiar to petrologists, although there are some possibly unique options such as calculation of the ferric/ferrous iron values from temperature and oxygen fugacity, and output normalized to an equimolar anion (oxygen) basis. Many petrology texts contain a thorough discussion of the CIPW norm and other indices. In the subroutine to calculate the CIPW norms (lines 1000-1520, Appendix A), REM statements key each step to the procedure outlined in detail in Barker (1983, p. 65-77). The CIPW norm calculation recasts the weight-percent oxide components of the analysis into weight percents of hypothetical mineral "molecules." These normative components may be more useful than oxide or elemental components for comparison with synthetic experiments and between texturally and mineralogically diverse rocks. Cation "mole" percent norms also have been used for many years and certain data for silicate liquids may be better represented in terms of components with equimolar amounts of oxygen anion (see Burnham, 1979). The cation and anion norms are produced by transforming the molecular proportions of the CIPW normative "molecules" into proportions normalized to a constant number of cations or anions.

The norm calculation is affected critically by the ratio of ferric/ferrous iron. Modern instrumental-analysis techniques provide only a determination of total iron content. Even when determined explicitly, values for ferric and ferrous iron may not represent magmatic values because this ratio is altered easily by later processes. An alternative approach is to utilize the relationship between temperature, oxygen fugacity, and magma composition established experimentally by Kilinc and others (1983) and Sack and others (1980). ROCALC incorporates this calculation as an option (subroutine in lines 5500-5799).

PROGRAMMING OBJECTIVES

In designing ROCALC there were several objectives which involve its interaction with the user (student users in particular). One of these was that the program should not require reference to external documentation. Some effort was made to design messages printed to the screen and prompts for input which lead the user through the program. This is illustrated in Figures 1A-1D. Anyone who is familiar with the petrological problems involved should be able to use this program successfully without additional written instructions. The printed output, Figures 2A, 2B, and 2C, also is labeled explicitly. A large number of comments are included in REM statements in order to increase the readability and to document internally the program for those users who may wish to understand and modify it.

Another objective was to make entering data as convenient and free of errors as possible. The norm calculation may take into account 25 oxide components (listed in DATA statements, lines 9040-9280). Not all are necessary for any particular calculation, and in any table of analyses as few as 10 or 12 will be listed. Some may be trace elements, not always determined; others represent alternatives such as H2O or LOI, and total Fe reported as FeO or Fe2O3. There also is no standard order in which the components are listed in a table of analyses. Rather than requesting input in some arbitrary format, ROCALC allows the user to set up the order of data input to correspond to the analysis table from which he will be transcribing the data. This is done in the subroutine in lines 7000-7270 (see the middle of Fig. 1B), where an array, EO(I), is loaded with a value J pointing to the Ith component to be entered. J is determined by searching the component list for a match between a series of string inputs and an internal array of oxide symbols (XN$). The EO-array values then are used in the data-input routine at line 5050 to print out the prompts and to store the data in the proper order (Fig. 1D, top). In fact, this arrangement is not only convenient, but it reduces significantly the errors that result when picking data from a table in some irregular pattern. These routines can be used easily in other programs with analytical data.

```
--------------------------------------
    - AMPERWARE -
FROM - SCIENTIFIC SOFTWARE PRODUCTS
   NOW BEING LOADED

  AMPERWARE MUST BE ON THE DISC

--------------------------------------


**************************************
*                                    *
*           PROGRAM ROCALC           *
*                                    *
* J. C. STORMER - GEOLOGY DEPT.      *
* RICE UNIVERSITY. P.O. BOX 1892     *
* HOUSTON  TX 77251                  *
* (713) 527-4054                     *
*                                    *
* FEO-FE2O3 CAN BE CALCULATED FROM   *
* T AND LOG F(O2) USING KILINC. ET AL *
* (1983) CONTR. MINERAL. PETROL..83.  *
* 136-140.   THE NORM CALCULATIONS   *
* FOLLOW THE SEQUENCE OF STEPS GIVEN *
* IN BARKER (1983) IGNEOUS ROCKS.    *
* PP. 65-71.                         *
**************************************
PROGRAM ROCALC - VERSION OF 27 JUNE 1984

ENTER THE SLOT # FOR THE PRINTER.

(1-7. USUALLY 1)?1

--------------------------------------
THE WRONG SLOT NUMBER WILL CAUSE THE
PROGRAM TO ''HANG''.

YOUR PRINTER SHOULD NOW PRINT A HEADER.

IF NOT. TURN OFF COMPUTER AND START OVER

--------------------------------------
SET PAPER TO THE TOP OF A PAGE.

HIT ANY KEY WHEN READY Q

--------------------------------------
```

A

Figure 1. Printed facsimile of output to monitor screen. Bell accompanies many of prompts. Dashed lines indicate points where screen is cleared for new input. Underlined portions indicate responses typed from keyboard. A, Amperware and printer setup. B, Data input setup. C, Input of calculation options. D, Analysis data input. Parts A, B, C, and D follow each other in sequence during running of program. For multiple calculations part D loops between top and "ARE YOU READY FOR THE NEXT SAMPLE?"

```
----------------------------------------
OXIDE COMPONENTS FOR NORM ---------
SYMBOL MOL.WT.  SYMBOL MOL.WT.
SIO2   60.09    TIO2   79.9
AL203  101.96   FE203  159.69
FEO    71.85    MNO    70.94
MGO    40.3     CAO    56.08
NA20   61.98    K20    94.2
P205   141.95   CO2    44.01
CR203  151.99   NIO    74.7
BAO    153.34   SRO    103.62
CL     35.45    F      19
S      32.06    SO3    80.06
ZRO2   123.22   H20    18
H20+   18       H20-   18
LOI

ENTER ONLY THE COMPONENT SYMBOLS IN
YOUR TABLE IN THE ORDER THAT THEY WILL
BE ENTERED.''END'' WILL END LIST

COMPONENT 1 IS?SIO2
COMPONENT 2 IS?TIO2
COMPONENT 3 IS?AL203
COMPONENT 4 IS?FE203
COMPONENT 5 IS?FEO
COMPONENT 6 IS?MNO
COMPONENT 7 IS?MGO
COMPONENT 8 IS?CAO
COMPONENT 9 IS?NA20
COMPONENT 10 IS?K20
COMPONENT 11 IS?H20+
COMPONENT 12 IS?H20-
COMPONENT 13 IS?CO2
COMPONENT 14 IS?P205
COMPONENT 15 IS?END

----------------------------------------
YOUR TABLE IS:
1. SIO2
2. TIO2
3. AL203
4. FE203
5. FEO
6. MNO
7. MGO
8. CAO
9. NA20
10. K20
11. H20+
12. H20-
13. CO2
14. P205

IS`THIS`CORRECT?Y
```

B

Figure 1. (continued)

```
----------------------------------------
DO YOU WANT ANALYSES  RECALCULATED
TO 100% IGNORING WATER (Y/N)?Y

DO YOU WANT FE2O3/FEO RECALCULATED
FROM TEMP. AND FUGACITY OF OXYGEN(Y/N)?Y

DO YOU WANT CO2 CALCULATED TO CALCITE?
(THE ALTERNATIVE IS SODIUM CARBONATE
FOR CANCRINITE BEARING ROCKS ONLY.) Y/N?Y


SAVE OUTPUT ON DISK (Y/N)?N

----------------------------------------
ARE YOU READY FOR NEXT SAMPLE (Y OR 1)
EXIT OR CHANGE OPTIONS OR TABLE (N)

 Y/N ?Y
```

C

Figure 1. (continued)

```
-----------------------------------------
SAMPLE NAME= PILOT KNOB BASALT

ENTER ANALYSIS
SIO2 = 42.80
TIO2 = 3.57
AL2O3= 10.15
FE2O3= 2.71
FEO  = 9.90
MNO  = .17
MGO  = 13.10
CAO  = 12.27
NA2O = 2.16
K2O  = .36
H2O+ = 2.26
H2O- = .58
CO2  = .11
P2O5 = .62
TOTAL  100.76


IS THIS CORRECT(Y=1/N)?Y


TEMPERATURE (DEG.C) = 1000

LOG FUGACITY OF OXYGEN=-11.0
FE2O3 1.98311041
FEO   10.554105
TOTAL  100.687215


IS THIS CORRECT(Y(=1)/N)?1

    AT THIS POINT THE THREE OUTPUT
    PAGES WILL BE PRINTED

-----------------------------------------
ARE YOU READY FOR NEXT SAMPLE (Y OR 1)2
EXIT OR CHANGE OPTIONS OR TABLE (N)

 Y/N ?N

-----------------------------------------
CHANGE CALCULATION OPTIONS (Y/N)?N

-----------------------------------------
RUN COMPLETE
PROGRAM ROCALC - VERSION OF  14 JAN 1985
```

D

Figure 1. (continued)

```
PILOT KNOB BASALT
OXIDE AND ELEMENTAL ANALYSIS                PROGRAM ROCALC - VERSION OF  14 JAN 1985

*** FEO-FE2O3 RECALCULATED USING - T(C)= 1000   LOG F(O2)= -11   ***
-----------------------------------------------------------------------------------
OXIDE WT.% RCAL.100%   OXIDE MOLE %      CATION %   CATIONS PER 100 ANIONS

SIO2   42.80  43.74    SIO2   44.31      SI  40.41      SI/O  27.31
TIO2    3.57   3.65    TIO2    2.78      TI   2.53      TI/O   1.71
AL2O3  10.15  10.37    AL2O3   6.19      AL  11.30      AL/O   7.63
FE2O3   1.98   2.03    FE2O3    .77      F3   1.41      F3/O    .95
FEO    10.55  10.79    FEO     9.14      F2   8.33      F2/O   5.63
MNO      .17    .17    MNO      .15      MN    .14      MN/O    .09
MGO    13.10  13.39    MGO    20.22      MG  18.44      MG/O  12.46
CAO    12.27  12.54    CAO    13.61      CA  12.41      CA/O   8.39
NA2O    2.16   2.21    NA2O    2.17      NA   3.95      NA/O   2.67
K2O      .36    .37    K2O      .24      K     .43      K /O    .29
P2O5     .62    .63    P2O5     .27      P     .50      P /O    .33
CO2      .11    .11    CO2      .16      C     .14      C /O    .10
CR2O3   0.00   0.00    CR2O3   0.00      CR   0.00      CR/O   0.00
NIO     0.00   0.00    NIO     0.00      NI   0.00      NI/O   0.00
BAO     0.00   0.00    BAO     0.00      BA   0.00      BA/O   0.00
SRO     0.00   0.00    SRO     0.00      SR   0.00      SR/O   0.00
CL      0.00   0.00    CL      0.00      CL   0.00      CL/O   0.00
F       0.00   0.00    F       0.00      F    0.00      F /O   0.00
S       0.00   0.00    S       0.00      S    0.00      S /O   0.00
SO3     0.00   0.00    SO3     0.00      S6   0.00      S6/O   0.00
ZRO2    0.00   0.00    ZRO2    0.00      ZR   0.00      ZR/O   0.00
-----------------------------------------------------------------------------------
       97.85  SUB-TOTAL
       97.85  TOTAL (LESS O=F+CL+S)
        2.26  H2O+
         .58  H2O-
     -------
      100.69   TOTAL
-----------------------------------------------------------------------------------

    WEIGHT   FORMULA    CATION
    ------   -------    ------

     .1879     .0845     .1691   FE2O3/FEO (FE3+/FE2+)

     .5538     .6888     .6888   MGO / (MGO + FEO)
     .5115     .6512     .6512   MGO / (MGO + FEO* + MNO)   FEO*=TOTAL FE AS FEO

      2.52      2.41      4.39   NA2O + K2O
     .0589     .0543     .1086   (NA2O + K2O)/ SIO2
     .2483     .3885     .3885   (NA2O + K2O)/ AL2O3
    1.4571    2.5863    1.4874   (CAO + NA2O + K2O)/ AL2O3

TERNARY AFM PLOTTING PERCENTAGES
      9.01      7.22     13.47   A - (NA2O + K2O)
     44.13     32.07     29.91   F - (FEO+.8998*FE2O3)(FEO+2*FE2O3)(FE2+ + FE3+
     46.85     60.71     56.62   M - (MGO)

TERNARY CA-NA-K PLOTTING PARAMETERS
     82.96     84.98     73.88   CAO /(CAO+NA2O+K2O)
     14.60     13.54     23.54   NA2O/(CAO+NA2O+K2O)
      2.43      1.48      2.58   K2O /(CAO+NA2O+K2O)
```

A

Figure 2. Printed output from ROCALC. Each rock analysis produces three pages of output. A, Page 1, Oxide and mole percent analyses, elemental ratios and indices. B, CIPW and other norm percentages and normative indices. C, Ternary ratios as percentages for plotting in several common ternary diagrams.

PILOT KNOB BASALT
NORMATIVE ANALYSES PROGRAM ROCALC - VERSION OF 14 JAN 1985

*** FEO-FE2O3 RECALCULATED USING - T(C)= 1000 LOG F(O2)= -11 ***
*** BASED ON ANALYSIS RECALCULATED TO 100% WATER FREE***

	CIPW WT.%	CATION PERCENT	ANION PERCENT	COMPONENT FORMULAS
Q	0.00	0.00	0.00	SIO2
C	0.00	0.00	0.00	AL2O3
OR	2.17	2.17	2.34	K2AL2SI6O16
AB	10.45	11.06	11.95	NA2AL2SI6O16
AN	17.31	17.27	18.66	CAAL2SI2O8
LC	0.00	0.00	0.00	K2AL2SI4O12
NE	4.46	5.23	4.71	NA2AL2SI2O8
KP	0.00	0.00	0.00	K2AL2SI2O8
AC	0.00	0.00	0.00	NA2FE2SI4O12
NS	0.00	0.00	0.00	NA2SIO3
KS	0.00	0.00	0.00	K2SIO3
WO	0.00	0.00	0.00	CASIO3
DI-DI	24.29	24.91	25.23	CAMGSI2O6
DI-HD	7.89	7.06	7.16	CAFESI2O6
DI	32.18	31.97	32.39	CA(MG,FE)SI2O6
HY-EN	0.00	0.00	0.00	MGSIO3
HY-FS	0.00	0.00	0.00	FESIO3
HY	0.00	0.00	0.00	(MG,FE)SIO3
OL-FO	15.48	18.32	16.50	MG2SIO4
OL-FA	6.36	5.20	4.68	FE2SIO4
OL	21.84	23.52	21.18	(MG,FE)2SIO4
CS	0.00	0.00	0.00	CA2SIO4
MT	2.94	2.11	1.90	(FE+2)(FE+3)2O4
IL	6.93	5.07	5.14	FETIO3
HM	0.00	0.00	0.00	FE2O3
NC	0.00	0.00	0.00	NA2CO3
TN	0.00	0.00	0.00	CATISIO5
PF	0.00	0.00	0.00	CATIO3
RU	0.00	0.00	0.00	TIO2
AP	1.50	1.32	1.45	1/3 (CA10P6O24F2)
CC	.26	.28	.29	CACO3
PR	0.00	0.00	0.00	FES2
TH	0.00	0.00	0.00	NA2SO4
FR	0.00	0.00	0.00	CAF2
ZR	0.00	0.00	0.00	ZRSIO4
HL	0.00	0.00	0.00	NACL
CM	0.00	0.00	0.00	FECR2O4

	100.04	SUB-TOTAL		
	2.84	H2O/L.O.I.		
	102.88	TOTAL		

17.08 THORNTON-TUTTLE DIFF. INDEX

PARAMETERS FOR STRECKEISEN-LEMATRE (1979) ROCK CLASSIFICATION PLOT

88.84 AN/(AN+OR)
0.00 Q/(Q+OR+AB+AN)
12.97 (NE+LC+KP)/(NE+LC+KP+OR+AB+AN)

B

Figure 2. (continued)

PILOT KNOB BASALT
NORMATIVE PLOTTING PARAMETERS PROGRAM ROCALC - VERSION OF 14 JAN 1985

*** FEO-FE2O3 RECALCULATED USING - T(C)= 1000 LOG F(O2)= -11 ***

	----------TERNARY-	------------		
	-WT %	-MOLE %	-EQIV.OXYGEN	'MOLE' FORMULA (EQUIVALENT OX.)
OR	7.26	7.11	7.11	KALSI3O8
AB	34.90	36.26	36.26	NAALSI3O8
AN	57.84	56.63	56.63	CAAL2SI2O8
Q	33.52	54.66	37.61	SIO2 (SI4O8)
NE	59.24	40.86	56.23	NAALSIO4 (NA2ASI2O8)
KP	7.23	4.48	6.17	KALSIO4 (K2AL2SI2O8)
NE	33.30	27.30	32.22	NAALSIO4 (NA2AL2SI2O8)
FO	50.94	42.17	49.77	MG2SIO4 (MG4SI2O8)
Q	15.75	30.53	18.02	SIO2 (SI4O8)
DI	54.92	44.23	54.33	CAMGSI2O6 (CA2MG2SI4O12)
FO	35.00	43.39	35.53	MG2SIO4 (MG6SI3O12)
NE	10.09	12.38	10.14	NAALSIO4 (NA3AL3SI3O12)
PL	33.94	26.35	36.36	CAAL2SI2O8 + NAALSI3O8
OL	26.71	36.47	25.16	(MG,FE)2SIO4 ((MG,FE)4SI2O8)
DH	39.35	37.18	38.48	CA(MG,FE)SI2O6 ([CA4(MG,FE)4SI8O24]/3
WO	51.97	50.00	50.00	CASIO3
EN	34.99	38.95	38.95	MGSIO3
FS	13.04	11.05	11.05	FESIO3

C

Figure 2. (continued)

Almost all of the inputs provide for some way of checking errors. After input of an array of data (such as an analysis), the array is printed to the screen, and the user is prompted to check its accuracy. Then the user is given the option of reentering the values (see lines 5095 and 7205-7260 and the top of Figure 1D and bottom of 1B). Simple "Y" or "N" and integer responses are checked for inappropriate entries (see lines 520-530 and 600-610).

The program is structured in such a way that the main program is relatively short (31 lines beginning at 500) the calculations are performed in several major subroutines called from this main program. In development, this allowed testing of individual segments. The selection of optional or alternative calculations also is facilitated. This structure also may make the program easier to read and understand, as well as to modify or use individual routines in other similar programs.

Most of these features are inefficient relatively with regard to speed of execution and use of memory. However, within the microcomputer environment, computer efficiency may be less important than optimizing the speed and accuracy of human interaction.

ACKNOWLEDGMENTS

Many petrology students at Rice University contributed to the testing of this program. This material is based on work supported by the National Science Foundation under Grant No. EAR-8341668.

REFERENCES

Barker, D.S., 1983, Igneous rocks: Prentice-Hall, Englewood Cliffs, New Jersey, 147 p.

Burnham, C.W., 1979, The importance of volatile constituents, in Yoder, H.S., ed., The evolution of igneous rocks: Princeton Univ. Press., p.439-482.

Kilinc, A., Carmichael, I.S.E., Rivers, M.L., and Sack, R.O. 1983. The ferric - ferrous ratio of natural liquids equilibrated in air: Contrib. Mineral. Petrol., v. 83, no. 1/4, p. 136-140.

Sack, R.O., Carmichael, I.S.E., Rivers, M.L., and Ghiorso, M.S. 1980. Ferric-ferrous equilibria in natural silicate liquids at 1 bar: Contrib. Mineral. Petrol., v. 75, no. 4, p. 369-376.

APPENDIX A

PROGRAM LISTING FOR ROCALC

Notes - All statements beginning with & are interpreted by the Amperware subroutines (copywrited by Scientific Software Products, Inc.) which provide extensions to normal Applesoft BASIC.
Commands to the Apple DOS 3.3 operating system are given by PRINT D9$;.... (D9$ contains ctrl-D, ASCII code #4)
Commands "PR#0" and "PR#";P% direct subsequent output to the screen or to the printer (interface slot# P%).

```
1  REM

** PROGRAM ROCALC -J. C. STORMER, DEPT. OF GEOLOGY, RICE UNIVERSITY, HOUSTON TX
     77251 - (713) 527-4054

2  REM   THE PROGRAM USES THE EXTENSIONS TO APPLESOFT BASIC THAT ARE PROVIDED
     BY THE "AMPERWARE" SUBROUTINES (STATEMENTS WITH &) FROM SCIENTIFIC
     SOFTWARE PRODUCTS, 5726 PROFESSIONAL CIRCLE, INDIANAPOLIS  IN  46241
3  HOME : PRINT "    - AMPERWARE -": PRINT "FROM - SCIENTIFIC SOFTWARE
     PRODUCTS": PRINT "   NOW BEING LOADED"
4  PRINT : PRINT "  AMPERWARE MUST BE ON THE DISC"

****** SEE AMPERWARE MANUAL FOR THE PROPER FORM OF LINE 5

5  PRINT  CHR$ (4);"BRUN XXX": REM RUN THE PROPER AMPERWARE LOADER
10  GOTO 200: REM  FORMATTING SUBROUTINES ARE PLACED FIRST TO IMPROVE THE SPEED
     - GO TO MAIN PROGRAM
49  REM

** SUBROUTINES FOR FORMATTING AND PRINTING.  THE F#$ ARE THE FORMAT STRINGS
     SIMILAR TO FORMATS WITH PRINT USING IN OTHER BASICS*

50  &  PRINT WITH - F1$,Z1$,Z2(1),Z2(0),Z2(2),Z3$: RETURN
52  &  PRINT WITH - F2$,Z1$,Z2(1),Z2(0),Z2(2),Z3$: RETURN
54  &  PRINT WITH - F3$,Z2(0),Z1$: RETURN
56  &  PRINT WITH - F4$,Z1$,Z2(0),Z2(1),Z2(2),Z3$: RETURN
58  &  PRINT WITH - F5$,Z2(0),Z2(1),Z2(2),Z1$: RETURN
60  &  PRINT WITH - F6$,Z2(0),Z2(1),Z2(2),Z1$: RETURN
62  &  PRINT WITH - F7$,Z1$,Z2(0),Z2(4),Z1$,Z2(1),Z3$,Z2(2),Z4$,Z2(3): RETURN
200  REM

** FORMATS FOR "&PRINT WITH-" STATEMENTS*

210 F1$ = "##  #####.##  #####.##  #####.##        #########################/"
220 F2$ = "#####  #####.##  #####.##  #####.##     #########################/"
230 F3$ = "    #####.##  ########################################/"
240 F4$ = "##  #####.##  #####.##  #####.##
     #######################################/"
250 F5$ = "  #####.##  #####.##  #####.##
     #################################################/"
260 F6$ = "  ###.####  ###.####  ###.####
     #################################################/"
270 F7$ = "##### ###.## ###.##     ##### ###.##     ## ###.##     #### ###.##/"
499 REM

** MAIN PROGRAM STARTS HERE**

500  HOME : GOSUB 9500: REM  CLEARS SCREEN AND PRINTS HEADER
510 D9$ =  CHR$ (4): REM          CTRL-D AFTER PRINT INDICATES COMMAND TO
     OPERATING SYSTEM
520  PRINT : PRINT "ENTER THE SLOT # FOR THE PRINTER.": PRINT  CHR$ (7): INPUT
     "(1-7, USUALLY 1)?";P%
```

```
530  IF (P% < 1) + (P% > 7) THEN  PRINT  CHR$ (7): GOTO 520
540  HOME : VTAB 5: PRINT "THE WRONG SLOT NUMBER WILL CAUSE THE ": PRINT
     "PROGRAM TO ''HANG''.": PRINT : PRINT "YOUR PRINTER SHOULD NOW PRINT A
     HEADER.": PRINT : PRINT "IF NOT, TURN OFF COMPUTER AND START OVER"
550  PRINT D9$;"PR#";P%: GOSUB 9500
560  PRINT  CHR$ (12)
570  PRINT D9$;"PR#0": PRINT  CHR$ (7): HOME : PRINT "SET PAPER TO THE TOP OF A
     PAGE."; CHR$ (7): PRINT : PRINT "HIT ANY KEY WHEN READY": GET A$: PRINT
580  GOSUB 9000: REM      READ CONSTANTS
590  GOSUB 7000: REM      SET UP INPUT TABLE
600  HOME : VTAB 5: PRINT "DO YOU WANT ANALYSES  RECALCULATED": PRINT "TO 100%
     IGNORING WATER "; CHR$ (7);: INPUT "(Y/N)?";RC$
610 RC$ =  LEFT$ (RC$,1): IF RC$ <  > "Y" THEN  IF RC$ <  > "N" THEN  GOTO 600:
     REM    RC$ IS A FLAG FOR RECALCULATED ANALYSES
620  PRINT : PRINT "DO YOU WANT FE2O3/FEO RECALCULATED": PRINT "FROM TEMP. AND
     FUGACITY OF OXYGEN"; CHR$ (7);: INPUT "(Y/N)?";FC$
630 FC$ =  LEFT$ (FC$,1): IF FC$ <  > "Y" THEN  IF FC$ <  > "N" THEN  GOTO 620:
     REM        FC$ IS A FLAG FOR FE2O3/FEO RECALCULATIONS FROM TEMPERATURE AND
     OXYGEN FUGACITY BY THE METHOD OF KILINC, ET AL. (1983) CONTR. MINERAL.
     PETROL.
640  PRINT : PRINT "DO YOU WANT CO2 CALCULATED TO CALCITE?": PRINT "(THE
     ALTERNATIVE IS SODIUM CARBONATE": PRINT "FOR CANCRINITE BEARING ROCKS
     ONLY.) Y/N";: INPUT A$
650 CC$ =  LEFT$ (A$,1): IF CC$ <  > "Y" THEN  IF CC$ <  > "N" GOTO 640
660  PRINT : PRINT  CHR$ (7): INPUT "SAVE OUTPUT ON DISK (Y/N)?";DK$:DK$ =
     LEFT$ (DK$,1): IF DK$ = "N" GOTO 690
670  IF DK$ <  > "Y" GOTO 660
680  GOTO 8000: REM  ROUTINE TO SET UP DISC FILE
690  HOME : VTAB 5: PRINT  CHR$ (7);"ARE YOU READY FOR NEXT SAMPLE (Y OR 1)":
     PRINT "EXIT OR CHANGE OPTIONS OR TABLE (N)": PRINT : INPUT " Y/N ?";A$
700 A$ =  LEFT$ (A$,1): IF A$ <  > "Y" THEN  IF A$ <  > "N" THEN  IF A$ <  >
     "1" GOTO 690
710  IF A$ = "N" GOTO 760
720  GOSUB 5000: REM          READS OXIDES CALCULATES MOLE AND CATION
     PROPORTIONS ETC.
730  IF DK$ = "Y" THEN  GOSUB 8300: REM        SAVES OUTPUT TO DISK
740  GOSUB 1000: REM        CALCULATES AND  PRINTS NORM VALUES
750  GOTO 690
760  IF DK$ = "Y" THEN  GOSUB 8360
770  HOME : VTAB 5: PRINT  CHR$ (7);"CHANGE CALCULATION OPTIONS (Y/N)";: INPUT
     A$
780 A$ =  LEFT$ (A$,1): IF A$ <  > "Y" THEN  IF A$ <  > "N" GOTO 770
790  IF A$ = "N" THEN  HOME : PRINT "RUN COMPLETE": GOSUB 9900: GOTO 999
800  GOTO 590
999  END
1000  REM

     **CALCULATE CIPW NORM**

1010  REM     CALCULATIONS ARE KEYED TO THE SERIES OF STEPS DESCRIBED IN DETAIL
     BY BARKER (1983) IGNEOUS ROCKS, PRENTICE HALL, SECTION 4.3, PAGES 65-71.
1020  REM      STEP 1 IS THE CALCULATION OF THE MOLECULAR PROPORTIONS OF THE
     OXIDES -MP(I)- BY SUBROUTINE 5000
1030 MP(5) = MP(5) + MP(6) + MP(14): REM     STEP 2
1040 MP(8) = MP(8) + MP(15) + MP(16): REM    STEP 3
1050 ZR = MP(21):Y = MP(21): REM    STEP 4
1060 AP = MP(11):MP(8) = MP(8) - 3.33 * MP(11):MP(18) = MP(18) - 2 * MP(11) /
     3: REM    STEP 5
1070 FR = 0: IF MP(18) > 0 THEN FR = .5 * MP(18):MP(8) = MP(8) - FR: REM  STEP
     6
1080 HL = MP(17):MP(9) = MP(9) - 0.5 * HL: REM    STEP 7
1090 TH = MP(20):MP(9) = MP(9) - TH: REM    STEP 8
1100 PR = 0.5 * MP(19):MP(5) = MP(5) - PR: REM    STEP 9
1110 CC = 0:NC = 0: IF MP(12) = 0 GOTO 1140
1120  IF CC$ = "Y" THEN CC = MP(12):MP(8) = MP(8) - CC: GOTO 1140: REM
```

```
     STEP10
1130  IF CC$ <  > "Y" THEN NC = MP(12):MP(9) = MP(9) - NC: GOTO 1140: REM
     STEP 10 ALT.
1140 CM = MP(13):MP(5) = MP(5) - CM: REM    STEP 11
1150  IF MP(5) > MP(2) THEN IL = MP(2):MP(5) = MP(5) - IL:MP(2) = 0: GOTO 1170
1160 IL = MP(5):MP(2) = MP(2) - IL:MP(5) = 0: REM    STEP 12
1170 KS = 0: IF MP(3) > MP(10) THEN OT = MP(10):MP(3) = MP(3) - OT:MP(10) = 0:Y
     = Y + 6 * OT: GOTO 1190: REM   OT IS PROVISIONAL ORTHOCLASE
1180 OT = MP(3):MP(3) = 0:MP(10) = MP(10) - OT:KS = MP(10):Y = Y + 6 * OT + KS:
     REM    STEP 13
1190 AC = 0:NS = 0: IF MP(3) > MP(9) THEN AB = MP(9):MP(3) = MP(3) - AB:MP(9) =
     0:Y = Y + 6 * AB: GOTO 1230: REM      STEP 14
1200 AB = MP(3):MP(9) = MP(9) - AB:MP(3) = 0:Y = Y + 6 * AB
1210 AC = 0:NS = 0: IF MP(9) > MP(4) THEN AC = MP(4):MP(9) = MP(9) - AC:MP(4) =
     0:NS = MP(9):Y = Y + 4 * AC + NS: GOTO 1230
1220 AC = MP(9):MP(9) = 0:MP(4) = MP(4) - AC:Y = Y + 4 * AC: REM    STEP 15
1230  IF MP(3) > MP(8) THEN AN = MP(8):MP(8) = 0:MP(3) = MP(3) - AN:Y = Y + 2 *
     AN: GOTO 1250
1240 AN = MP(3):MP(3) = 0:MP(8) = MP(8) - AN:Y = Y + 2 * AN: REM  STEP 16
1250 RU = 0:TN = 0: IF MP(8) > MP(2) THEN TN = MP(2):MP(2) = 0:MP(8) = MP(8) -
     TN:Y = Y + TN: GOTO 1270
1260 TN = MP(8):MP(2) = MP(2) - TN:Y = Y + TN:RU = MP(2):MP(2) = 0: REM    STEP
     17
1270 HM = 0: IF MP(4) > MP(5) THEN MT = MP(5):MP(5) = 0:MP(4) = MP(4) - MT:HM =
     MP(4):MP(4) = 0: GOTO 1290
1280 MT = MP(4):MP(4) = 0:MP(5) = MP(5) - MT: REM    STEP 18
1290 M1 = MP(7) + MP(5): IF M1 <  = 0 THEN M = 0:M2 = 1: GOTO 1320
1300 M = MP(7) / M1:M2 = M - 1: REM    STEP 19
1310 WO = 0:HY = 0
1320  IF MP(8) > M1 THEN DH = M1:M1 = 0:MP(7) = 0:MP(5) = 0:MP(8) = MP(8) -
     DH:WO = MP(8):MP(8) = 0:Y = Y + 2 * DH + WO: GOTO 1340: REM      DH IS
     DIOPSIDE-HEDENBERGITE COMBINED
1330 DH = MP(8):MP(8) = 0:M1 = M1 - DH:HY = M1:MP(5) = 0:MP(7) = 0:Y = Y + 2 *
     DH + HY: REM    STEP 20
1340 Q = MP(1) - Y:OL = 0:FO = 0:FA = 0:PF = 0:PF = 0:NE = 0:KP = 0:LC = 0:CS =
     0: IF Q >  = 0 GOTO 1490: REM      CALCULATE Q    -SET ALL SILICA
     DEFICIENT MINERALS TO 0 -IF Q IS  NEGATIVE BEGIN DESILICATION STEPS
1350 D = Y - MP(1):Q = 0
1360  IF 2 * D < HY THEN OL = D:HY = HY - 2 * D: GOTO 1490
1370 OL = HY / 2:D = D - OL:HY = 0
1380  IF TN <  = 0 GOTO 1410
1390  IF D < TN THEN TN = TN - D:PF = D: GOTO 1490
1400 PF = TN:TN = 0:D = D - PF: REM    STEP 23
1410  IF D < 4 * AB THEN NE = D / 4:AB = AB - NE:D = 0: GOTO 1490
1415 NE = AB:AB = 0:D = D - 4 * NE: REM  STEP 24
1420  IF D < 2 * OT THEN LC = D / 2:OT = OT - LC: GOTO 1490
1430 LC = OT:D = D - 2 * OT:OT = 0: REM    STEP 25
1440 X = WO / 2: IF D < X THEN CS = D:WO = WO - D: GOTO 1490
1450 CS = WO:D = D - X: REM   STEP 26
1460 X = D / 2: IF D < DH THEN CS = CS + X:OL = OL + X:DH = DH - D: GOTO 1490
1470 X = DH / 2:D = D - DH:DH = 0:CS = CS + X:OL = OL + X: REM    STEP 27
1480 X = D / 2:KP = X:LC = LC - X
1490 DI = DH * M:HD = DH - DI: REM      CALCULATE DIOPSIDE AND HEDENBERGITE
1500 EN = HY * M:FS = HY - EN: REM    ENSTATITE AND FERROSILITE
1510 FO = OL * M:FA = OL - FO: REM    FORSTERITE/FAYALITE
1520 C = MP(3): REM  CORUNDUM IS THE REMAINING ALUMINA AFTER STEP 16
1530 D = Q + C * 2 + OT * 10 + AB * 10 + AN * 5 + LC * 8 + NE * 6 + KP * 6 + AC
     * 8 + NS * 3 + KS * 3 + WO * 2 + DI * 4 + EN * 2 + HD * 4 + FS * 2 + FO *
     3 + FA * 3 + CS * 3 + MT * 3 + IL * 2 + HM * 2 + NC * 3 + TN * 3 + PF * 2
     + RU + AP * 16 / 3 + CC * 2 + PR + TH * 3 + FR + ZR * 2 + HL + CM * 3
1540 D1 = Q * 2 + C * 3 + OT * 16 + AB * 16 + AN * 8 + LC * 12 + NE * 8 + KP *
     8 + AC * 12 + NS * 3 + KS * 3 + WO * 3 + DI * 6 + EN * 3 + HD * 6 + FS * 3
     + FO * 4 + FA * 4 + CS * 4 + MT * 4 + IL * 3 + HM * 3 + NC * 3 + TN * 5 +
     PF * 3 + RU * 2 + AP * 26 / 3 + CC * 3 + PR * 2 + TH * 4 + FR * 2 + ZR * 4
     + HL + CM * 4
```

```
1550 D = D / 100:D1 = D1 / 100: REM     D IS THE SUM OF THE CATION PROPORTIONS
     OF THE MINERALS DIVIDED BY 100 TO GIVE % WHEN DIVIDED INTO THE INDIVIDUAL
     MOLECULAR PROPORTIONS. D1 IS THE SUM OF ANION PROP. /100
1560  REM
```

**SECTION TO CALCULATE AND PRINT THE NORMATIVE MINERALS*

```
1561  REM    THE CATION NORM IS OBTAINED FROM THE MOLECULAR PROPORTIONS TIMES
     THE NUMBER OF CATIONS PER FORMULA DIVIDED BY D
1562  REM   THE ANION NORM IS CALCULATED BY MULTIPLYING THE MOL. PROP. BY THE
     NO. OF ANIONS IN THE FORMULA AND DIVIDING BY D1
1570  REM      THE CIPW NORM (WT.%) IS OBTAINED BY MULTIPLYING BY THE MOLECULAR
     WEIGHT OF THE NORMATIVE MINERAL, AND THE ANION NORM IS CALCULATED BY
     MULTIPLYING BY THE NUMBER OF ANIONS AND DIVIDING BY D1 (SUM OF ANIONS).
1580  PRINT D9$;"PR#";P%
1590  PRINT : PRINT SN$: PRINT "NORMATIVE ANALYSES                          ";:
     GOSUB 9900: IF FC$ = "Y" THEN  PRINT "*** FEO-FE2O3 RECALCULATED USING -
     T(C)= ";CT;"  LOG F(O2)= ";CX;"  ***"
1592  IF RC$ = "Y" THEN  PRINT "*** BASED ON ANALYSIS RECALCULATED TO 100%
     WATER FREE***"
1594  PRINT
     "
     "------------------------------------------------------------------------
1600  PRINT "        CIPW   CATION    ANION           COMPONENT ": PRINT "
     WT.%   PERCENT   PERCENT         FORMULAS": PRINT
1610 Q(1) = Q * 60.09:Q(2) = Q * 2 / D1:Q(0) = Q / D
1620 Z1$ = "Q ":Z2(1) = Q(1):Z2(0) = Q(0):Z2(2) = Q(2):Z3$ = " SIO2": GOSUB 50
1630 C(1) = C * 101.96:C(2) = C * 3 / D1:C(0) = C * 2 / D
1640 Z1$ = "C ":Z2(1) = C(1):Z2(0) = C(0):Z2(2) = C(2):Z3$ = " AL2O3": GOSUB 50
1650 OT(1) = OT * 556.70:OT(2) = OT * 16 / D1:OT(0) = OT * 10 / D
1660 Z1$ = "OR":Z2(1) = OT(1):Z2(0) = OT(0):Z2(2) = OT(2):Z3$ = " K2AL2SI6O16":
     GOSUB 50
1670 AB(1) = AB * 524.48:AB(2) = AB * 16 / D1:AB(0) = AB * 10 / D
1680 Z1$ = "AB":Z2(1) = AB(1):Z2(0) = AB(0):Z2(2) = AB(2):Z3$ = "
     NA2AL2SI6O16": GOSUB 50
1690 AN(1) = AN * 278.22:AN(2) = AN * 8 / D1:AN(0) = AN * 5 / D
1700 Z1$ = "AN":Z2(1) = AN(1):Z2(0) = AN(0):Z2(2) = AN(2):Z3$ = " CAAL2SI2O8":
     GOSUB 50
1710 LC(1) = LC * 436.52:LC(2) = LC * 12 / D1:LC(0) = LC * 8 / D
1720 Z1$ = "LC":Z2(1) = LC(1):Z2(0) = LC(0):Z2(2) = LC(2)::Z3$ = "
     K2AL2SI4O12": GOSUB 50
1730 NE(1) = NE * 284.12:NE(2) = NE * 8 / D1:NE(0) = NE * 6 / D
1740 Z1$ = "NE":Z2(1) = NE(1):Z2(0) = NE(0):Z2(2) = NE(2):Z3$ = " NA2AL2SI2O8":
     GOSUB 50
1750 KP(1) = KP * 316.34:KP(2) = KP * 8 / D1:KP(0) = KP * 6 / D
1760 Z1$ = "KP":Z2(1) = KP(1):Z2(0) = KP(0):Z2(2) = KP(2):Z3$ = " K2AL2SI2O8":
     GOSUB 50
1780 AC(1) = AC * 462.03:AC(2) = AC * 12 / D1:AC(0) = AC * 8 / D
1790 Z1$ = "AC":Z2(1) = AC(1):Z2(0) = AC(0):Z2(2) = AC(2):Z3$ = "
     NA2FE2SI4O12": GOSUB 50
1800 NS(1) = NS * 122.07:NS(2) = NS * 3 / D1:NS(0) = NS * 3 / D
1810 Z1$ = "NS":Z2(1) = NS(1):Z2(0) = NS(0):Z2(2) = NS(2):Z3$ = " NA2SIO3":
     GOSUB 50
1820 KS(1) = KS * 154.29:KS(2) = KS * 3 / D1:KS(0) = KS * 3 / D
1830 Z1$ = "KS":Z2(1) = KS(1):Z2(0) = KS(0):Z2(2) = KS(2):Z3$ = " K2SIO3":
     GOSUB 50
1840 WO(1) = WO * 116.17:WO(2) = WO * 3 / D1:WO(0) = WO * 2 / D
1850 Z1$ = "WO":Z2(1) = WO(1):Z2(0) = WO(0):Z2(2) = WO(2):Z3$ = " CASIO3":
     GOSUB 50
1860 DI(1) = DI * 216.56:DI(2) = DI * 6 / D1:DI(0) = DI * 4 / D
1870 Z1$ = "DI-DI":Z2(1) = DI(1):Z2(0) = DI(0):Z2(2) = DI(2):Z3$ = "
     CAMGSI2O6": GOSUB 52
1880 HD(1) = HD * 248.11:HD(2) = HD * 6 / D1:HD(0) = HD * 4 / D
1890 Z1$ = "DI-HD":Z2(1) = HD(1):Z2(0) = HD(0):Z2(2) = HD(2):Z3$ = "
```

```
     CAFESI2O6": GOSUB 52
1900 Z1$ = "DI":Z2(1) = DI(1) + HD(1):Z2(0) = DI(0) + HD(0):Z2(2) = DI(2) +
     HD(2):Z3$ = " CA(MG,FE)SI2O6": GOSUB 50
1910 EN(1) = EN * 100.39:EN(2) = EN * 3 / D1:EN(0) = EN * 2 / D
1920 Z1$ = "HY-EN":Z2(1) = EN(1):Z2(0) = EN(0):Z2(2) = EN(2):Z3$ = " MGSIO3":
     GOSUB 52
1930 FS(1) = FS * 131.94:FS(2) = FS * 3 / D1:FS(0) = FS * 2 / D
1940 Z1$ = "HY-FS":Z2(1) = FS(1):Z2(0) = FS(0):Z2(2) = FS(2):Z3$ = " FESIO3":
     GOSUB 52
1950 Z1$ = "HY":Z2(1) = EN(1) + FS(1):Z2(0) = EN(0) + FS(0):Z2(2) = EN(2) +
     FS(2):Z3$ = " (MG,FE)SIO3": GOSUB 50
1960 FO(1) = FO * 140.69:FO(2) = FO * 4 / D1:FO(0) = FO * 3 / D
1970 Z1$ = "OL-FO":Z2(1) = FO(1):Z2(0) = FO(0):Z2(2) = FO(2):Z3$ = " MG2SIO4":
     GOSUB 52
1980 FA(1) = FA * 203.79:FA(2) = FA * 4 / D1:FA(0) = FA * 3 / D
1990 Z1$ = "OL-FA":Z2(1) = FA(1):Z2(0) = FA(0):Z2(2) = FA(2):Z3$ = " FE2SIO4":
     GOSUB 52
2000 Z1$ = "OL":Z2(1) = FO(1) + FA(1):Z2(0) = FO(0) + FA(0):Z2(2) = FO(2) +
     FA(2):Z3$ = " (MG,FE)2SIO4": GOSUB 50
2020 CS(1) = CS * 172.25:CS(2) = CS * 4 / D1:CS(0) = CS * 3 / D
2030 Z1$ = "CS":Z2(1) = CS(1):Z2(0) = CS(0):Z2(2) = CS(2):Z3$ = " CA2SIO4":
     GOSUB 50
2040 MT(1) = MT * 231.54:MT(2) = MT * 4 / D1:MT(0) = MT * 3 / D
2050 Z1$ = "MT":Z2(1) = MT(1):Z2(0) = MT(0):Z2(2) = MT(2):Z3$ = "
     (FE+2)(FE+3)2O4": GOSUB 50
2060 IL(1) = IL * 151.75:IL(2) = IL * 3 / D1:IL(0) = IL * 2 / D
2070 Z1$ = "IL":Z2(1) = IL(1):Z2(0) = IL(0):Z2(2) = IL(2):Z3$ = " FETIO3":
     GOSUB 50
2080 HM(1) = HM * 159.69:HM(2) = HM * 3 / D1:HM(0) = HM * 2 / D
2090 Z1$ = "HM":Z2(1) = HM(1):Z2(0) = HM(0):Z2(2) = HM(2):Z3$ = " FE2O3": GOSUB
     50
2100 NC(1) = NC * 105.99:NC(2) = NC * 3 / D:NC(0) = NC * 3 / D
2110 Z1$ = "NC":Z2(1) = NC(1):Z2(0) = NC(0):Z2(2) = NC(2):Z3$ = " NA2CO3":
     GOSUB 50
2120 TN(1) = TN * 196.07:TN(2) = TN * 5 / D1:TN(0) = TN * 3 / D
2130 Z1$ = "TN":Z2(1) = TN(1):Z2(0) = TN(0):Z2(2) = TN(2):Z3$ = " CATISIO5":
     GOSUB 50
2140 PF(1) = PF * 135.98:PF(2) = PF * 3 / D1:PF(0) = PF * 2 / D
2150 Z1$ = "PF":Z2(1) = PF(1):Z2(0) = PF(0):Z2(2) = PF(2):Z3$ = " CATIO3":
     GOSUB 50
2160 RU(1) = RU * 79.90:RU(2) = RU * 2 / D1:RU(0) = RU / D
2165 Z1$ = "RU":Z2(1) = RU(1):Z2(0) = RU(0):Z2(2) = RU(2):Z3$ = " TIO2": GOSUB
     50
2180 AP(1) = AP * 336.32:AP(2) = AP * 26 / 3 / D1:AP(0) = (AP * 16 / 3) / D
2190 Z1$ = "AP":Z2(1) = AP(1):Z2(0) = AP(0):Z2(2) = AP(2):Z3$ = " 1/3
     (CA10P6O24F2)": GOSUB 50
2200 CC(1) = CC * 100.09:CC(2) = CC * 3 / D1:CC(0) = CC * 2 / D
2210 Z1$ = "CC":Z2(1) = CC(1):Z2(0) = CC(0):Z2(2) = CC(2):Z3$ = " CACO3": GOSUB
     50
2220 PR(1) = PR * 119.97:PR(2) = PR * 2 / D1:PR(0) = PR / D
2230 Z1$ = "PR":Z2(1) = PR(1):Z2(0) = PR(0):Z2(2) = PR(2):Z3$ = " FES2": GOSUB
     50
2240 TH(1) = TH * 142.04:TH(2) = TH * 4 / D1:TH(0) = TH * 3 / D
2250 Z1$ = "TH":Z2(1) = TH(1):Z2(0) = TH(0):Z2(2) = TH(2):Z3$ = " NA2SO4":
     GOSUB 50
2260 FR(1) = FR * 78.08:FR(2) = FR * 2 / D1:FR(0) = FR / D
2270 Z1$ = "FR":Z2(1) = FR(1):Z2(0) = FR(0):Z2(2) = FR(2):Z3$ = " CAF2": GOSUB
     50
2280 ZR(1) = ZR * 183.31:ZR(2) = ZR * 4 / D1:ZR(0) = ZR * 2 / D
2290 Z1$ = "ZR":Z2(1) = ZR(1):Z2(0) = ZR(0):Z2(2) = ZR(2):Z3$ = " ZRSIO4":
     GOSUB 50
2300 HL(1) = HL * 58.44:HL(2) = HL / D1:HL(0) = HL / D
2310 Z1$ = "HL":Z2(1) = HL(1):Z2(0) = HL(0):Z2(2) = HL(2):Z3$ = " NACL": GOSUB
     50
2320 CM(1) = CM * 223.84:CM(2) = CM * 4 / D1:CM(0) = CM * 3 / D
```

2330 Z1$ = "CM":Z2(1) = CM(1):Z2(0) = CM(0):Z2(2) = CM(2):Z3$ = " FECR2O4": GOSUB 50
2340 T = Q(1) + C(1) + OT(1) + AB(1) + AN(1) + LC(1) + NE(1) + KP(1) + AC(1) + NS(1) + KS(1) + WO(1) + DI(1) + HD(1) + EN(1) + FS(1) + FO(1) + FA(1) + CS(1) + MT(1) + IL(1) + HM(1) + NC(1) + TN(1) + PF(1) + RU(1) + AP(1) + CC(1) + PR(1) + TH(1) + FR(1) + ZR(1) + HL(1) + CM(1)
2350 PRINT "   -------":Z2(0) = T:Z1$ = " SUB-TOTAL": GOSUB 54:Z2(0) = OP(22) + OP(23) + OP(24) + OP(25):Z1$ = " H2O/L.O.I.": GOSUB 54:Z2(0) = T + Z2(0):Z1$ = " TOTAL": GOSUB 54
2361 PRINT
"
"--

2362 PRINT : REM

**CALCULATE THE THORNTON-TUTTLE DIFFERENTIATION INDEX*

2363 TT = Q(1) + AB(1) + NE(1) + OT(1) + LC(1) + KP(1)
2364 Z2(0) = TT:Z1$ = " THORNTON-TUTTLE DIFF. INDEX": GOSUB 54
2365 PRINT : REM

** CALCULATE QPRIME, FPRIME , AND PPRIME FOR THE NORMATIVE CLASSIFICATION OF STRECKEISEN AND LEMATRE (1979) NEUES JAHRBUCH FUR MINERALOGIE AB., 134,1-14 *

2366 QP = 0:FP = 0:PP = 0:X = (AN(1) + OT(1)) / 100:Y = (AN(1) + OT(1) + AB(1) + Q(1) + NE(1) + KP(1) + LC(1)) / 100: IF X > 0 THEN IF Y > 0 THEN PP = AN(1) / X:QP = Q(1) / Y:FP = (NE(1) + LC(1) + KP(1)) / Y
2367 PRINT "PARAMETERS FOR STRECKEISEN-LEMATRE (1979) ROCK CLASSIFICATION PLOT": PRINT :Z2(0) = PP:Z1$ = " AN/(AN+OR) ": GOSUB 54
2368 Z2(0) = QP:Z1$ = " Q/(Q+OR+AB+AN)": GOSUB 54
2369 Z2(0) = FP:Z1$ = " (NE+LC+KP)/(NE+LC+KP+OR+AB+AN)": GOSUB 54
2380 PRINT CHR$ (12)
3000 REM

**CALCULATE NORMATIVE PARAMETERS FOR PLOTTING VARIOUS DIAGRAMS*

3010 PRINT : PRINT SN$: PRINT "NORMATIVE PLOTTING PARAMETERS   ";: GOSUB 9900: IF FC$ = "Y" THEN PRINT "*** FEO-FE2O3 RECALCULATED USING - T(C)= ";CT;" LOG F(O2)= ";CX;" ***"
3020 PRINT
"
"--

3040 PRINT : PRINT "   ----------TERNARY-------------": PRINT "   -WT % -MOLE % -EQIV.OXYGEN   'MOLE' FORMULA (EQUIVALENT OX.)"
3050 GOSUB 3600: REM TERNARY FELDSPARS
3060 IF Q(1) > 0 THEN GOSUB 3800: REM Q-AB-OR
3070 IF Q(1) < = 0 THEN GOSUB 4200: REM Q-NE-KP
3080 GOSUB 4600: REM Q-NE-FO
3090 IF (Q(1) + EN(1)) > 0 THEN GOSUB 4000: REM DI-FO-Q
3100 IF (Q(1) + EN(1)) < = 0 THEN GOSUB 4400: REM DI-FO-NE
3110 GOSUB 4800: REM DI-OL-PLAG
3120 GOSUB 4500: REM QUAD PYROXENES
3150 PRINT CHR$ (12): PRINT D9$;"PR#0"
3190 RETURN
3600 PRINT : REM

**COMPUTE FACTORS FOR TERNARY FELDSPARS*

3610 X = OT * 2 + AB * 2 + AN:X = X / 100:Y = OT(1) + AB(1) + AN(1):Y = Y / 100
3615 IF X = 0 GOTO 3650
3620 Z1$ = "OR":Z2(0) = OT(1) / Y:Z2(1) = OT * 2 / X:Z2(2) = Z2(1):Z3$ = " KALSI3O8": GOSUB 56
3630 Z1$ = "AB":Z2(0) = AB(1) / Y:Z2(1) = AB * 2 / X:Z2(2) = Z2(1):Z3$ = " NAALSI3O8": GOSUB 56

```
3640 Z1$ = "AN":Z2(0) = AN(1) / Y:Z2(1) = AN / X:Z2(2) = Z2(1):Z3$ = "
     CAAL2SI2O8": GOSUB 56
3650  RETURN
3800  PRINT : REM

**CALCULATE FACTORS FOR AB-OR-Q SYSTEM*

3810 X = OT * 2 + AB * 2 + Q:X = X / 100
3820 Y = OT(1) + AB(1) + Q(1):Y = Y / 100
3830 Z = OT * 2 + AB * 2 + Q / 4:Z = Z / 100: REM     CALCULATE TO CONSTANT 8
     ANIONS (IE. KALSI3O8-SI4O8)
3840  IF X = 0 THEN X = 1:Y = 1:Z = 1: REM   IF SUMS=0 THEN COMPONENTS=0
     SETTING SUMS=1 WILL AVOID DIV BY 0 ERROR IN CALCULATING %
3850 Z1$ = "Q ":Z2(0) = Q(1) / Y:Z2(1) = Q / X:Z2(2) = (Q / 4) / Z:Z3$ = " SIO2
     (SI4O8)": GOSUB 56
3860 Z1$ = "OR":Z2(0) = OT(1) / Y:Z2(1) = OT * 2 / X:Z2(2) = OT * 2 / Z:Z3$ = "
     KALSI3O8": GOSUB 56
3870 Z1$ = "AB":Z2(0) = AB(1) / Y:Z2(1) = AB * 2 / X:Z2(2) = AB * 2 / Z:Z3$ = "
     NAALSI3O8": GOSUB 56
3880  RETURN
4000  PRINT : REM

**CALCULATE FACTORS FOR DI-FO-Q DIAGRAM*

4020 X = DI(1) + FO(1) + EN(1) + Q(1):Y = DI + FO + EN + Q:Z = DI / 2 + FO / 3
     + Q / 6 + 3 * EN / 12: REM      FOR Z WE FORM 'MOLES' WITH 12 OXYGENS EACH
4030 A = 0:B = 0:C = 0: IF X > 0 THEN A = 100 * DI(1) / X:B = 100 * DI / Y:C =
     100 * (DI / 2) / Z
4040 Z1$ = "DI":Z2(0) = A:Z2(1) = B:Z2(2) = C:Z3$ = " CAMGSI2O6
     (CA2MG2SI4O12)": GOSUB 56
4050 A = 0:B = 0:C = 0: IF X > 0 THEN A = 100 * (FO(1) + .70 * EN(1)) / X:B =
     100 * (FO + EN / 2) / Y:C = 100 * (FO / 3 + EN / 6) / Z
4060 Z1$ = "FO":Z2(0) = A:Z2(1) = B:Z2(2) = C:Z3$ = " MG2SIO4 (MG6SI3O12)":
     GOSUB 56
4070 A = 0:B = 0:C = 0: IF X > 0 THEN A = 100 * (Q(1) + .30 * EN(1)) / X:B =
     100 * (Q + EN / 2) / Y:C = 100 * (Q / 6 + EN / 12) / Z
4080 Z1$ = "Q ":Z2(0) = A:Z2(1) = B:Z2(2) = C:Z3$ = " SIO2 (SI6O12)": GOSUB 56
4090  RETURN
4200  PRINT : REM

**CALCULATE POSITION IN SIO2-KALSIO4-NAALSIO4 SYSTEM*

4202 X = NE(1) + AB(1) + OT(1) + LC(1) + KP(1):Y = 2 * NE + 6 * AB + 2 * KP + 6
     * OT + 4 * LC:Z = NE + KP + 12 * LC / 8 + 2 * AB + 2 * OT: REM
     FOR Z WE FORM 'MOLES' WITH  8 OXYGENS EACH
4204 A = 0:B = 0:C = 0: IF X > 0 THEN A = 100 * (.4583 * AB(1) + .4318 * OT(1)
     + .2753 * LC(1)) / X:B = 100 * (4 * AB + 4 * OT + 2 * LC) / Y:C = 100 *
     (AB + OT + LC / 2) / Z
4206 Z1$ = "Q ":Z2(0) = A:Z2(1) = B:Z2(2) = C:Z3$ = " SIO2 (SI4O8)": GOSUB 56
4208 A = 0:B = 0:C = 0: IF X > 0 THEN A = 100 * (NE(1) + .5417 * AB(1)) / X:B =
     100 * (2 * NE + 2 * AB) / Y:C = 100 * (NE + AB) / Z
4210 Z1$ = "NE":Z2(0) = A:Z2(1) = B:Z2(2) = C:Z3$ = " NAALSIO4 (NA2ASI2O8)":
     GOSUB 56
4212 A = 0:B = 0:C = 0: IF X > 0 THEN A = 100 * (KP(1) + .5682 * OT(1) + .7247
     * LC(1)) / X:B = 100 * (2 * KP + 2 * OT + 2 * LC) / Y:C = 100 * (KP + OT +
     LC) / Z
4214 Z1$ = "KP":Z2(0) = A:Z2(1) = B:Z2(2) = C:Z3$ = " KALSIO4 (K2AL2SI2O8)":
     GOSUB 56
4218  RETURN
4400  PRINT : REM

**CALCULATE FACTORS FOR DI-FO-NE DIAGRAM*

4404 X = DI(1) + FO(1) + NE(1):Y = DI + FO + NE * 2:Z = DI / 2 + FO / 3 + 8 *
     NE / 12: REM            FOR Z WE FORM 'MOLES' WITH 12 OXYGENS EACH
```

```
5280 Z2(0) = A:Z2(1) = B:Z2(2) = C:Z1$ = "  FE2O3/FEO (FE3+/FE2+)": GOSUB 60:
     PRINT
5290 A = 0:B = 0:C = 0: IF OP(7) > 0 THEN A = OP(7) / (OP(7) + OP(5)):B = MP(7)
     / (MP(7) + MP(5)):C = CP(7) / (CP(7) + CP(5))
5300 Z2(0) = A:Z2(1) = B:Z2(2) = C:Z1$ = "  MGO / (MGO + FEO)": GOSUB 60
5310 A = 0:B = 0:C = 0: IF OP(7) > 0 THEN A = OP(7) / (OP(7) + OP(5) + OP(6) +
     .8998 * OP(4)):B = MP(7) / (MP(7) + MP(5) + MP(6) + MP(4) * 2):C = CP(7) /
     (CP(7) + CP(5) + CP(6) + CP(4))
5320 Z2(0) = A:Z2(1) = B:Z2(2) = C:Z1$ = "  MGO / (MGO + FEO* + MNO)
     FEO*=TOTAL FE AS FEO": GOSUB 60: PRINT
5330 A = OP(9) + OP(10):B = 100 * (MP(9) + MP(10)) / MP(0):C = 100 * (CP(9) +
     CP(10))
5340 Z2(0) = A:Z2(1) = B:Z2(2) = C:Z1$ = "  NA2O + K2O": GOSUB 58
5350 A = 0:B = 0:C = 0: IF OP(1) > 0 THEN A = (OP(9) + OP(10)) / OP(1):B =
     (MP(9) + MP(10)) / MP(1):C = (CP(9) + CP(10)) / CP(1)
5360 Z2(0) = A:Z2(1) = B:Z2(2) = C:Z1$ = "  (NA2O + K2O)/ SIO2": GOSUB 60
5370 A = 0:B = 0:C = 0: IF OP(3) > 0 THEN A = (OP(9) + OP(10)) / OP(3):B =
     (MP(9) + MP(10)) / MP(3):C = (CP(9) + CP(10)) / CP(3)
5380 Z2(0) = A:Z2(1) = B:Z2(2) = C:Z1$ = "  (NA2O + K2O)/ AL2O3": GOSUB 60
5390 A = 0:B = 0:C = 0: IF OP(3) > 0 THEN A = (OP(8) + OP(9) + OP(10)) /
     OP(3):B = (MP(8) + MP(9) + MP(10)) / MP(3):C = (CP(8) + CP(9) + CP(10)) /
     CP(3)
5400 Z2(0) = A:Z2(1) = B:Z2(2) = C:Z1$ = "  (CAO + NA2O + K2O)/ AL2O3": GOSUB
     60: PRINT
5408  REM

**AFM DIAGRAM CALCULATIONS *

5409  PRINT "TERNARY AFM PLOTTING PERCENTAGES"
5410 X = OP(9) + OP(10) + OP(5) + OP(4) / 1.1113 + OP(7):Y = MP(9) + MP(10) +
     MP(5) + MP(4) * 2 + MP(7):Z = CP(9) + CP(10) + CP(5) + CP(4) + CP(7):X = X
     / 100:Y = Y / 100:Z = Z / 100: REM    SUMS FOR AFM DIAGRAM
5420 A = 0:B = 0:C = 0: IF X > 0 THEN A = (OP(9) + OP(10)) / X:B = (MP(9) +
     MP(10)) / Y:C = (CP(9) + CP(10)) / Z
5425 Z2(0) = A:Z2(1) = B:Z2(2) = C:Z1$ = "  A - (NA2O + K2O) ": GOSUB 58
5430 A = 0:B = 0:C = 0: IF X > 0 THEN A = (OP(5) + OP(4) / 1.1113) / X:B =
     (MP(5) + MP(4) * 2) / Y:C = (CP(5) + CP(4)) / Z
5435 Z2(0) = A:Z2(1) = B:Z2(2) = C:Z1$ = "  F -
     (FEO+.8998*FE2O3)(FEO+2*FE2O3)(FE2+ + FE3+)": GOSUB 58
5440 A = 0:B = 0:C = 0: IF X > 0 THEN A = OP(7) / X:B = MP(7) / Y:C = CP(7) / Z
5445 Z2(0) = A:Z2(1) = B:Z2(2) = C:Z1$ = "  M - (MGO)": GOSUB 58: PRINT
5448  REM

**CA-NA-K TERNARY DIAGRAM CALCULATIONS*

5449  PRINT "TERNARY CA-NA-K PLOTTING PARAMETERS"
5450 X = OP(9) + OP(10) + OP(8):Y = MP(9) + MP(10) + MP(8):Z = CP(9) + CP(10) +
     CP(8):X = X / 100:Y = Y / 100:Z = Z / 100: REM     SUMS FOR CA-NA-K
     DIAGRAM
5460 A = 0:B = 0:C = 0: IF X > 0 THEN A = OP(8) / X:B = MP(8) / Y:C = CP(8) / Z
5465 Z2(0) = A:Z2(1) = B:Z2(2) = C:Z1$ = "  CAO /(CAO+NA2O+K2O)": GOSUB 58
5470 A = 0:B = 0:C = 0: IF X > 0 THEN A = OP(9) / X:B = MP(9) / Y:C = CP(9) / Z
5475 Z2(0) = A:Z2(1) = B:Z2(2) = C:Z1$ = "  NA2O/(CAO+NA2O+K2O)": GOSUB 58
5480 A = 0:B = 0:C = 0: IF X > 0 THEN A = OP(10) / X:B = MP(10) / Y:C = CP(10)
     / Z
5485 Z2(0) = A:Z2(1) = B:Z2(2) = C:Z1$ = "  K2O /(CAO+NA2O+K2O)": GOSUB 58
5490  PRINT  CHR$ (12): PRINT D9$;"PR#0"
5495  RETURN
5500  REM

**SUBROUTINE TO RECALCULATE FE2O3/FEO USING THE RELATIONSHIPS IN KILINC,
     CARMICHAEL, RIVERS AND SACK (1983) CONTR. MINERAL. PETROL., 83:136-140*

5510  PRINT : PRINT  CHR$ (7): INPUT "TEMPERATURE (DEG.C) = ";CT
5520  PRINT  CHR$ (7): INPUT "LOG FUGACITY OF OXYGEN=";CX
```

```
5530 OP(0) = OP(0) - OP(4) - OP(5): REM  TAKE FE2O3 AND FEO FROM THE TOTALS
5535 MP(0) = MP(0) - MP(4) - MP(5)
5540 CP(0) = CP(0) - CP(4) - CP(5)
5545 TX = TX - MP(4) * XN(4) - MP(5) * XN(5)
5550 MP(5) = MP(5) + 2 * MP(4): REM    COMBINE FE2O3 AS FEO WITH FEO
5560 MT = MP(1) + MP(3) + MP(5) + MP(7) + MP(8) + MP(9) + MP(10): REM  TOTAL
     THE MOLECULAR PROPORTIONS FOR MOLE FRACTION CALCULATION
5570 SUM =  - 2.24 * MP(3) / MT + 1.55 * MP(5) / MT + 2.96 * MP(8) / MT + 8.42
     * MP(9) / MT + 9.59 * MP(10) / MT: REM   SUM THE COEFICIENTS TIMES THE
     MOLE FRACTIONS (MP/MT) IN EQUATION 1 OF KILINC,ET AL.
5580 KT = CT + 273:LX = 2.303 * CX: REM  COMPUTE THE KELVIN TEMPERATURE AND
     NATURAL LOG OF THE OXYGEN FUGACITY
5590 R = .2185 * LX + 12670 / KT - 7.54 + SUM: REM   COMPUTE THE RIGHT SIDE OF
     EQUATION 1 OF KILINC, ET AL.
5600 R =  EXP (R): REM   TAKE THE ANTILOG TO OBTAIN THE MOLE RATIO OF FE2O3 TO
     FEO
5610 X = MP(5)
5620 MP(5) = X / (2 * R + 1): REM   CALCULATE MOLAR PROPORTION OF FEO
5630 MP(4) = (X - MP(5)) / 2: REM  CALCULATE MOLAR PROPORTION OF FE2O3
5640 OP(4) = MP(4) * MW(4):CP(4) = MP(4) * CN(4)
5650 OP(5) = MP(5) * MW(5):CP(5) = MP(5) * CN(5)
5660 OP(0) = OP(0) + OP(4) + OP(5)
5670 MP(0) = MP(0) + MP(4) + MP(5)
5680 CP(0) = CP(0) + CP(4) + CP(5)
5690 TX = TX + MP(4) * XN(4) + MP(5) * XN(5)
5700  PRINT "FE2O3 ";OP(4)
5710  PRINT "FEO   ";OP(5)
5720  PRINT "TOTAL  ";OP(0): PRINT : PRINT  CHR$ (7): INPUT "IS THIS
     CORRECT(Y<=1>/N)?";A$:A$ =  LEFT$ (A$,1): IF (A$ <  > "Y") AND (A$ <  >
     "1") GOTO 5500
5799  RETURN
5900  REM

**SUBROUTINE TO RECALCULATE MOLE PROPORTIONS FROM THE ANALYSIS RECALCULATED TO
     100% WATER FREE *

5905 MP(0) = 0
5910  FOR I = 1 TO 21
5920 MP(I) = NX(I) / MW(I):MP(0) = MP(0) + MP(I)
5930  NEXT I
5950  FOR I = 22 TO 25:MP(I) = 0:CP(I) = 0: NEXT I
5960  RETURN
7000  REM

**SUBROUTINE TO SET UP INPUT**

7010  HOME : PRINT "OXIDE COMPONENTS FOR NORM ---------": PRINT "SYMBOL
     MOL.WT.","SYMBOL MOL.WT.": FOR I = 1 TO 24 STEP 2: PRINT XN$(I); SPC(
     2);MW(I),XN$(I + 1); SPC( 2);MW(I + 1): NEXT I: PRINT XN$(25)
7020  PRINT : PRINT "ENTER ONLY THE COMPONENT SYMBOLS IN ": PRINT "YOUR TABLE
     IN THE ORDER THAT THEY WILL": PRINT "BE ENTERED.''END'' WILL END LIST":
     PRINT
7030  FOR I = 1 TO 25:EO(I) = 0: NEXT I
7040  FOR I = 1 TO 25
7050  PRINT "COMPONENT ";I;" IS";: INPUT A$:A$ = A$ + "      ":A$ =  LEFT$
     (A$,5)
7060  IF A$ = "END  " THEN EI = I - 1: GOTO 7200
7130  FOR J = 1 TO 25: IF A$ = XN$(J) THEN EO(I) = J:J = 25: NEXT J: GOTO 7190
7140  NEXT J
7150  PRINT  CHR$ (7);"SYMBOL -";A$;"- NOT RECOGNIZED": PRINT  CHR$ (7): PRINT
     : GOTO 7050
7190  NEXT I
7200  IF EI = 0 THEN  PRINT "TRY THE TABLE AGAIN": GOTO 7010
7205  HOME : PRINT "YOUR TABLE IS:"
7210  FOR I = 1 TO EI
```
````
````

```
7220  IF EO(I) = 0 GOTO 7240
7230 J =  INT (EO(I)): PRINT I;". ";XN$(J)
7240  NEXT I
7250  PRINT : FLASH : PRINT  CHR$ (7);"IS THIS CORRECT?";: NORMAL : INPUT A$:A$
     =  LEFT$ (A$,1): IF A$ = "N" GOTO 7010
7260  IF A$ <  > "Y" GOTO 7250
7270  RETURN
8000  REM

**ROUTINE TO SET UP DISK FILE FOR OUTPUT*

8010  PRINT : PRINT  CHR$ (7): INPUT "FILE NAME?";F$: PRINT  CHR$ (7)
8020  PRINT  CHR$ (7): INPUT "DRIVE 1 OR 2?";A$:DD$ =  LEFT$ (A$,1): IF DD$ <
     > "1" THEN  IF DD$ <  > "2" GOTO 8020
8030  HOME :F$ = F$ + ".MPC":DD$ = ",D" + DD$: PRINT : PRINT "INSERT PROPER
     DISC IN DRIVE ";DD$: PRINT : PRINT "FOR FILE ";F$;DD$: PRINT : PRINT "HIT
     ANY KEY WHEN READY.": GET A$: PRINT
8040  ONERR  GOTO 8180
8050  PRINT  CHR$ (13);D9$;"VERIFY ";F$;DD$: POKE 216,0: REM    RESETS ERROR
     HANDLING
8060  PRINT  CHR$ (7): PRINT : PRINT
8070  PRINT "FILE ALREADY EXISTS - WRITE OVER";: INPUT "(Y/N)?";A$:A$ =  LEFT$
     (A$,1): IF A$ = "N" GOTO 8010
8080  IF A$ = "Y" GOTO 8110
8090  GOTO 8060
8100  IF  PEEK (222) <  > 6 THEN  GOTO 8130
8110  ONERR  GOTO 8130
8120  PRINT : PRINT D9$;"OPEN";F$;DD$: PRINT D9$;"DELETE";F$;DD$: POKE 216,0:
     GOTO 8190
8130  PRINT : PRINT  CHR$ (7): PRINT "CHECK DISC AND START OVER":X =  PEEK
     (222):Y =  PEEK (218) + 256 *  PEEK (219): POKE 216,0: IF X = 10 THEN
     PRINT "******FILE LOCKED******": GOTO 8010
8140  IF X = 4 THEN  PRINT "******DISC WRITE PROTECTED******": GOTO 8010
8150  IF X = 9 THEN  PRINT "******DISC FULL******": GOTO 8010
8160  IF X = 8 THEN  PRINT "******I/O ERROR******": GOTO 8010
8170  PRINT "****************************************": PRINT "ERROR #";X;" AT
     LINE #";Y: STOP
8180  IF  PEEK (222) <  > 6 THEN  GOTO 8130
8190  & SPACE - B = X:X = X / 2: PRINT : PRINT "DISC HAS SPACE FOR ABOUT ";X;"
     ANALYSES.": PRINT : INPUT "IS THIS ENOUGH (Y/N)?";A$:A$ =  LEFT$ (A$,1)
8200  IF A$ = "N" GOTO 660
8210  IF A$ <  > "Y" GOTO 8190
8220  PRINT  CHR$ (7): INPUT "FILE TITLE (MAX. 80 CHARACTERS)?";FI$
8230  PRINT  CHR$ (13);D9$;"OPEN";F$;DD$: PRINT D9$;"WRITE";F$
8240 NN = 25: PRINT FI$: PRINT NN: FOR I = 1 TO 25: PRINT XN$(I): NEXT I
8250  PRINT D9$;"CLOSE";F$
8260  GOTO 690
```

```
8300  REM

**SUBROUTINES TO WRITE ANALYSIS DATA TO DISC FILE*

8310 SN$ = SN$ + "                                        ":SN$ =  LEFT$
     (SN$,40)
8320  PRINT D9$;"APPEND";F$;DD$: PRINT D9$;"WRITE";F$
8330  PRINT SN$: FOR I = 1 TO 25: PRINT MX(I): NEXT I
8340  PRINT D9$;"CLOSE";F$
8350  RETURN
8360  PRINT D9$;"APPEND";F$;DD$: PRINT D9$;"WRITE";F$
8370  PRINT "END OF DATA"
8380  PRINT D9$;"CLOSE";F$
8390  RETURN
9000  REM

**ROUTINE TO SET CONSTANTS AND DIM ARRAYS**

9010  DIM EN$(25),XN$(25),MW(25),CN(25),XN(25),Z2(4)
9011  DIM OP(25),MP(25),MX(25),CP(25),CX(25),NX(25),EO(25)
9012  DIM
     Q(2),C(2),OT(2),AB(2),AN(2),LC(2),NE(2),KP(2),AC(2),NS(2),KS(2),WO(2),DI(2)
     ,HD(2),EN(2),FS(2)
9013  DIM
     FO(2),FA(2),CS(2),MT(2),IL(2),HM(2),NC(2),TN(2),PF(2),RU(2),AP(2),CC(2),PR(
     2),TH(2),FR(2),ZR(2),HL(2),CM(2)
9030  FOR I = 1 TO 25: READ EN$(I),XN$(I),MW(I),CN(I),XN(I): NEXT I
9040  DATA   "SI","SIO2 ",60.09,1,2
9050  DATA   "TI","TIO2 ",79.90,1,2
9060  DATA   "AL","AL2O3",101.96,2,3
9070  DATA   "F3","FE2O3",159.69,2,3
9080  DATA   "F2","FEO  ",71.85,1,1
9090  DATA   "MN","MNO  ",70.94,1,1
9100  DATA   "MG","MGO  ",40.30,1,1
9110  DATA  "CA","CAO  ",56.08,1,1
9120  DATA  "NA","NA2O ",61.98,2,1
9130  DATA  "K ","K2O  ",94.20,2,1
9140  DATA  "P ","P2O5 ",141.95,2,5
9150  DATA  "C ","CO2  ",44.01,1,2
9160  DATA  "CR","CR2O3",151.99,2,3
9170  DATA  "NI","NIO  ",74.70,1,1
9180  DATA  "BA","BAO  ",153.34,1,1
9190  DATA  "SR","SRO  ",103.62,1,1
9200  DATA  "CL","CL   ",35.45,1,-0.5
9210  DATA  "F ","F    ",19.00,1,-0.5
9220  DATA  "S ","S    ",32.06,1,-1
9230  DATA  "S6","SO3  ",80.06,1,3
9240  DATA  "ZR","ZRO2 ",123.22,1,2
9250  DATA   "H ","H2O  ",18.00,2,1
9260  DATA  "H ","H2O+ ",18.00,2,1
9270  DATA   "H ","H2O- ",18.00,2,1
9280  DATA    "H ","LOI  ",9E+20,0,0
9290  RETURN
```

```
9500  REM

**SUBROUTINE TO PRINT HEADING**

9510  PRINT : PRINT : PRINT "***************************************"
9520  PRINT "*                                    *"
9530  PRINT "*           PROGRAM ROCALC           *"
9540  PRINT "*                                    *"
9550  PRINT "* J. C. STORMER - GEOLOGY DEPT.      *"
9560  PRINT "* RICE UNIVERSITY, P.O. BOX 1892     *"
9570  PRINT "* HOUSTON  TX 77251                  *"
9580  PRINT "* (713) 527-4054                     *"
9581  PRINT "*                                    *"
9582  PRINT "* FEO-FE2O3 CAN BE CALCULATED FROM   *"
9583  PRINT "* T AND LOG F(O2) USING KILINC, ET AL *"
9584  PRINT "* (1983) CONTR. MINERAL. PETROL.,83,  *"
9585  PRINT "* 136-140.   THE NORM CALCULATIONS   *"
9586  PRINT "* FOLLOW THE SEQUENCE OF STEPS GIVEN  *"
9587  PRINT "* IN BARKER (1983) IGNEOUS ROCKS,     *"
9588  PRINT "* PP. 65-71.                          *"
9610  PRINT "**************************************"
9620  GOSUB 9900
9670  RETURN
9900  REM

**SUBROUTINE TO PRINT NAME AND VERSION*

9910  PRINT "PROGRAM ROCALC - VERSION OF  14 JAN 1985             "
9920  RETURN
```

Sketching a Cross Section of Folded Terrain with a Microcomputer

M. M. Kimberly

North Carolina State University

ABSTRACT

TRAVERSE is a Pascal program which interpolates observations of bedding-plane orientations along a traverse using an Apple II microcomputer. Initial input to TRAVERSE is a set of elevations which then are interpolated to illustrate topography. To plot geologic contacts, the user specifies location by a distance along the cross section, corresponding elevation, bearing of the cross section, strike of the contact, angle of dip, and direction of dip (either forward or backward relative to direction of traverse). TRAVERSE calculates vertical exaggeration, offers the option of decreasing vertical exaggeration, and then calculates apparent dips as they should appear along the bearing of the cross section. TRAVERSE forces interpolation between outcrops to follow the observed dip for part of the distance toward the closest adjacent contact and then allows an interpolation routine to calculate the remaining intervening region. Interpolation options are cubic spline or Lagrangian polynomials. In the situation of a single observed contact, for example a fault plane, TRAVERSE extends the contact straight along the dip of the fault. Multiple sets of interpolated contacts may be overprinted to produce a compete cross section.

INTRODUCTION

A structural geologist usually measures bedding orientations exposed in scattered outcrops and then deduces geometric relationships in the intervening areas. An experienced structural geologist is able to perform this interpolation mentally. Interpolation while in the field helps direct mapping toward the most important areas. However, many geologists cannot perform this mental interpolation in highly deformed areas and the program described herein, TRAVERSE, is offered to assist with interpolation (Fig. 1, Appendix Tables 1 and 2).

TRAVERSE is a Pascal program which runs on the Apple II family of microcomputers (II+, IIe, IIc). The portable, battery-operated Apple IIc would be most appropriate for field applications. The Apple IIc could be back-packed into a remote area, given a CPU weight of only 3.4 kg and a comparably light printer or monitor. Minimum field gear would be an Apple IIc, a DC power supply, and a thin, liquid-crystal-type monitor. Collectively, this could fit into a briefcase.

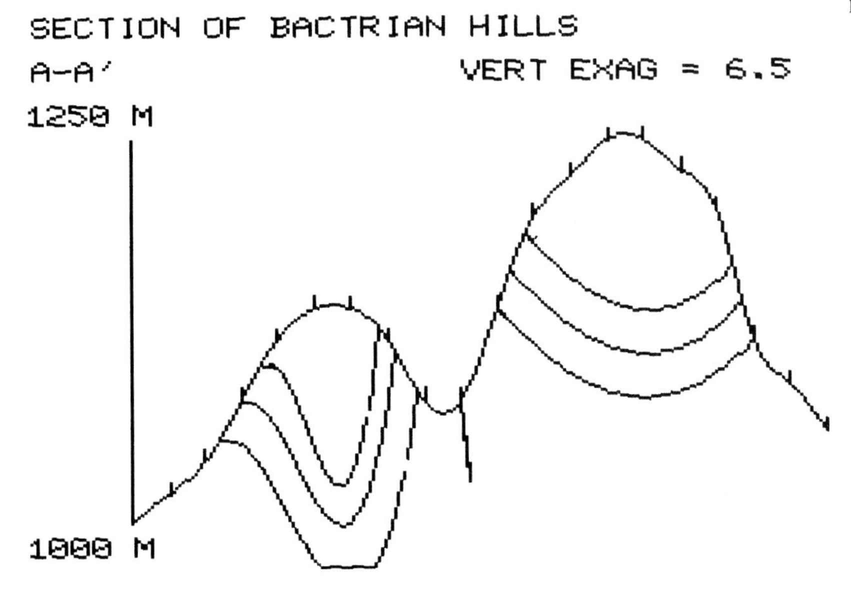

Figure 1. Example of output from program TRAVERSE, given input data of Appendix Table 2. Fictitious Bactrian Hills section (named for type of camel) consists of two synclinal fault blocks separated by illustrated fault. Lowermost syncline has been truncated at lower limit of Apple-generated image.

The Apple IIc runs Pascal Version 1.1 similar to an Apple IIe. The slight differences noted herein between the Apple II+ and the IIe therefore also apply to the IIc. Pascal has been selected for TRAVERSE instead of BASIC because Pascal permits the usage of variable names which have obvious meaning. TRAVERSE has been designed for easy modification by diverse users and only a readable language like Pascal lends itself to such modification. Even programmers unfamiliar with Pascal should have little trouble understanding the code in the programs listed in the Appendix.

MICROCOMPUTING IN THE FIELD

Microcomputers offer several field applications, for example, direct collection of geophysical data, surveying calculations, and rapid plotting to locate trends which may influence data collection. One of the most promising uses of microcomputers in the field is the direct collection and initial interpretation of geophysical data. A battery-powered Apple II+ has been used by Sowerbutts and Mason (1984) for direct data collection in various land-based surveys, that is, with a magnetic vertical gradiometer, ground conductivity meter, and proton magnetometer. We have used an Apple II+ and IIe on a research vessel, the R/V Cape Hatteras, to digitize seismic-profiler data so that multiple reflections could be filtered. An inexpensive analog-to-digital converter and large-volume tape drive recorded digitized data simultaneously with the usual strip-chart recording of analog profiling.

The most extensive field application of microcomputers to date probably has been surveying for the petroleum industry. Commercial surveying packages based upon an Apple II are available, for example, the package from Petty-Ray Geophysical. Kimberley (in press) provides four Apple Pascal programs for geochemistry and structural geology. These four calculate correlation coefficients, evaluate an ore deposit, plot compositional data, and plot unfolded bedding.

The type of application discussed in this paper is the graphical presentation of folds while they are being mapped. Most plotting of geologic data is done after leaving the field, given that field work is expensive and usually seasonally limited. Inevitably, plotting helps identify crucial areas which may not have been studied thoroughly. Initial plotting of data by a microcomputer could help with recognition of important areas and guide data collection. Plotting by a microcomputer is faster than plotting by hand because scaling and interpolation are automatic. The program described here (TRAVERSE) interpolates strike-and-dip data to postulate fold patterns and thus predict where fold axes may be located (Fig. 1). TRAVERSE also is useful for displaying faults but a simpler program is adequate for this purpose (Kimberley, 1985).

To avoid excess weight in the field, TRAVERSE may be modified readily for use without a printer. Alternatively, the monitor may be eliminated in favor of the printer but this would require major modification of the program and would hinder data entry. The code for this alternative is not supplied. A cross section displayed on the monitor could be traced from it by hand almost as quickly as a printer would print it.

To modify TRAVERSE (Appendix Table 1) to eliminate the printer, the user should erase ten lines of code, starting sixteen lines from the end of the program, that is, from CHOICE:='Y' through END; (* WHILE CHOICE=Y*). Two other lines should be changed; the third line of procedure PRINTDATA should become REWRITE (PRINTFILE,'CONSOLE:'). This would direct the complete set of input data to the screen where it could be checked for typographical errors. A large data set would fill more than a single screen and further modification of TRAVERSE therefore may be required to segment the data set prior to transfer to the monitor. The other program line to be changed is the third line from the bottom of procedure PLOT. This one-line DO-loop controls the duration of cross-section display on the screen. To facilitate tracing, this duration either should be increased substantially by increasing the number of loops or else the DO-loop should be replaced with READ (KEYBOARD,ENTRY);. This statement would hold the cross section on the screen until the user pressed any key.

If TRAVERSE is modified to wait for a keyboard entry, a companion statement should be added to inform the user of the necessity of pressing a key to continue. However, this companion statement could not occur beside the READ (KEYBOARD,ENTRY) statement in procedure PLOT because the computer would be in graphics mode at this point and would not display the message on the monitor. The companion statement could be inserted near the beginning of the main block at the end of the program.

Retention of the displayed cross section on the screen with an increase in the number of loops would be a convenient program modification but would not be convenient during execution. Moreover, the maximum delay per DO-loop statement is limited to about 20 seconds because the maximum ordinary integer with an Apple II is 32,767. To achieve a longer delay, the user could declare a LONG INTEGER variable which could exceed this limit or else stack one delaying DO-loop after another.

PROGRAM TRAVERSE: INTERPOLATION OF STRIKE-AND-DIP DATA

To run program TRAVERSE in its standard version (Appendix Table 1), one must have a 64K byte Apple II with Apple Pascal version 1.1, a dot-matrix printer, and a Grappler (Orange Micro, Inc., Anaheim, CA) or Grappler-compatible interface to the printer. Two disk drives are needed to compile TRAVERSE but the compiled code will run with just a single drive. TRAVERSE has been compiled on an Apple II+ because the IIe generally limits the number of Pascal procedures to that which can be listed on a single screen. Once compiled on an Apple II+, modification and recompilation of TRAVERSE on a IIe is, however, usually successful.

Interpolation of both topography and bedding in the standard version of TRAVERSE (Appendix Table 1) employs a cubic-spline algorithm (for example, Conte and de Boor, 1980). Alternative Pascal code for interpolation of bedding based on Lagrangian polynomials also is listed (Appendix Table 3). Lagrangian polynomial interpolation generally produces more and larger oscillations than does spline interpolation (Figs. 1,2). These oscillations generally are most severe at both ends of the calculated curve and typically the oscillations at the ends are absurdly large for large data sets. However, in the central region of the data set, polynomial interpolation may produce a smoothly changing and visually appealing curve. If the user is pressed for time, slow Lagrangian interpolation should be avoided. Otherwise, it is recommended that both cubic-spline and Lagrangian interpolations be plotted routinely just to remind the user that there is no unique solution to subsurface deformation based on outcrop data. Lagrangian interpolation of topography is not recommended and no Pascal code is provided herein to perform it.

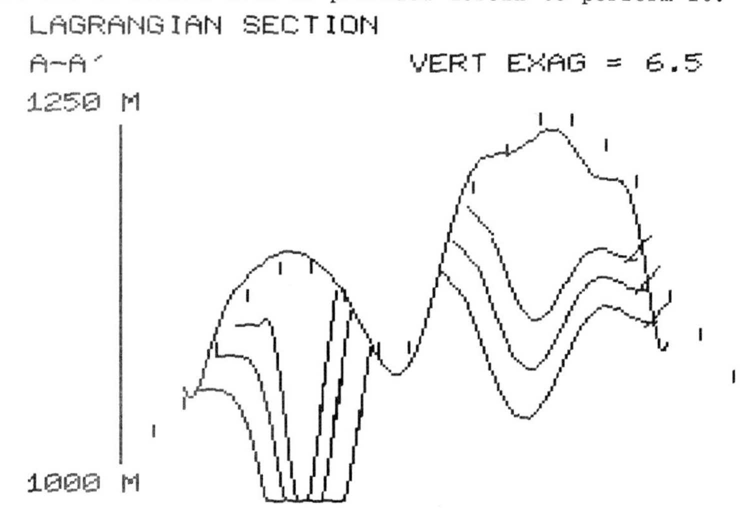

Figure 2. Example of output from Lagrangian version of program TRAVERSE, using same input data (Appendix Table 2) as was used to generate Figure 1. Fault contact of Figure 1 has been omitted. All three synclinal contacts on left side of Figure 2 have been truncated at lower limit of Apple generated graphics.

In the Lagrangian code of this paper, plotting near each end of the curve is omitted to avoid the typically absurd oscillations near each end. To limit the omitted region and constrain Lagrangian oscillatory tendency, three data points are calculated straight along each dip into the subsurface. By contrast, a single subsurface point adequately guides cubic-spline interpolation to reflect observed dip. Even the additional pair of Lagrangian data points may be inadequate, however, in areas of tight folding because of the inherent tendency for exaggeration by polynomial interpolation.

Spline interpolations consistently are conservative and reliable. Given any real geologic contact observed at N outcrops, a cubic spline calculated from the N data points theoretically always will contain more conservative oscillations than the real contact (Cheney and Kincaid, 1980, p.180). Cubic spline probably is used most widely for interpolation and is the highest order of spline which presently fits into a user-friendly microcomputer program. However, new generations of microcomputers will be capable of handling more sophisticated and more reliable algorithms, for example, quadratic splines (McAllister and Roulier, 1981; McLaughlin, 1983).

The Lagrangian interpolation used herein is calculated at the highest possible order, that is, an order equal to the number of data points. This simplifies the algorithm but slows execution. Guarding against the effects of unreasonable input data also slows execution. The Lagrangian version of TRAVERSE executes about fifty times slower than the cubic-spline version, that is, more than 20 minutes versus 25 seconds for a twenty-point data set. A reduction in the number of data points substantially increases the speed of Lagrangian interpolation. Assignment of more code for the calculation of lower order Lagrangian polynomials also would speed execution but might leave insufficient room for the user-friendly aspects of TRAVERSE.

TRAVERSE is limited to 60 topographic points and either 10 (Lagrangian) or 15 (cubic spline) outcrops per folded contact. The cubic-spline version creates an additional data point along the dip under each observed outcrop of the folded surface and so the number of permissible outcrops is half the dimension of XSPLINE and YSPLINE in the declaration of variables. In the Lagrangian situation, the permissible number is one-quarter. The user is encouraged to increase any of these limits if necessary; the maximum capacity of the Apple II for larger dimensions has not been determined.

TRAVERSE automatically prints all entered data and then allows for corrections before execution. TRAVERSE is completely menu-driven. The user need not know anything about computer programming to use it. Virtually the only fatal user error would be the typing of a nonnumeric character if a number is expected. The user may backspace and correct any entered titles but backspacing is inoperable during entry of numbers. The user must wait until all numbers have been entered before corrections may be made.

As with any other graphics program, TRAVERSE must contend with the aspect ratios of the computer, monitor and printer. TRAVERSE automatically modifies all values to reflect the horizontal range of 280 pixels versus vertical range of 192 on an Apple II. This ratio is 192/280 = 0.6857 in Procedure NORMALIZE. If TRAVERSE is modified to work on a different microcomputer, the aspect ratio should be changed to that of the alternative computer. Multiplication by this aspect ratio does not solve the problem of using various monitors and printers which are characterized by their own aspect ratios. The user, therefore, cannot expect the displayed dip angles to correspond exactly with the input angles without some experimentation with the aspect ratio in Procedure NORMALIZE. In general, vertical exaggeration and divergence of line-of-section from local strike combine to affect dip angles so markedly that minor errors in aspect ratio are unnoticeable.

TRAVERSE effectively assumes that only folds, not faults, exist because it uses an interpolation algorithm which avoids discontinuities when connecting contacts between outcrops. This tacit assumption may be overridden by entering contacts in each fault block separately. With each overprinting of new sets of contacts, TRAVERSE juxtaposes fault blocks on the same cross section. In Figure 1, TRAVERSE has plotted a pair of fault blocks, each containing a syncline, separated by a steeply dipping fault plane. The data which generated Figure 1 are listed in the Appendix. A vertical exaggeration of 6.5 has been calculated by TRAVERSE from these data. The option of changing exaggeration during execution was not exercised and so this value became printed as a subtitle in Figure 1.

DETAILED DESCRIPTION OF PROGRAM TRAVERSE

Declarations and Main Block of Program

TRAVERSE is not well documented internally because internal comments would consume too much memory and not allow full development of the program (Appendix Table 1). Each END in TRAVERSE is followed by a comment because it would be tedious otherwise to determine which statement or loop it is ending. Given the lack of internal documentation, a detailed description of TRAVERSE follows.

Given its size, TRAVERSE begins with a command, (*$S+*), which reduces the amount of memory needed for compilation by swapping with the disk, the portions of the compiler needed for declarations versus statements. Two Apple library units are used, a two-dimensional graphics package termed TURTLEGRAPHICS and and a package to calculate transcendental functions termed TRANSCEND. Although a three-dimensional graphics package is available for the Apple II (APPLEGRAPHICS), it requires too much memory to be used with a program long enough to solve practical structural problems.

Three margins of the graph (XLOW, XHIGH, YHIGH) are specified as constants at the beginning of TRAVERSE so that the user readily may locate and change them, if desired. The default position of the origin is (35,15) and the diagonally opposite corner of the graph (255,145) also is well inward from the edge of the screen. Given the aspect ratio of the Apple II (180/192), an origin of (35,15) is not as skewed as it might seem. The Y value of the origin, YLOW, is not declared as a constant because the user is allowed to modify YLOW to control vertical exaggeration. The constant, RADIAN, is used to convert degrees to radians in the main block of TRAVERSE.

After examining the aforementioned declarations, the reader should skip to the action portion of the main block, at the end of the program. Here TRAVERSE commences by asking for a title on the cross section, for example, the name of the map area. Then a subtitle is requested and it is suggested that it record the ends of the cross section, as labeled on an accompanying map, for example A-A'. The length of the section could be entered on the same line if desired; A-A' IS A DISTANCE OF 2000 M. Both the title and subtitle are limited to 40 printed characters, but the user may type any number of characters without aborting the program. Typing mistakes may be corrected simply by backspacing. The user is prompted for the lowest and highest elevations to label the graph. TRAVERSE could calculate these automatically but manual entry allows the user to specify the unit of measurement, for example 500 METERS.

Two boolean variables, PLOTFOLD and LINEARSECTION, are initialized to the value of FALSE. These and other boolean variables subsequently record the user's responses which control program action. User response also is recorded in single-character (CHAR) variables such as ENTRY. ENTRY controls looping through the remainder of the main block of TRAVERSE. Each passage through the loop begins with procedure READDATA for the entry of points to be interpolated. These points are printed and, if necessary, corrected by procedure PRINTDATA. All angles then are converted to radian measure to conform to the TRANSCEND library.

The length of the topographic profile is adjusted to fit the screen width of XHIGH-XLOW and the topographic relief is adjusted to fit YHIGH-YLOW. Minima and maxima of the entered X and Y values must be determined to perform this fitting. During the initial loop controlled by ENTRY, PLOTFOLD is FALSE and so procedure NORMALIZE is called to calculate these minima and maxima. NORMALIZE also determines the vertical exaggeration. NORMALIZE never is recalled and so all geologic contacts subsequently become plotted on the same base profile. The actual fitting of X and Y values to the screen dimensions is performed in the main block because this fitting is required during each loop, whether the loop is for topographic or geologic data points.

Boolean PLOTFOLD directs TRAVERSE into the procedure for topographic plotting (PLOTRELIEF) during the initial loop and subsequently into the pair of procedures for plotting geologic contacts (STRUCTURE and PLOTCONTACTS). PLOTRELIEF labels the graph with user-supplied titles and then controls interpolation of topography through the elevation-distance points. STRUCTURE creates additional subsurface data points along the dip of each geologic contact. PLOTCONTACTS controls interpolation through these subsurface points and the corresponding outcrop points. The resultant plots are displayed by the monitor.

An inner loop, controlled by boolean CHOICE, prints the image displayed on the monitor. Communication with the printer is established with the REWRITE command and the paper is advanced to a clean sheet by PAGE (PRINTFILE). PRINTFILE is an arbitrary name assigned to the printer in the initial declaration of variables. Actual printing is performed by WRITELN (PRINTFILE,CHR(25),'GDR'). It is assumed that the printer-interface card is a standard Grappler or Grappler-compatible card which expects CHR(25) to dump graphics from Apple Pascal. 'GDR' is the Grappler-compatible code for dumping the first page of graphics memory (G) at double size (D) and rotated ninety degrees on the page (R). Any number of copies may be obtained, as controlled by boolean CHOICE.

After plotting a simple topographic profile, TRAVERSE prompts the user for data on geologic contacts. By responding negatively to the prompt, one may use TRAVERSE as a general-purpose interpolation program for such activities as plotting borehole chemical data. One may enter nothing in response to each of the prompts for a title and subsequently add titles by high-quality drafting onto the dot-matrix interpolation. Dot-matrix titles never approach drafting quality (Fig. 1).

With each affirmative response to the question of plotting contacts, the user adds another interpolated curve to the existing graph. Eight loops with the data in Appendix Table 2 produced Figure 1. Following the question about contacts, the bearing of the line-of-section is converted back to degrees because this is reprinted in degrees each time that the main block loops through its call to procedure PRINTDATA. At the end of TRAVERSE, PLOTFOLD automatically becomes TRUE because only the initial loop plots topography.

Procedures in TRAVERSE

The procedures in TRAVERSE are discussed in the order that they are called by the main block. These procedures subsequently call other procedures which also are discussed in detail. This detail is provided so that the user may modify readily any part of TRAVERSE to meet specialized needs.

Procedure READDATA starts with a message to the user about how to signal an end to data input. The alternative to this approach is to demand that the user specify at the outset the number of data points to be entered. The displayed signal is the entry of a distance of 1E20 or greater. This distance would be absurd in anything other than celestial data sets. Although the displayed message calls for signal numbers greater than 1E19, there is an upper limit which is characteristic of the computer being used. The largest number which an Apple II will accept without fatal error is 1E38.

The message about signalling data termination should be displayed for as long as necessary for comprehension by the user. For an experienced user, it need not be displayed for more than an instant. To permit variable duration of display, the user is instructed to depress any key when ready to proceed. The following statement, READ (KEYBOARD,ENTRY), speeds execution by not requiring the usual depression of the RETURN key to accept keyboard entries.

The number of data points in each set to be interpolated is labeled MAXSIZE. This is initialized to zero and augmented by one with each passage through the data-entry loop. This loop is controlled by boolean STOP. STOP becomes TRUE whenever the entered distance exceeds 1E19. Distances and corresponding elevations may be entered either as real or integer numbers. Numbers are displayed on the screen, of course, during entry. The screen is refreshed by PAGE (OUTPUT) with each passage through the loop because the Apple IIe usually aborts Pascal if it tries to write beyond the bottom of the screen. This statement is unnecessary with an Apple II+.

If READDATA has been called to enter geologic contacts, boolean PLOTFOLD would be TRUE and the user would be prompted for outcrop data. The first prompt concerns the bearing of the line-of-section. This is required because the angle between the line-of-section and the strike of the contact influences the apparent dip. In general, the dip of virtually any contact in any cross section is an apparent dip rather than the true dip measured at the outcrop.

Some cross sections are straight lines on maps and some are not. After entry of the first line-of-section, the user is asked if the cross section is linear. User response is recorded in boolean LINEARSECTION. Once LINEARSECTION becomes TRUE, the user no longer is asked to enter the section bearing on subsequent calls to READDATA.

Procedure READDEGREES is called by READDATA to handle alternative compass formats. The two alternatives are readings on the full scale of 0 to 360 degrees or in a quadrant-based format, for example N45W. If the user previously has indicated that all bearings are in the same format, whichever format, then UNIFORMITY will have become 'Y' and the initial prompt of READDEGREES would be bypassed. If the user selects the quadrant-based format, boolean MODIFICATION becomes TRUE and compass readings become modified to 360-degree format in the latter half of READDEGREES. Only the northern two quadrants are acceptable input. The user initially specifies whether the bearing is in the eastern or western quadrant and then enters the number of degrees. A bearing in the eastern quadrant need not be modified because it is equivalent to the 360-degree value. A bearing in the eastern quadrant need not be modified because it is equivalent to the 360-degree value. A bearing in the western quadrant is subtracted from 360 to become equivalent to the 360-degree value. Returning to Procedure READDATA, the bearing is recorded as the line-of-section through the corresponding outcrop.

If LINEARSECTION previously has become TRUE in a call to procedure READDATA, the line-of-section at a given outcrop automatically may be equated to that of the first outcrop, SECTION[1]. SECTION[1] must be in degrees instead of radians at this point, hence the aforementioned reconversion from radians to degrees at the very end of program TRAVERSE.

If LINEARSECTION previously has become TRUE or if the user has progressed beyond entering the first strike-and-dip observation, the user need not be queried if the cross section is a straight line. Otherwise, the question is asked and the response is recorded in boolean LINEARSECTION. Similarly, the user is asked if all strikes are to be entered in the same format, 0-360 degrees versus quadrant format. Given uniformity of format, the user need not be queried about format when entering each strike. The user's response is recorded in the variable UNIFORMITY which controls procedure READDEGREES.

Input of the dip angle is followed by a query about dip direction. Dip direction generally is recorded relative to the strike and TRAVERSE could be modified to accept this type of input. In its present form, however, TRAVERSE asks if the dip is to the left on the cross section (backward along the traverse) or to the right. To circumvent any typing mistake, TRAVERSE will accept only the letter R (right) or L (left) from the keyboard.

With the termination of data entry (distance > 1E19), the number of data points (MAXSIZE) is reduced by one to remove the absurdly large distance used to signal termination. If READDATA has been called for the entry of just topographic data, the three numeric variables which record strike-and-dip data are each initialized to zero and DIPDIRECTION is initialized to an asterisk. This initialization is useful during subsequent printing because procedure PRINTDATA attempts to print these four variables. PRINTDATA would execute printing even if the variables were undefined, but a printer-control character may be transferred randomly as an undefined variable and induce an undesirable effect, for example a radical change in the size of all subsequently printed characters.

Procedure PRINTDATA starts by establishing communication with the printer through the REWRITE command. A heading then is printed for the data table. The user may wish to alter the spacing of column titles in this heading to conform to the column widths of numbers normally entered and printed. Numbers larger than a single digit are printed automatically in exponentiation format and require a greater column width. All angles in the table are printed in degrees as entered, but subsequently must be converted to radians prior to computation.

PRINTDATA prompts the user to correct any mistakes spotted in the printed table. The leftmost column of the table incrementally numbers the position of each entered observation. This position number is used to identify each entry requiring correction. Any mistake in a row of data requires reentry of the entire row, with the possible exception of the line-of-section. It is assumed that the user will enter the first line-of-section correctly if the cross section is linear. If LINEARSECTION is TRUE, the line-of-section never again is entered and therefore could not require correction. Procedure PRINTDATA ends when the user indicates that no more corrections are needed.

Procedure NORMALIZE identifies the minima and maxima in the input distance (X) and elevation (Y) data. The user is asked if elevation and distance data are in the same units. Usually they are not and this would need to be rectified before vertical exaggeration could be calculated. The user may rectify nonconformity by entering the distance between extreme topographic points in the same units as are used for elevation data.

Vertical exaggeration is a function of the aspect ratio of the Apple II (192/280=0.6857), the ratio of the graph dimensions (YHIGH-YLOW)/(XHIGH-XLOW), and the ratio of the data ranges (XMAX-XMIN)/(YMAX-YMIN). A value less than one would indicate horizontal exaggeration but this rarely is encountered. The calculated exaggeration is displayed and the user is asked if a smaller exaggeration is desired. A smaller exaggeration is obtained by decreasing the vertical graph dimension. The lower edge of the graph (YLOW) is raised proportionately. A greater exaggeration could be obtained by decreasing the horizontal graph dimension but greater exaggeration generally is undesirable and is not offered herein as an option. The user could modify readily procedure NORMALIZE to provide this option if desired. Once vertical exaggeration is determined, the user is prompted to type a subtitle to record exaggeration on the printed section. The printing is not handled automatically because it is expected that the user will want to round the calculated exaggeration to a one-digit or two-digit number for printing (Fig. 1). In a routine application, NORMALIZE may be modified to round and print vertical exaggeration uniformly.

Procedure PLOTRELIEF controls plotting of topography. It prints titles, plots topographic points, and calls Procedure PLOT to interpolate through the topographic points. PLOTRELIEF starts by switching into GRAFMODE and erasing any preexisting graphics image with FILLSCREEN (BLACK). Plotting is performed by an imaginary turtle which is directed around the screen, dragging its tail to make lines or lifting its tail to move without drawing. Initially, the tail is lifted with PENCOLOR (NONE). The turtle moves toward the top of the graph and prints the user-supplied title (SECTIONNAME). This is followed by a subtitle (ENDSSECTION) and information about vertical exaggeration (EXAGGERATION). The turtle is directed to the screen position of the bottom, left corner of each initial letter before printing. The WSTRING command induces printing even while the turtle's tail remains elevated. The user readily may change the location of any title by changing the position of the initial letter.

Plotting of original topographic points occurs in a loop which first rounds each real topographic point to become an integral screen position. The turtle moves to that position, lowers its tail, and draws a vertical tick mark four pixels high, TURNTO(90); MOVE(4). The tail is raised before moving to the next data point. Following data plotting, the turtle moves invisibly to the top of the Y axis and draws downward to the bottom of that axis. The Y axis then is labeled with the user-supplied titles, HIGHEST and LOWEST. Procedure PLOTRELIEF ends with a call to procedure SPLINE to calculate cubic-spline parameters and then procedure PLOT to plot the interpolated curve. PLOT ends by reverting to TEXTMODE so that the screen will display geologic-contact data while it is being entered by the user.

Procedures STRUCTURE and PLOTCONTACTS control plotting of geologic contacts. STRUCTURE starts by calculating the apparent dip of each contact projected onto the plane of the cross section. As noted by Ragan (1973, p.6), the tangent of the angle of apparent dip may be calculated from the tangent of the true dip and the sine of the angle between the line-of-section and the strike. The library unit, TRANSCEND, provides sine and cosine values which are divided to provide the required tangent.

An extension of the dip observed in outcrop into the subsurface forces the subsequent interpolation to reflect observed bedding near each outcrop. The lower end of this dip line becomes a specific number of pixels horizontally right or left of the top end. This number is labeled EXTENSION. It is difficult to calculate EXTENSION in a formula which is ideal for all data sets. The user is encouraged to experiment with alternative formulae. As a minimum, EXTENSION should be one screen unit to avoid subsequent division by zero.

EXTENSION has been made inversely proportional to the vertical exaggeration because great exaggeration produces steep apparent dips which are difficult to connect smoothly by interpolation. Shortening steep dips ameliorates interpolation. For a vertical exaggeration of five, the horizontal component of the dip-line vector would become one-fiftieth of the horizontal length of the graph (XHIGH-XLOW).

The length of the dip vector is a function of both its horizontal and vertical components. The ratio of these components depends not only on vertical exaggeration but on the angle between the cross section and the strike, as previously used to calculate apparent dip. When this angle approaches zero degrees, all dips appear to be shallow. The angle of apparent dip is obtained with the library function, ATAN, operating on the calculated tangent. The cosine of the angle of apparent dip decreases with a decrease in the ratio of horizontal to vertical vector components. This cosine term therefore moderates the situation of a small ratio of horizontal to vertical components. Otherwise, the dip line could become so long that it routinely would extend to the lower limit of Apple Graphics. An alternative method of calculating EXTENSION would be equating it to some fixed proportion of the distance to the next data point. However, this method fails if there is just a single observation of the contact, as for an unfolded bed or a fault plane. The user could test for this condition and calculate EXTENSION as a proportional distance only for multiple-contact situations.

STRUCTURE creates a new array of points (XSPLINE,YSPLINE) which represents the ends of each of the dip lines. Every other (XSPLINE,YSPLINE) point is a geologic contact observed in outcrop. The extension is guided beneath the topographic surface by the user-supplied dip direction. If DIPDIRECTION is to the right on the cross section, the X coordinate for the lower end of the dip line will be greater than that of the upper end, and vice versa. The corresponding Y coordinate depends on the vertical exaggeration and the tangent of the apparent dip angle, as well as on the X coordinate. Calculation of the full set of X and Y coordinates ends procedure STRUCTURE.

Procedure PLOTCONTACTS switches back to GRAFMODE and the topographic profile reappears on the monitor. The turtle moves invisibly between each geologic contact and draws each dip line downward to the corresponding coordinates calculated by STRUCTURE. Prior to drawing, each dip end must be checked to determine if it lies within the vertical bounds of Apple graphics. Any extraneous position is truncated at the closest boundary. An attempt to draw outside the boundary would abort the program on an Apple II. PLOTCONTACTS does not check on the horizontal boundary and so the unlikely, but possible, situation of exceeding this boundary without exceeding the vertical boundary would be fatal.

To prepare for interpolation through both ends of each dip line, the (X,Y) coordinates of each pair of ends are transferred to the same arrays as were used for entry of just the upper-end data. The number of points in the new arrays is double that of the old ones and so MAXSIZE is doubled. Procedure SPLINE then is called to calculate cubic-spline parameters for the new arrays and procedure PLOT is called to perform the plotting.

Procedure SPLINE performs most of the mathematical activity in program TRAVERSE by calculating second derivatives at each of the data points for a cubic spline. A cubic spline is a series of polynomials joined at the data points so that both the first and second derivatives are continuous through those points. Interpolation between data points is based on the assumption that the second derivatives differ linearly between them (for example, Cheney and Kincaid, 1980). SPLINE constitutes less than 5% of the code in TRAVERSE. This proportion is typical of user-friendly programs like TRAVERSE and much of the abundant input-output in TRAVERSE may be used in programs with different mathematical functions. The calculated second derivatives are stored in array A for use by procedure PLOT.

PLOT calculates the cubic-spline interpolation for each horizontal screen position along the curve. In the initial plotting of topography, this ranges virtually the full width of the graph from XLOW to XHIGH, about 220 pixels. However, most geologic contacts extend for only a portion of the graph (Fig. 1). The first and last screen positions are omitted because cubic-spline interpolation usually calculates absurd values at these extremities; this omission is not noticeable.

Calculation by PLOT of the cubic-spline interpolation for any horizontal screen position requires only second derivatives (array A) at the two adjacent data points. PLOT determines which pair of data points encloses a given screen position (XGIVEN) by comparing XGIVEN with each data point. It is assumed that the X data have been entered as monotonously increasing values. The first X value which XGIVEN exceeds in its increasing DO-loop therefore must be the adjacent, lower value and the succeeding X value must be the adjacent, higher value. If the X data are to be entered as monotonously decreasing values, only a single inequality symbol in PLOT need be changed. The seventh line would become, IF XGIVEN <= X[N] THEN K:=N;. If TRAVERSE is to be used routinely with alternately increasing or decreasing distance values, the user should be prompted to select one mode or the other and the response should be stored in a boolean variable which controls the inequality sign in PLOT. Note that the question could not be asked from PLOT itself because PLOT is in GRAFMODE and the question would not appear on the monitor. The question could be asked from the main block at the end of the program.

If distances are to be entered randomly, TRAVERSE must be modified to order them before proceeding further. Within PLOT, the proximity of XGIVEN to the lower adjacent X value is sorted in SPLIT. SPLIT determines the relative contribution of the second derivative on this lower side to the weighted average of the pair of adjacent second derivatives. This algorithm works for the situation of a single traversed contact, for example a fault, because a pair of points is created at each end of the dip extension. The algorithm would fail, however, in the absurd situation of interpolating a single topographic point due to attempted division by zero.

Each calculated interpolation in PLOT is rounded to an integral screen position and then tested against the vertical bounds of Apple II graphics. Extraneous values are truncated and plotted at the boundary. The turtle moves invisibly to the first plotting point before drawing commences. Upon completion of the curve, the image is held on the screen for a few seconds by a do-nothing DO-loop. TEXTMODE then resumes and action reverts to the main block which prints the image.

Lagrangian Alternative

Lagrangian interpolation requires substitution of procedure SPLINE with an alternative mathematical algorithm, function PCALC (Appendix Table 3). A function can return only a single value whereas procedure SPLINE creates an array. PCALC requires even less code than SPLINE but several ancillary changes in TRAVERSE are required to compensate for the inherent instability of a continuous polynomial interpolation such as the Lagrangian algorithm. Fortunately, Pascal is a convenient language for describing such modifications because each Pascal statement ends with an explicit semicolon rather than an invisible carriage return as in FORTRAN and BASIC.

To constrain Lagrangian oscillation, it is necessary to add more points along the dip line at each contact (Fig. 2). A cubic spline interpolates well with a single point at each end of the dip line but Lagrangian interpolation requires an additional pair of points intermediate between the ends. All four points of all contacts are used in Lagrangian interpolation. However, plotting is omitted for the first two and last two points along the interpolation curve because oscillation is most severe in these extremities and the calculated curve obviously is unrealistic. Omission is effected by replacing the third line of procedure PLOT with the following, "FOR J:=ROUND (X[3]) TO TRUNC (X[MAXSIZE-2]) DO BEGIN". Omission is apparent in the topographic profile in Figure 2 but not in the interpolated geologic contacts because straight dip lines routinely are drawn even where the interpolation curve is not.

To accommodate additional dip-line points, the declaration of variables at the beginning of TRAVERSE must be modified (Appendix Table 1). Integers K, L, and M must be added, along with boolean variables FLAG1 and FLAG2. The dimension of the XSPLINE and YSPLINE arrays should be augmented to 40. Despite this augmentation, Lagrangian TRAVERSE can handle only 10 geologic contacts in each interpolation curve whereas cubic-spline TRAVERSE can handle 15.

Two procedures in Appendix Table 1 must be replaced by code in Appendix Table 3. Procedure SPLINE must be replaced by function PCALC and procedure STRUCTURE must be replaced by a new procedure with the same name. Function PCALC includes its own declaration of variables to minimize alteration of the original program. PCALC expects to receive three parameters whenever it is called. Internally, these are named FLAG, P, and XK. PCALC returns a real number which is labeled PCALC.

One-half of PCALC is devoted to avoiding division by zero if the user mistakenly enters identical distances for different elevations or geologic contacts. This type of mistake is difficult to discover if the program terminates prematurely because the error message simply indicates division by zero. To speed execution, this error checking is invoked only when the calculated PCALC will be returned to become a denominator. Boolean FLAG controls error checking. A zero value for PCALC is avoided by adding a small proportion of the distance range to one of the identical distances.

Calculation in PCALC is trivial compared to that in procedure SPLINE. For any given value within the range of the entered data, PCALC is the product of the differences between that value and all other entered values (Sokolnikoff and Sokolnikoff, 1941). Despite this simplicity, Lagrangian calculations are slow because they occur repetitively within loops. PCALC contains a loop and PCALC is used within another loop by Procedure PLOT which calls it. Both loops proceed incrementally through all data points and so large data sets execute slowly.

Procedure PLOT requires not only the aforementioned change in its third line but lines six through twelve should be replaced by the following two Pascal statements which may be concatenated or spread out over five lines as listed here:

```
YCALC:=0.0;
FOR N:=1 TO MAXSIZE DO
 YCALC:=YCALC +
        Y[N]*PCALC(FLAG1,N,XGIVEN)/
        PCALC(FLAG2,N,X[N]);
```

Lagrangian procedure STRUCTURE (Appendix Table 3) is similar to the cubic-spline STRUCTURE but it calculates three subsurface points along the dip instead of just one. The algorithm for assigning dip direction in cubic-spline STRUCTURE is not used in the Lagrangian procedure because it would require a complicated series of IF-THEN-ELSE statements. In Appendix Table 3, a single IF-THEN-ELSE statement used the dip direction to assign a value of +1 or -1 to the factor MINUS.

Lagrangian STRUCTURE differs from cubic-spline STRUCTURE in its calculation of EXTENSION. EXTENSION is the horizontal distance of a subsurface point on a dip line measured from the intersection of the dip line with the Earth's surface. One would expect Lagrangian EXTENSION to be shorter than cubic-spline EXTENSION because it is added three times instead of just once to produce the complete dip line. However, it is important to make EXTENSION as long as possible because a single short EXTENSION may induce severe oscillation throughout the length of the Lagrangian curve. In cubic-spline STRUCTURE, EXTENSION is proportional inversely to vertical exaggeration but this dependence is omitted from Lagrangian STRUCTURE to avoid decreasing EXTENSION excessively. If the apparent dip is zero degrees, Lagrangian EXTENSION becomes one-fiftieth of the width of the cross section. EXTENSION decreases with increasing apparent dip, as in the cubic-spline situation.

Other small modifications are required for Lagrangian TRAVERSE. Three of these occur in procedure PLOTCONTACTS. Line six must be replaced with, "K:=4*N-3;" and line seven becomes "L:=4*N-1;". These two statements create a series of integers which represent the first and third array positions, respectively, in XSPLINE and YSPLINE for each four-point data set associated with a given dip line. This permits drawing from the topmost point on each dip line directly to the third point. Lagrangian plotting starts at this third point and continues to the third-from-the-last point on the ultimate interpolated dip. The eighth-from-the-last line in PLOTCONTACTS should become, "MAXSIZE:=4*MAXSIZE:". This reflects the extra dip-line points. Lastly, two new statements must be added at or near the beginning of the main block at the end of the program, "FLAG1:=FALSE; FLAG2:=TRUE;".

CONCLUSION

TRAVERSE conveniently plots accurate geologic cross sections on any type of Apple II computer. It clearly demonstrates to students the difference between true and apparent dip due to vertical exaggeration or an angular difference between strike and line-of-section.

According to a poll by the American Association of Petroleum Geologists (Anonymous, 1984), the Apple II comprises about 16% of all microcomputers used by petroleum geologists at home or work versus 18% (at home) to 27% (at work) for the IBM-PC. Globally, almost a thousand professional geologists have daily access to an Apple II and those interested in cross sections routinely could use TRAVERSE or some modification of it.

Interactive plotting programs like TRAVERSE generally require modification to conform to individual preferences for data entry and graphical display. The detailed description of TRAVERSE herein and the suggestions for possible changes should facilitate this modification. With the advent of more powerful microcomputers, TRAVERSE may be expanded to become three dimensional.

ACKNOWLEDGMENTS

The Exxon Foundation has supported generously development and field testing of TRAVERSE on a mapping project in New Mexico.

REFERENCES

Anonymous, 1984, Microcomputers gain in popularity: Am. Assoc. Petroleum Geologists Explorer, v. 5, no. 16, p. 38-39.

Cheney, W., and Kincaid, E., 1980, Numerical mathematics and computing: Brooks/Cole Publishing, Monterey, California, 362 p.

Conte, S.D., and DeBoor, C., 1980, Elementary numerical analysis: McGraw-Hill Book Co., New York, 432 p.

Kimberley, M.M., 1985, Geochemistry and structure of stratiform deposits with a portable microcomputer: Ore-Geology Reviews, v. 1, no. 1, in press.

McAllister, D.F., and Roulier, J.A., 1981, An algorithm for computing a shape preserving osculating quadric spline: Trans. Mathematical Software, v. 7, p. 331-347.

McLaughlin, H.W., 1983, Shape-preserving planar interpolation: an algorithm: IEEE Computer Graphics and Applications, v. 3, p. 58-67.

Ragan, D.M., 1973, Structural geology, an introduction to geometrical techniques: John Wiley & Sons, New York, 208 p.

Sokolnikoff, I.S., and Sokolnikoff, E.S., 1941, Higher mathematics for engineers and physicists: McGraw-Hill Book Co., New York, 587 p.

Sowerbutts, W.T.C., and Mason, R.W.I., 1984, A microcomputer based system for small-scale geophysical surveys: Geophysics, v. 49, no. 2, p. 189-193.

APPENDIX

Table 1. Pascal program TRAVERSE for Apple II. TRAVERSE runs on single-drive or dual-drive Apple II+, IIe, or IIc, but is best compiled on dual-drive Apple II+. TRAVERSE interpolates folded cross section from strike-and-dip data in outcrops (Fig. 1). Apparent dips are corrected for vertical exaggeration and for any angle of less than ninety degrees between strike and line-of-section. TRAVERSE interpolates with cubic spline but alternative code for Lagrangian interpolation is given in Table 3 and text.

```
(*$S+*)
PROGRAM TRAVERSE;
USES TURTLEGRAPHICS,TRANSCEND;
CONST
 RADIAN=57.296; YHIGH=145;
 XLOW=35; XHIGH=255;
VAR
 YLOW,PLOTX,PLOTY,
 N,J,K,L,MAXSIZE: INTEGER;
 EXTENSION,VERTICAL,SPLIT,TAN,
 XGIVEN,YCALC,XMIN,XMAX,YMIN,YMAX,
 XRANGE,BEARING: REAL;
 X,Y,STRIKE,DIP,SECTION,
 A,B,C: ARRAY [1..60] OF REAL;
 XSPLINE,YSPLINE: ARRAY [1..30] OF REAL;
 SECTIONNAME,ENDSSECTION,LOWEST,
 HIGHEST,EXAGGERATION: STRING[40];
 STOP,PLOTFOLD,MODIFICATION,
 LINEARSECTION: BOOLEAN;
 ENTRY,CHOICE,UNIFORMITY,EASTWEST: CHAR;
 DIPDIRECTION: ARRAY [1..60] OF CHAR;
 PRINTFILE: TEXT;

 PROCEDURE READDEGREES;
 BEGIN
  IF (UNIFORMITY='N') OR (MAXSIZE=1) THEN BEGIN
   WRITELN('DEGREES 0-360 OR N5W FORMAT? (3/N) ->');
   REPEAT READ (KEYBOARD,ENTRY) UNTIL
    ENTRY IN ['3','N'];
   IF ENTRY='N' THEN MODIFICATION:=TRUE
    ELSE MODIFICATION:=FALSE
  END; (* IF MAXSIZE=1 *)
  IF MODIFICATION=FALSE THEN BEGIN
   WRITELN('ENTER BEARING IN 360 DEG NOTATION->');
   READLN (BEARING)
  END
  ELSE BEGIN
   WRITELN('BEARING E OR W OF N? (E/W)->');
   REPEAT READ (KEYBOARD,EASTWEST) UNTIL
    EASTWEST IN ['E','W'];
   WRITELN('ENTER NUMBER OF DEGREES ->');
   READLN (BEARING);
   IF EASTWEST='W' THEN
    BEARING:=360.0 - BEARING
  END (* ELSE IF MODIFICATION=TRUE *)
 END; (* PROCEDURE READDEGREES *)
```

```
PROCEDURE READDATA;
BEGIN
 WRITELN('ENTER DISTANCE OF 1E2Ø OR MORE TO END DATA.');
 WRITELN('PRESS ANY KEY TO CONTINUE->');
 READ (KEYBOARD,ENTRY);
 MAXSIZE:=Ø;
 STOP:=FALSE;
 WHILE STOP=FALSE DO BEGIN
  MAXSIZE:=MAXSIZE+1;
  PAGE(OUTPUT);
  WRITELN;
  WRITELN('ENTER DISTANCE ',MAXSIZE,' ALONG TRAVERSE -> ');
  READLN (X [MAXSIZE]);
  IF (X[MAXSIZE]>1E19) THEN STOP:=TRUE
   ELSE BEGIN
   WRITELN('ENTER CORRESPONDING ELEVATION ',MAXSIZE,' -> ');
   READLN (Y [MAXSIZE]);
   IF PLOTFOLD=TRUE THEN BEGIN
    IF LINEARSECTION=FALSE THEN BEGIN
     WRITELN('ENTER BEARING OF LINE-OF-SECTION ->');
     READDEGREES;
     SECTION [MAXSIZE]:=BEARING
     END (* IF LINEARSECTION=FALSE *)
     ELSE SECTION [MAXSIZE]:=SECTION[1];
    IF (LINEARSECTION=FALSE) AND (MAXSIZE=1) THEN BEGIN
     WRITELN('ENTIRE SECTION ALONG THIS LINE? (Y/N) ->');
     REPEAT READ (KEYBOARD,ENTRY)
      UNTIL ENTRY IN ['Y','N'];
     IF ENTRY='Y' THEN LINEARSECTION:=TRUE;
     WRITELN('ONE STRIKE FORMAT? Ø-36Ø OR N5W (Y/N)->');
     REPEAT READ (KEYBOARD,UNIFORMITY)
       UNTIL UNIFORMITY IN ['Y','N']
    END; (* IF MAXSIZE=1 *)
    WRITELN('ENTER STRIKE FOR CONTACT ',MAXSIZE);
    READDEGREES;
    STRIKE [MAXSIZE]:=BEARING;
    WRITELN('ENTER DIP ANGLE OF CONTACT ->');
    READLN (DIP [MAXSIZE]);
    WRITELN('DIP TO LEFT OR RIGHT ALONG X-SECTN? (L/R)->');
    REPEAT READ (KEYBOARD,DIPDIRECTION [MAXSIZE])
     UNTIL DIPDIRECTION [MAXSIZE] IN ['L','R']
    END (* IF PLOTFOLD=TRUE *)
  END (* IF X LESS THAN 1E19 *)
 END; (* WHILE STOP=FALSE *)
 MAXSIZE:=MAXSIZE-1;
 IF PLOTFOLD=FALSE THEN
  FOR N:=1 TO MAXSIZE DO BEGIN
   STRIKE[N]:=Ø.Ø;
   DIP[N]:=Ø.Ø;
   DIPDIRECTION[N]:='*';
   SECTION[N]:=Ø.Ø
   END (* FOR N=1 TO MAXSIZE *)
 END; (* PROCEDURE READDATA *)
```

```
PROCEDURE PRINTDATA;
BEGIN
 REWRITE (PRINTFILE,'PRINTER:');
 WRITELN(PRINTFILE,
   'POSITION DISTANCE ELEVATION   DIP  DIRECTION   X-SECTION');
 WRITELN (PRINTFILE);
 FOR J:=1 TO MAXSIZE DO
 WRITELN(PRINTFILE,'    ',J,'   ',X[J],'   ',
  Y[J],'   ',DIP[J],'     ',DIPDIRECTION[J],
  '   ',SECTION[J]);
 WRITELN('ANY CORRECTIONS NEEDED? (Y/N) ->');
 REPEAT READ (KEYBOARD,ENTRY)
  UNTIL ENTRY IN ['Y','N'];
  WHILE ENTRY='Y' DO BEGIN
  PAGE (OUTPUT);
  WRITELN('ENTER POSITION OF INCORRECT DATA ->');
  READLN (J);
  WRITELN('ENTER CORRECT DISTANCE ->');
  READLN (X[J]);
  WRITELN('ENTER CORRECT ELEVATION ->');
  READLN (Y[J]);
  IF PLOTFOLD=TRUE THEN BEGIN
   WRITELN('ENTER CORRECT STRIKE IN DEG 360 FORMAT ->');
   READLN (STRIKE[J]);
   WRITELN('ENTER CORRECT DIP ANGLE ->');
   READLN (DIP[J]);
   WRITELN('ENTER DIP DIRECTION (L/R) ->');
   REPEAT READ (DIPDIRECTION[J])
    UNTIL DIPDIRECTION[J] IN ['L','R'];
   IF LINEARSECTION=FALSE THEN BEGIN
    WRITELN('ENTER X-SECTION DIRECTION ->');
    READLN (SECTION[J])
   END (*IF LINEARSECTION=FALSE*)
  END; (* IF PLOTFOLD=TRUE *)
  WRITELN;WRITELN('MORE CORRECTIONS NEEDED? (Y/N)');
  REPEAT READ (KEYBOARD,ENTRY)
  UNTIL ENTRY IN ['Y','N']
 END; (* WHILE ENTRY=Y *)
 CLOSE (PRINTFILE,LOCK)
END; (* PROCEDURE PRINTDATA *)
```

```
PROCEDURE SPLINE;
BEGIN
 C[2]:=2*((X[2]-X[1])+(X[3]-X[2]));
 B[2]:=6*((Y[3]-Y[2])/(X[3]-X[2])
        -(Y[2]-Y[1])/(X[2]-X[1]));
 FOR J:=3 TO MAXSIZE-1 DO BEGIN
  C[J]:=2*((X[J+1]-X[J])+(X[J]-X[J-1]))
         -SQR(X[J]-X[J-1])/C[J-1];
  B[J]:=6*((Y[J+1]-Y[J])/(X[J+1]-X[J])
         -(Y[J]-Y[J-1])/(X[J]-X[J-1]))
         -(X[J]-X[J-1])*B[J-1]/C[J-1]
 END; (* FOR J=3 TO MAXSIZE-1 *)
 A[1]:=Ø.Ø;
 A[MAXSIZE]:=Ø.Ø;
 FOR J:=(MAXSIZE-1) DOWNTO 2 DO
  A[J]:=(B[J]-(X[J+1]-X[J])*A[J+1])/C[J]
END; (* PROCEDURE SPLINE *)

PROCEDURE NORMALIZE;
BEGIN
 XMIN:=X[1]; XMAX:=X[1];
 YMIN:=Y[1]; YMAX:=Y[1];
 FOR N:= 2 TO MAXSIZE DO BEGIN
  IF X[N] > XMAX THEN XMAX:=X[N];
  IF X[N] < XMIN THEN XMIN:=X[N];
  IF Y[N] > YMAX THEN YMAX:=Y[N];
  IF Y[N] < YMIN THEN YMIN:=Y[N]
 END; (*FOR DO LOOP*)
 WRITELN;
 WRITELN('SAME UNITS FOR ELEVATION AND DIST? (Y/N) ->');
 READ (KEYBOARD,ENTRY);
 IF ENTRY='Y' THEN
  XRANGE:=XMAX-XMIN
  ELSE BEGIN WRITELN('ENTER DISTANCE RANGE IN ELEV UNITS ->');
  READLN (XRANGE)
  END; (* ELSE ENTER XRANGE *)
 VERTICAL:=((YHIGH-YLOW)/(Ø.6857*(XHIGH
      -XLOW)))*(XRANGE/(YMAX-YMIN));
 WRITELN;
 WRITELN('VERTICAL EXAGGERATION = ',VERTICAL);
 WRITELN('PREFER DIFFERENT EXAGGERATION? (Y/N) ->');
 READ (KEYBOARD,CHOICE);
 IF CHOICE = 'Y' THEN BEGIN
  WRITELN('ENTER SMALLER EXAGGERATION ->');
 READLN (VERTICAL);
 YLOW:=ROUND(YHIGH-VERTICAL*(Ø.6857*(XHIGH
       -XLOW))/(XRANGE/(YMAX-YMIN)))
 END; (* IF CHOICE = Y *)
 WRITELN('TYPE -> VERT EXAG= NUMBER ABOVE ->');
 READLN (EXAGGERATION)
END; (* PROCEDURE NORMALIZE *)
```

```
PROCEDURE PLOT;
BEGIN
 FOR J:=(ROUND(X[1])+1) TO TRUNC(X[MAXSIZE]) DO BEGIN
  PLOTX:=J;
  XGIVEN:=J;
  FOR N:=1 TO MAXSIZE-1 DO
   IF XGIVEN >= X[N] THEN K:=N;
  SPLIT:=XGIVEN-X[K];
  YCALC:=Y[K]+SPLIT*(SPLIT*(SPLIT
   *(A[K+1]-A[K])/(6.Ø*(X[K+1]-X[K]))+Ø.5*A[K])
   +(Y[K+1]-Y[K])/(X[K+1]-X[K])-(X[K+1]
   -X[K])*(2.Ø*A[K]+A[K+1])/6.Ø);
  PLOTY:=ROUND(YCALC);
  IF PLOTY >= 191 THEN PLOTY :=191;
  IF PLOTY <= Ø THEN PLOTY := Ø;
  MOVETO (PLOTX,PLOTY);
  PENCOLOR (WHITE)
 END; (* FOR J= XLOW TO XHIGH *)
 FOR N:=1 TO 4ØØØ DO;
 TEXTMODE
END; (* PROCEDURE PLOT *)

PROCEDURE PLOTRELIEF;
BEGIN
 GRAFMODE;
 FILLSCREEN(BLACK);
 PENCOLOR (NONE);
 MOVETO(3,18Ø);
 WSTRING (SECTIONNAME);
 MOVETO(3,165);
 WSTRING (ENDSSECTION);
 MOVETO(14Ø,165);
 WSTRING (EXAGGERATION);
 FOR J:=1 TO MAXSIZE DO BEGIN
  PLOTX:=ROUND(X[J]);
  PLOTY:=ROUND(Y[J]);
  MOVETO (PLOTX,PLOTY);
  PENCOLOR (WHITE);
  TURNTO (9Ø);
  MOVE (4);
  PENCOLOR (NONE)
  END; (*FOR J=1 TO MAXSIZE*)
 MOVETO (XLOW,YHIGH);
 PENCOLOR (WHITE);
 MOVETO (XLOW,YLOW);
 PENCOLOR (NONE);
 N:= YHIGH+5;
 MOVETO (1,N);
 WSTRING (HIGHEST);
 N:= YLOW-12;
 MOVETO (1,N);
 WSTRING (LOWEST);
 SPLINE;
 PLOT
END; (* PROCEDURE PLOTRELIEF *)
```

```
PROCEDURE STRUCTURE;
BEGIN
 K:=-1;
 L:=Ø;
 FOR N:=1 TO MAXSIZE DO BEGIN
  TAN:=SIN(DIP[N])/COS(DIP[N])
       *ABS(SIN(SECTION[N]-STRIKE[N]));
  EXTENSION:=1.Ø+COS(ATAN(TAN))
             *(XHIGH-XLOW)/(1Ø.Ø*VERTICAL);
  K:=K+2;
  L:=L+2;
  XSPLINE[K]:=X[N];
  YSPLINE[K]:=Y[N];
  IF DIPDIRECTION[N]='R' THEN
   XSPLINE[L]:=XSPLINE[K]+EXTENSION
   ELSE XSPLINE[L]:=XSPLINE[K]-EXTENSION;
   YSPLINE[L]:=YSPLINE[K]-VERTICAL
              *EXTENSION*TAN
  END (* FOR N=1 TO MAXSIZE *)
 END; (*PROCEDURE STRUCTURE*)

PROCEDURE PLOTCONTACTS;
BEGIN
 GRAFMODE;
 FOR N:=1 TO MAXSIZE DO BEGIN
  PENCOLOR (NONE);
  K:=N+N-1;
  L:=N+N;
  PLOTX:=ROUND(XSPLINE[K]);
  PLOTY:=ROUND(YSPLINE[K]);
  IF PLOTY>=191 THEN PLOTY:=191;
  IF PLOTY<=Ø THEN PLOTY:=Ø;
  MOVETO (PLOTX,PLOTY);
  PENCOLOR (WHITE);
  PLOTX:=ROUND(XSPLINE[L]);
  PLOTY:=ROUND(YSPLINE[L]);
  IF PLOTY>=191 THEN PLOTY:=191;
  IF PLOTY<=Ø THEN PLOTY:=Ø;
  MOVETO (PLOTX,PLOTY)
 END; (* FOR N=1 TO MAXSIZE *)
 MAXSIZE:=2*MAXSIZE;
 FOR N:=1 TO MAXSIZE DO BEGIN
  X[N]:=XSPLINE[N];
  Y[N]:=YSPLINE[N]
 END;
 SPLINE;
 PENCOLOR (NONE);
 PLOT
 END; (* PROCEDURE PLOTCONTACTS *)
```

```
BEGIN (* MAIN BLOCK OF TRAVERSE *)
YLOW:=15;
WRITELN('ENTER NAME OF X-SECTION ->');
READLN (SECTIONNAME);
WRITELN('ENTER ENDS OF SECTION, E.G. A-A" ->');
READLN (ENDSSECTION);
WRITELN ('ENTER LOWEST POINT, E.G. 500 M ->');
READLN (LOWEST);
WRITELN ('ENTER HIGHEST POINT, E.G. 2100 M ->');
READLN (HIGHEST);
PLOTFOLD:=FALSE;
LINEARSECTION:=FALSE;
ENTRY:='Y';
WHILE ENTRY='Y' DO BEGIN
 READDATA;
 PRINTDATA;
 FOR N:=1 TO MAXSIZE DO BEGIN
  DIP[N]:=DIP[N]/RADIAN;
  SECTION[N]:=SECTION[N]/RADIAN;
  STRIKE[N]:=STRIKE[N]/RADIAN
 END;(* CONVERSION TO RADIANS *)
IF PLOTFOLD=FALSE THEN NORMALIZE;
 FOR N:=1 TO MAXSIZE DO BEGIN
  X[N]:=XLOW+((X[N]-XMIN)/(XMAX-XMIN))
        *(XHIGH-XLOW);
  Y[N]:=YLOW+((Y[N]-YMIN)/(YMAX-YMIN))
        *(YHIGH-YLOW)
 END; (* FOR N=1 TO MAXSIZE *)
 IF PLOTFOLD=TRUE THEN STRUCTURE;
 IF PLOTFOLD=FALSE THEN PLOTRELIEF
  ELSE PLOTCONTACTS;
 CHOICE:='Y';
 WHILE CHOICE='Y' DO BEGIN
  PAGE (OUTPUT);
  REWRITE (PRINTFILE,'PRINTER:');
  PAGE (PRINTFILE);
  WRITELN (PRINTFILE,CHR(25),'GDR');
  WRITELN ('PRINT ANOTHER COPY? (Y/N)');
  READ (KEYBOARD,CHOICE);
  CLOSE (PRINTFILE,LOCK)
 END; (* WHILE CHOICE=Y *)
 WRITELN('WANT TO PLOT A SET OF CONTACTS? (Y/N)');
 READ (KEYBOARD,ENTRY);
 SECTION[1]:=SECTION[1]*RADIAN;
 PLOTFOLD:=TRUE
END (* WHILE ENTRY=Y *)
END. (* PROGRAM TRAVERSE *)
```

Table 2. Sample Data Used to Generate Figures 1 and 2.

Data Set #1. Topographic Data. Format is (distance, elevation)

(100,1000)	(200,1020)	(300,1040)	(400,1080)	(500,1120)
(600,1140)	(700,1140)	(800,1120)	(900,1080)	(1000,1080)
(1100,1140)	(1200,1200)	(1300,1226)	(1400,1250)	(1500,1250)
(1600,1230)	(1700,1204)	(1800,1120)	(1900,1090)	(2000,1060)

Data Sets #2 to #8. Geologic Contacts. All contacts have a strike which is ninety degrees from the line-of-section, e.g., strike N45W and section N45E. Dip direction is toward the left (L) or right (R) on the cross section. Data sets #2 to #4 are folds on left side of Figs. 1 & 2. Data sets #5 to #7 are folds on right side of Figs. 1 & 2. Data point #8 is a fault which appears only on Figure 1. Format is (distance, elevation, dip angle, dip direction).

#2	(340,1054,0,R)	(880,1084,45,L)
#3	(400,1078,0,R)	(820,1112,48,L)
#4	(460,1102,0,R)	(770,1126,54,L)
#5	(1100,1140,10,R)	(1800,1120,10,L)
#6	(1140,1164,10,R)	(1770,1144,10,L)
#7	(1180,1188,10,R)	(1740,1168,10,L)
#8	(1000,1080,55,R)	

Table 3. Pascal code to convert program TRAVERSE from cubic-spline interpolation to Lagrangian interpolation. Function PCALC and Procedure STRUCTURE of this table should replace Procedure SPLINE and Procedure STRUCTURE, respectively, in Table 1. Additional minor changes for Table 1 are listed in text. Output from Lagrangian interpolation is illustrated in Figure 2.

```
FUNCTION PCALC
(VAR FLAG:BOOLEAN;P:INTEGER;XK:REAL):REAL;
VAR PRODUCT: REAL;
    R: INTEGER;
BEGIN
 PRODUCT:=1.Ø;
 (*FLAG HANDLES CASE OF EQUAL X VALUES*)
 IF (FLAG=TRUE) THEN
  FOR R:=1 TO MAXSIZE DO
  IF (ABS(XK-X[R])<1E-1Ø) THEN
   XK:=X[R] + Ø.ØØ5*(X[MAXSIZE]-X[1]);
  FOR R:=1 TO MAXSIZE DO
   IF (R<>P) THEN
   PRODUCT:= PRODUCT*(XK-X[R]);
 PCALC:=PRODUCT
END; (* FUNCTION PCALC *)

PROCEDURE STRUCTURE;
VAR MINUS:REAL;
BEGIN
 J:=-3;
 K:=-2;
 L:=-1;
 M:=Ø;
 FOR N:=1 TO MAXSIZE DO BEGIN
  TAN:=SIN(DIP[N])/COS(DIP[N])
       *ABS(SIN(SECTION[N]-STRIKE[N]));
  EXTENSION:=COS(ATAN(TAN))
             *(XHIGH-XLOW)/5Ø.Ø;
  J:=J+4;
  K:=K+4;
  L:=L+4;
  M:=M+4;
  XSPLINE[J]:=X[N];
  YSPLINE[J]:=Y[N];
  IF DIPDIRECTION[N]='R' THEN MINUS:=+1.Ø
     ELSE MINUS:=-1.Ø;
  XSPLINE[K]:=XSPLINE[J]+MINUS*EXTENSION;
  XSPLINE[L]:=XSPLINE[J]+MINUS*2.Ø*EXTENSION;
  XSPLINE[M]:=XSPLINE[J]+MINUS*3.Ø*EXTENSION;
  YSPLINE[K]:=YSPLINE[J]-RATIO
              *EXTENSION*TAN;
  YSPLINE[L]:=YSPLINE[J]-RATIO
              *2.Ø*EXTENSION*TAN;
  YSPLINE[M]:=YSPLINE[J]-RATIO
              *3.Ø*EXTENSION*TAN
  END (* FOR N=1 TO MAXSIZE *)
 END; (*PROCEDURE STRUCTURE*)
```

A Computation Method for Drawing Mineral Stability Diagrams on a Microcomputer

E. Deloule and J–F. Gaillard

University of Paris

ABSTRACT

Stability diagrams depict chemical equilibria between an aqueous solution and minerals. They summarize and illustrate the thermodynamic data related to one element and its associated mineral phases in various geochemical systems.

Presented here is a new computation method for constructing such diagrams. The calculation is divided in three parts: (1) thermodynamic data reduction to plane equations in three-dimensional space, (2) computation of plane intersection with elimination of unstable minerals, and (3) diagram drawing according to a logical scheme selecting the boundary of each solid phase.

This method presents the advantages of being efficient and requiring a small core memory which permits programming and implementation on microcomputers.

INTRODUCTION

The thermodynamic study of mineral-phase formation at equilibrium with aqueous solution is aided by the construction of stability diagrams: Phourbais Eh-pH diagrams (Garrels and Christ, 1965) and activity-activity diagrams (Helgeson, Brown, and Leeper, 1969). Such diagrams enable the geochemist to predict the solid phase limiting, at equilibrium, the solubility of an element, and they facilitate interpretation of mineral paragenesis at low or high temperature.

Eh-pH diagrams however, are not convenient in studying natural environments because most of the processes are irreversible, for example

SO_4^{2-} <-> S^{2-} (Michard, 1967; Michard and Allegre, 1969). Activity

diagrams depicting chemical equilibrium among mineral and aqueous solutions are limited in their interpretation mostly because activity generally is accessible only through chemical speciation of the aqueous solution. But these diagrams can be constructed if no steady state in redox processes is observed - Eh having no meaning in this situation.

In Figure 1 we present a stability diagram for copper in seawater. Stability fields of different copper minerals at equilibrium are reported versus log(H+), the logarithm of H+ activity on the abscissa and log[SIIt], logarithm of total sulphur concentration on the ordinate; the composition of the aqueous solution is given in the left part of the diagram.

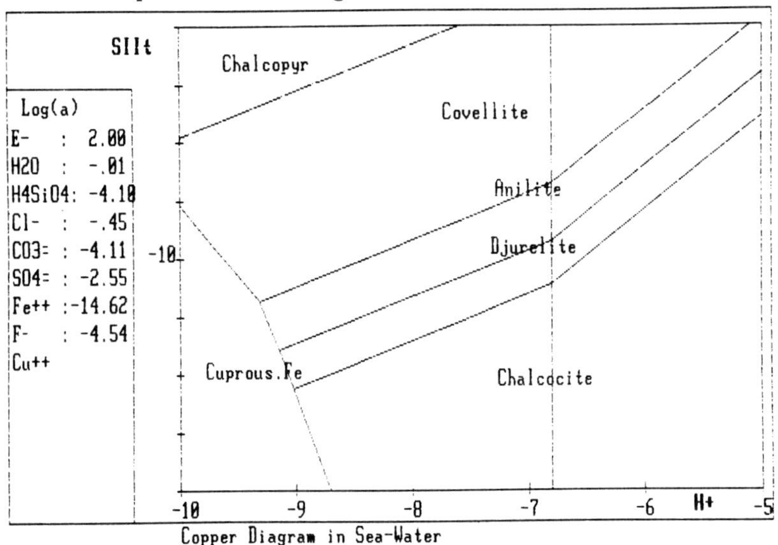

Figure 1. Mineral stability diagram for copper in seawater.

T. H. Brown (Helgeson, Brown, and Leeper, 1969) wrote a FORTRAN computer program for a CDC 6400 using a peripheral Calcomp plotter to draw activity-activity diagrams. Their computation method relies on the definition of a grid covering the entire diagram and computation at every grid point where the mineral is stable. The mineral boundaries then are constructed from this digital image. The algorithm requires large core memory and CPU time to run.

Proposed here is a new algorithm to run such calculations by directly defining the straight line segments to be drawn. This method demands less core memory and CPU time and the diagrams can be computed and drawn on a microcomputer.

THE KNOWLEDGE BASE

Chemical Speciation

In working with the chemical composition of an aqueous solution, the N species is in ionic form. One of the N species is the free species Nwf entering in all the solid phase compositions. Its activity is minimized by the minerals formed according to the composition of the studied natural water. Two other variables of the N species, Nwx and Nwy, define the X and Y axis of the diagram on the log-activity scale. The N-3 components have fixed activities and generally represent the chemical "inactive ions" of the aqueous solution.

Thermodynamic Equations of Mineral Stability

With the N chemical species present in the solution, P different minerals can be constructed. The precipitation of a mineral P1 is dependent upon its solubility product Ks according to the following equations:

-The stoichiometry of P1 formation being

$$xA + yB + zC \rightarrow P1$$

where A, B, and C are aqueous chemical species, one of them being Nwf. Then the ionic activity product (IAP) is:

$$IAP = (A)^x (B)^y (C)^z$$

where () stands for the activity of the chemical species.
At equilibrium:

IAP = Ks and P1 may precipitate if no inhibition factor is present.

Otherwise, when

IAP < Ks : P1 is undersaturated and no precipitation occurs;

IAP > Ks : P1 is oversaturated.

If in equilibrium then for each mineral phase is a linear expression taking the logarithm of activities and Ks:

$$\log Ks = x \log(A) + y \log(B) + z \log(C)$$

Given Nw-3 activities, the construction of the diagram relies on the Np linear equations defining Np planes in the three-dimensional space: Nwx, Nwy, and Nwf. For each point of the diagram area, the mineral whose composition gives the smallest activity for the free species is sought. Once these solid phases have been selected, the thermodynamic system is defined entirely.

In order to read the thermodynamic characteristic of each mineral, we employed the framework defined in PHREEQE (Parkhurst, Thorstenson, and Plummer, 1980).

COMPUTATION METHOD

Data Reduction

First of all the data set has to be simplified for considering only the true variables: Nwx, Nwy, and Nwf. For this, a new apparent solubility product Ks' is computed taking into account all activities of the fixed species. For simplification the equations are normalized by setting all Nwf stoichiometric coefficients to 1.0.

Hence, P-plane equations are obtained in three-dimensional space: Nwx, Nwy, and Nwf (Fig. 2A). Now, the phase diagram can be defined as the projection on the plane Nwx, Nwy of the lowest surface minimizing Nwf, using all the possible planes.

If the planes of two minerals have an intersection which projects in the Nwx, Nwy area, hereafter termed the couple line, such as the situation in Figure 2B, the two minerals coexist on the diagram, and we save the coordinates of the intersection points of the straight line with diagram boundaries. In the other situation (Fig. 2C) only one of the two minerals minimizes Nwf on the diagram area, and the other one will never be stable thermodynamically and we no longer consider it in the computation.

By looking at all possible couples, we have to select every mineral to be present on the diagram and to save all the segments defining their intersections in between them on the diagram frontiers. With this segment set, we now are able to draw the diagram.

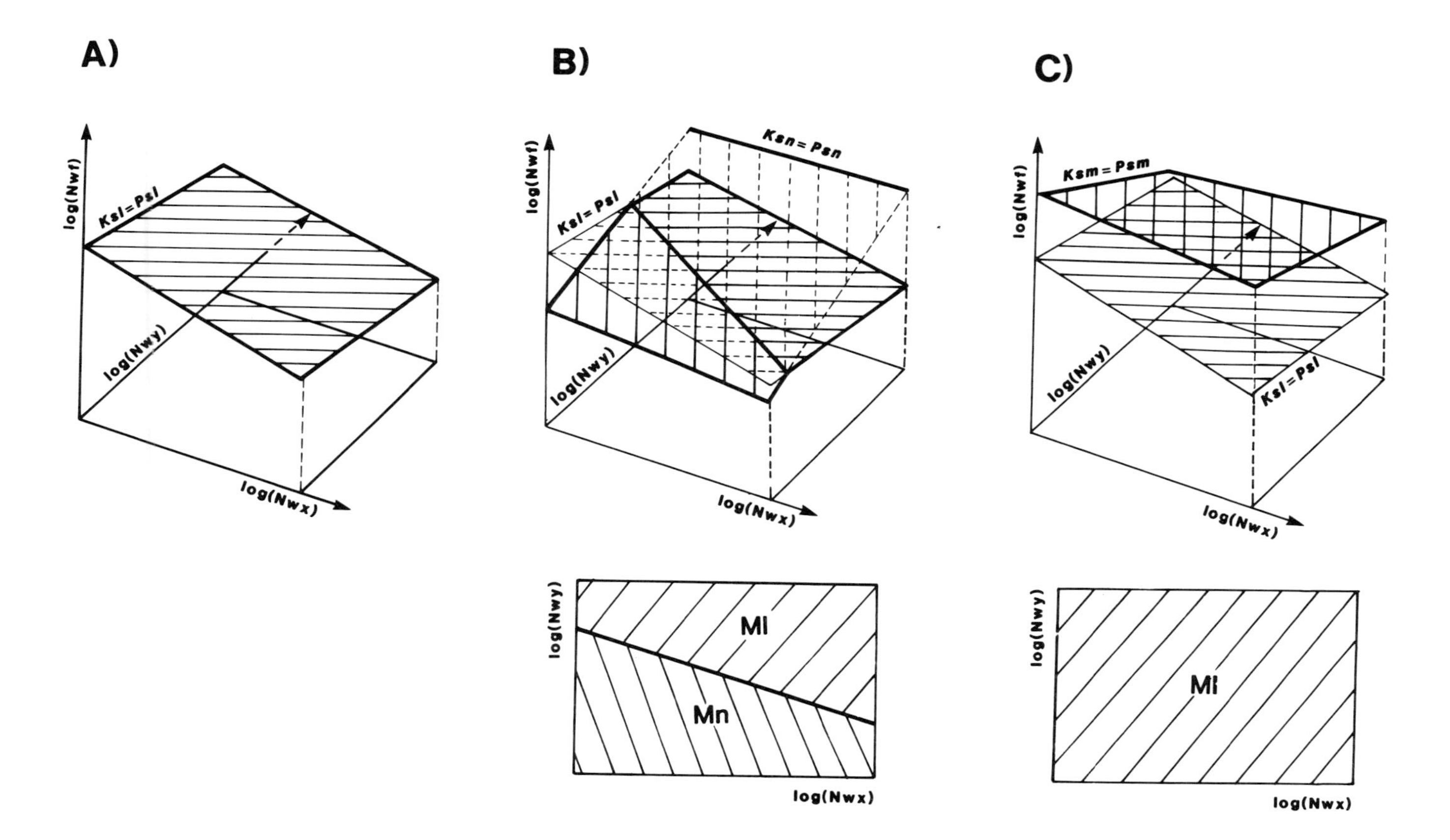

Figure 2. Selecting mineral stability fields. A, Plane representation for mineral L; B, planes intersection for mineral L and M defining two stability fields; and C, mineral L is minimizing Log(Nwf) on entire diagram frame.

Drawing the Diagram

Initialization. Drawing the diagram consists in selecting among all the segments the ones representing a real phase boundary and determining their limits. For this purpose, we investigate the boundaries of each mineral one after the other, starting at the lower-left corner and proceeding counterclockwise around the mineral and diagram area.

The first mineral to be drawn is the one which minimizes Nwf at the lower-left angle, and the first segment computed starts at this point and has a Nwx direction. The drawing procedure is the same for each mineral, if we know the starting point and the direction.

Determining the field of the ith mineral. The outline of the ith mineral is defined totally when all the segments indicating the boundaries with other minerals or diagram axes are selected. For drawing the S segment, whose origin P1 and direction are known, its end, P2, has to be determined. Initially, P2 is defined as the intersection of S with the diagram frame. Among all couples, including I, the one whose line intercepts S at the nearest point from the origin P1 is determined.

If a couple satisfying this condition is located then a new P2 is determined, so the origin and direction of the next segment (S+1) is respectively P2 and the direction of this couple line, otherwise the direction for (S+1) is the diagram frame. Segment-by-segment we outline the ith mineral back to the starting point, and save the coordinates.

Determining the next mineral to be drawn. In order to determine the next mineral stability field to be drawn (I+1), a test has to be made after each segment definition:

- if (I+1) is known already; and
- if (S+1) is defined by the (I,J) couple line, then (I+1) is the mineral J, unless J already has been outlined. The origin of the first segment for (I+1) is the intersection of S and (S+1) segments. The direction either will be the same as S, if S is on the diagram frame, or the one defined by the couple line (I+1, J-1), with S being on the couple line (I+1, J-1).

All this information is saved during the definition of the ith mineral contour. The program flowchart is presented in Figure 3.

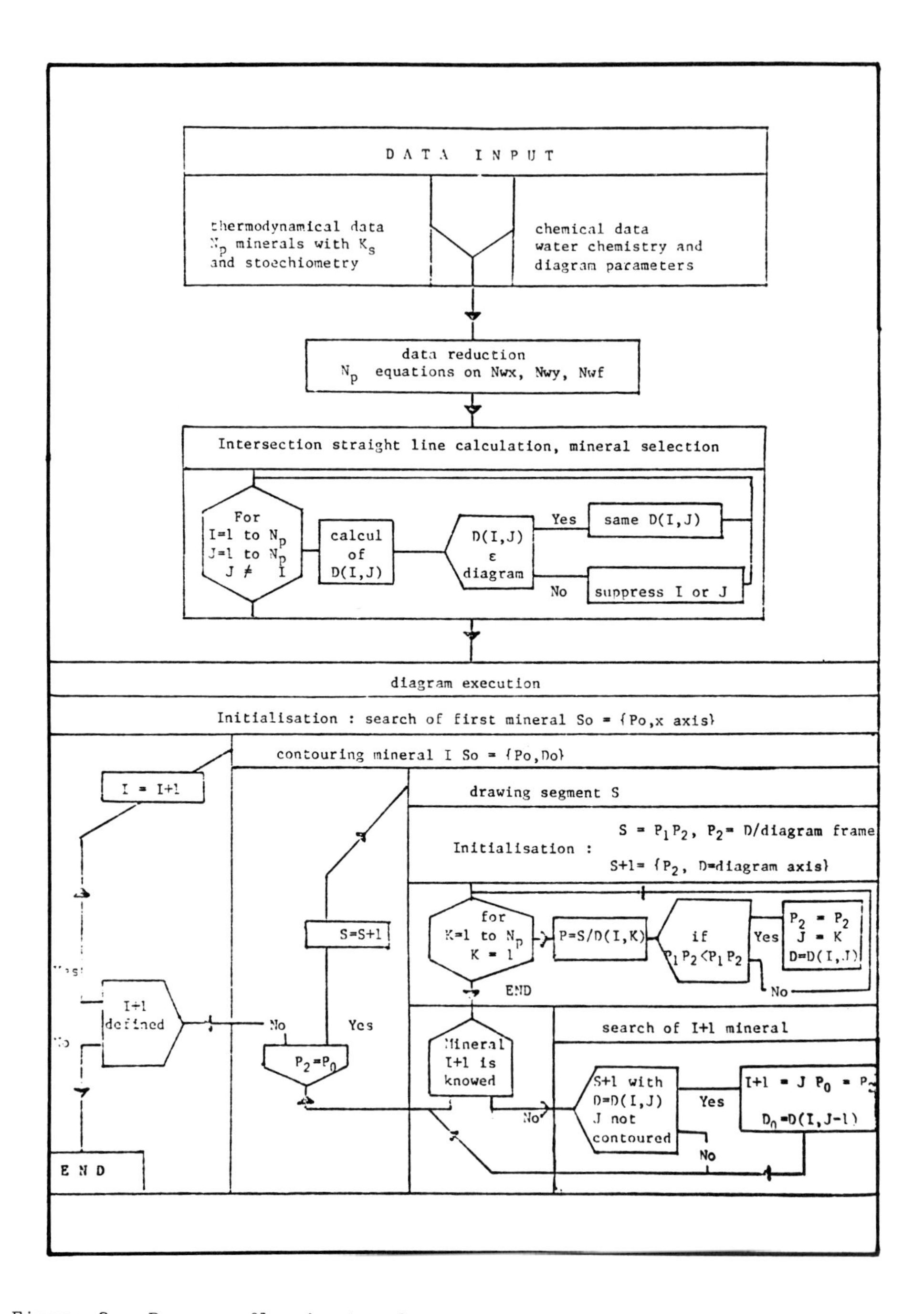

Figure 3. Program flowchart. Source program is written in FORTRAN and available on Victor S1 floppy disk.

Total Concentration-Activity Transformation

In constructing the diagram, the activity of the species involved in the solubility equations has to be known. For example, considering covellite in the Cu diagram (Fig. 1):

$$Cu^{2+} + HS^{-} - H^{+} \longleftrightarrow \underline{CuS}$$

(HS^{-}) should be known. Generally for a given aqueous solution, the total concentration of sulphur, SIIt, rather than (HS-) is known. Further, if (H+) is not constant through the diagram, (HS-) will change according to the equations:

$$\sum S_{II} = [H_2S] + [HS^{-}] + [S^{2-}]$$

with [] standing for the concentrations; and

$$K_1 = \frac{(HS^{-})(H^{+})}{(H_2S)} \quad ; \quad K_2 = \frac{(S^{2-})(H^{+})}{(HS^{-})}$$

It will be the same for all the acid-basic couples, or redox couples. By taking an acid-basic equation:

$$AH \longleftrightarrow A^{-} + H^{+}$$

with

$$\sum A = [A^{-}] + [AH]$$

$$Ka = \frac{[A^{-}]\ (H^{+})}{[AH]}$$

and

$$(AH) = \gamma * [AH]$$

then

$$\sum A = \frac{(AH)}{\gamma} * \left(1 + \frac{Ka}{(H^{+})} \right).$$

According to the value of (H+), this equation can be approximated by:

-if (H+) > Ka : $(AH) = \sum A * \gamma$ and $\log(AH) = \log\sum A + \log \gamma$

-if (H+) < Ka : $(AH) = \sum A * \gamma * (H^{+}) / Ka$ and $\log(AH) = \log\sum A + \log \gamma + \log(H^{+}) + \log Ka$.

During the calculation the mineral equations have to be changed, log(AH) by its log(H+) and log $\sum A$ dependent expression. Doing that preserves the linearity of the equation set. To construct the diagram with values of log(H+) both upper and lower than the log Ka value, the calculation has to be separated in two parts, with two different equation sets, and then the two obtained pictures drawn side by side.

EXAMPLE: MANGANESE DIAGENESIS

We are presenting here, an illustration, through mineral stability diagrams, of manganese geochemical behavior during early diagenesis in marine sediments. As shown by Michard (1971), Robbins and Callendar (1975), Holdren, Bricker, and Matisoff (1975), manganese undergoes various dissolutions or precipitations according to the chemical composition of interstitial waters. During diagenesis, the oxidation of organic matter, via a bacteria-mediated process, consumes oxygen and different electron acceptors. This results in a more reducing environment during burial, an increase of log(e^-) and a production of CO_2 buffering pH pore water at equilibrium with calcite.

So, in order to describe manganese diagenesis, we constructed three diagrams giving mineral stability versus log(H+) and log(CO_2t) at different oxydo-reduction states. The chemical speciation of the aqueous solutions and minerals considered are given in Table 1, the interstitial water evolution during diagenesis is presented in Figure 4A.

TABLE 1. Chemical speciation and minerals considered for manganese diagenesis diagrams.

Pore-water chemical composition in Log(activity).

Cl-:	-0.0453	H20:	-0.985	S04=:	-2.255	HS-:	-4.000	P04-3:	10.130

Chemical species and minerals investigated.

Mn3+	
Mn04-	
Mn04=	
Pyrolusite:	Mn02
Bixbyite:	Mn203
Hausmanite:	Mn304
Pyrochroite:	Mn(OH)2
Manganite:	MnOOH
Rhodocrosite:	MnC03
MnCl2.4H20	
MnS.green	
Mn3(P04)2	
Mn2(S04)3	
MnS04	

Figure 4B, emphasizing a log(e^-) = -8.22, represents an overlying water having the typical seawater pE. Pyrolusite (MnO_2) is the mineral limiting Mn solubility in such a situation.

After burial, the oxidation of organic matter produces more reducing conditions allowing MnO_2 reduction and concomittant solubilization of manganese as Mn^{2+} in pore waters and no solid phase are limiting. So, Mn^{2+} can diffuse up or down through the sedimentary column (Fig. 4C).

This example illustrates the way one can handle geochemical processes, for example, early diagenesis with mineral-stability diagrams.

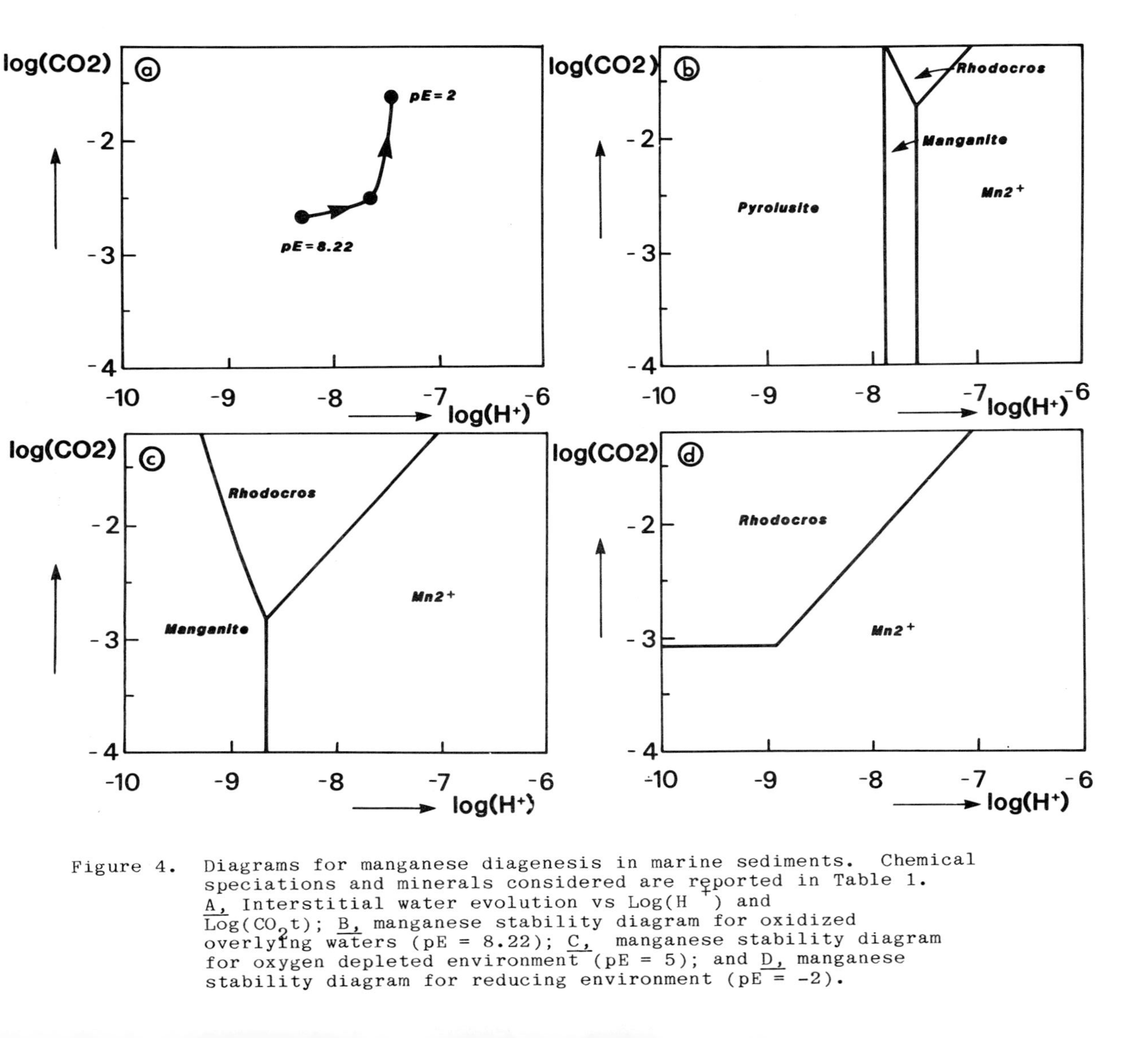

Figure 4. Diagrams for manganese diagenesis in marine sediments. Chemical speciations and minerals considered are reported in Table 1. A, Interstitial water evolution vs Log(H^+) and Log(CO_2t); B, manganese stability diagram for oxidized overlying waters (pE = 8.22); C, manganese stability diagram for oxygen depleted environment (pE = 5); and D, manganese stability diagram for reducing environment (pE = -2).

CONCLUSIONS

Computing time for drawing the diagrams on the CRT as presented in Figure 4 is only about 8 minutes on a standard Victor S1. This includes accessing data files stored or created on floppy disk and loading the graphics interface. It really is satisfying in regard to the time needed for preparing data sets for each diagram.

Program organization enables easy transformation to diagrams. Switching from an activity vs activity diagram to a total concentration vs pH diagram is illustrated by including some new chemical equations. Currently, we are looking at diagrams concerned with redox mass balances, on one hand, and temperature variations, on the other, in order to study mineral paragenesis in hydrothermal systems.

ACKNOWLEDGMENTS

We are indebted to C. J. Allegre who initiated this work, G. Michard for helpful theoretical discussions, and E. Levin for his know-how on the Victor 9000 system. We are grateful to C. Mercier for typing the manuscript. Musical support was provided by SAGA.

REFERENCES

Garrels, R.M., and Christ, C.L., 1965, Solutions, minerals and equilibria: Harper and Row, New York, 450 p.

Helgeson, H.C., Brown, T.H., and Leeper, R.H., 1969, Handbook of theoretical activity diagrams depicting chemical equilibria in geologic systems involving an aqueous phase at one atm. and 0 to 300 C.: Freeman, Cooper & Co., San Francisco, 253 p.

Holdren, G.R., Bricker, O.P., and Matisoff, G., 1975, A model for the control of dissolved manganese in the interstitial waters of Chesapeake Bay, in Church, T.M., ed., Marine chemistry in the coastal environment: Am. Chem. Soc. Series No. 18, Washington, p. 364-381.

Michard, G., 1967, Signification du potential redox dans les eaux naturelles: Mineral Deposita, v. 2, no. 1, p. 34-36.

Michard, G., 1971, Theoretical model for manganese distribution in calcareous sediments cores: Jour. Geophys. Res., v. 76, no. 9, p. 2179-2186.

Michard, G., and Allegre, C.J., 1969, Comportement geochimique des elements metalliques en milieu reducteur et diagrammes (log S, pH): Mineral Deposita, v. 4, no. 1, p. 1-17.

Parkhurst, D.L., Thorstenson, D.C., and Plummer, L.N., 1980, PHREEQE - A computer program for geochemical calculations: U.S. Geol. Survey, WRI 80-96, 210 p. (NTIS Techn. Rep. PB81-167801, Springfield, VA 22161).

Robbins, J.A., and Callender, E., 1975, Diagenesis of manganese in Lake Michigan sediments: Am. Jour. Sci., v. 275, no. 5, p. 512-533.

Promising Aspects of Geological Image Analysis

A. G. Fabbri

Instituto di Geologia Marina

ABSTRACT

Spatially distributed data are defined to introduce digital image processing. A review is made of the roots of geological image analysis. Recent developments of importance for image analysis and the geosciences are: (1) availability of new image analyzers, (2) design of new computer architectures, (3) programming of image-processing software, (4) development of languages and image algebra, and (5) new applications in economic geology, environmental geology, the study of textures in polycrystalline aggregates, and in interpreting the new remotely sensed imagery. Representative applications are reviewed in this paper. The character of the applications matches well J.C. Griffiths' views on quantification and the future of geoscience.

SPATIALLY DISTRIBUTED DATA

Geology is a science in which particular attention is devoted to information that has a distribution in space. Mathematical geology is concerned with the quantitative aspects in the geosciences and probably has contributed to the analysis of spatially distributed data more than most other disciplines. The usual way to represent quantitatively such data is the digital form, used, for example in remote sensing. Data are distributed as light-intensity readings, taken at regular intervals according to a grid or raster so that they cover an entire area on the ground. Because of the point-to-point correspondence between the readings and positions on the ground, obtained are "digital images."

How the geologist and the computer can compliment each other's abilities by establishing an interactive visual dialogue was demonstrated with examples of computerized vision by Kasvand (1983). What a computer "sees" are digital images, that is large matrices of numbers termed pixels. Each pixel is shaded a level of gray with white being the lightest and black being the darkest. The different pixel values (gray levels) and their spatial arrangement capture visual information. Computer processing extracts and characterizes such information quantitatively and also returns it visually. The family of computer techniques that are concerned with the systematic treatment of these data has been termed "image processing."

ROOTS OF GEOLOGICAL IMAGE ANALYSIS

Although the idea of capturing and analyzing rock fabrics and map patterns may be older, the first experiments in optical-data processing are exemplified by the work of Pincus and Dobrin (1966). Porosity in oil reservoir sandstones was studied by Davis (1971) using a laser optical bench. Data also were digitized by a flying spot scanner to produce 1024 lines of 1024 data points (pixels) arranged as a square array. Fourier optics and optical data processing of microporous fabrics later were refined and formalized by Preston and Davis (1976).

Serra (1966) and Agterberg (1967) were among the first to analyze thin sections of rocks by coding the occurrence of the different grain profiles by overlying regular grids on microphotographs. Their statistical analyses of the coded arrays of data was to illustrate new methods of studying spatial relationships between minerals or mineral groups and possibly contribute to petrological interpretations. Serra's study consisted in the analysis of a thin section of Lorraine oolitic iron ore, by comparing experimental "variograms" computed in different directions for three minerals. Agterberg worked instead on thin sections of a gabbro from the Muskox layered intrusion, District of MacKenzie in the Northwest Territories of Canada. He computed the two-dimensional "covariance function" and a two-dimensional "power spectrum" of the thin-section coded data.

Whereas the theory and applications of "mathematical morphology" also originated from the problem of characterizing porous media, in the work of Matheron (1967) they have developed a field of applicability. The concepts of "erosion" and "dilatation", and of "opening" and "closing" of sets by "structuring elements", are capable of describing in probabilistic terms the information in images of many different types as demonstrated by Serra (1972).

The work of Switzer (1975, 1976) who developed a method for estimating algebraically the spatial dependence of map patterns and the accuracy of digital maps, also has contributed to the problem of spatially distributed data digitization and map-data integration.

Srivastava (1977) performed optical processing of structural contour maps in the form of "zebra maps" in which every other interval is black.

Agterberg and Fabbri (1978a and 1978b) proposed to use image processing for spatial correlation of stratigraphic units digitized from maps for the statistical treatment of tectonic and mineral-deposit data.

NEW DIRECTION FOR IMAGE ANALYSIS AND GEOLOGY

The previous section scanned through some of the first representative approaches in geological image analysis. Currently, much activity can be envisaged from which it is predicted that geosciences will benefit in the future. The following are arguments worth considering in that they bridge analysis and geology:

(1) availability of new image analyzers;

(2) design of new computer architectures;

(3) programming of image-processing software;

(4) development of image algebra and languages; and

(5) new applications in economic geology, environmental geology, the study of textures in polycrystalline aggregates, and in interpreting the new remotely sensed imagery.

New Image Analyzers

Several specialized instruments, known as image-analyzing systems or image analyzers, have been built for the automatic detection of binary images (black and white) obtained from various types of picture material (ore microscopy, metallography, cytology, etc.) and for precision morphological measurements of the detected patterns. Until recently, the repertoire of measurement built as hardware in these instruments was limited. Then, in collaboration with the School of Mathematical Morphology of Fontainebleau, France, the Leitz Company in West Germany built an image analyzer that could satisfy the measurements involved in the theory of mathematical morphology. It was termed TAS (Texture Analysing System) and was described by Muller (1974) and Serra (1974).

Since then, most image analyzers have undergone continuous developments and expansions that required redesigning of the component modules for specific tasks and addition of new options for added programmability.

Some of the more recent developments are programmable systems such as the "AT4" (Digabel, 1976) and the "TAS second model" used by Chermant and Coster (1978) in France. Aside from having a rather extended programmability and being able to interface with large computers, these instruments have several memories for storage of the detected images and their transformations.

New examples of image analyzing systems are: the OMNICON ALPHA 500 (Bausch & Lomb), the POLYPROCESSOR C-1285 (Hamamatsu), the QUANTIMET 900 (Imanco, Cambridge Instruments), and IBAS (Kontron, Zeiss).

A new technique was developed by Miller, Reid, and Zuiderwyk (1982) and Reid and others (1984), that uses a QEM*SEM image analyzer (Quantitative Evaluation of Minerals by Scanning Electron Microscopy)> This fully automated instrumentation can determine the type of phases present in a polished section by an energy dispersive (multielement) x-ray detector and back-scattered electron detector. A motor-driven, moving stage brings successive fields into position, and at each sample point, at regularly spaced intervals, a signal generated by the back-scattered electrons is used to determine the average atomic number of the small area of material irradiated by the beam. "Mineral maps" can be computed, within 1-2 hours, in a digital form that is similar to the one of phase-labeled images (each pixel has a numeric value that indicates the membership to a given crystalline phase). The capture of textural data, that is a complex problem for crystalline material due to the variability of the optical characteristics of the crystal grains, thus is facilitated greatly.

Mainwaring and Petruk (1984) reviewed automatic image analysis based on signal inputs from scanning electron microscopes as used in metallurgy and mineral beneficiation. They describe a JEOL 733 electron microscope-based image analysis system with energy dispersive spectrometer and an interfaced KONTRON image analyzer with 16 image memories accessible by a microprocessor.

The Design of New Computer Architectures

In an analysis of the computational aspects of image processing for geoscience data, Kasvand and Fabbri (1984) subdivided them into three classes: (I) primarily interactive, (S) sequential, and (P) "pipelineable". Some of those aspects are summarized in Figure 1A in terms of types of processing and form of input and output data; (P) indicates the pipelineable processes. The term pipeline refers to a particular hardware realization of a processing structure

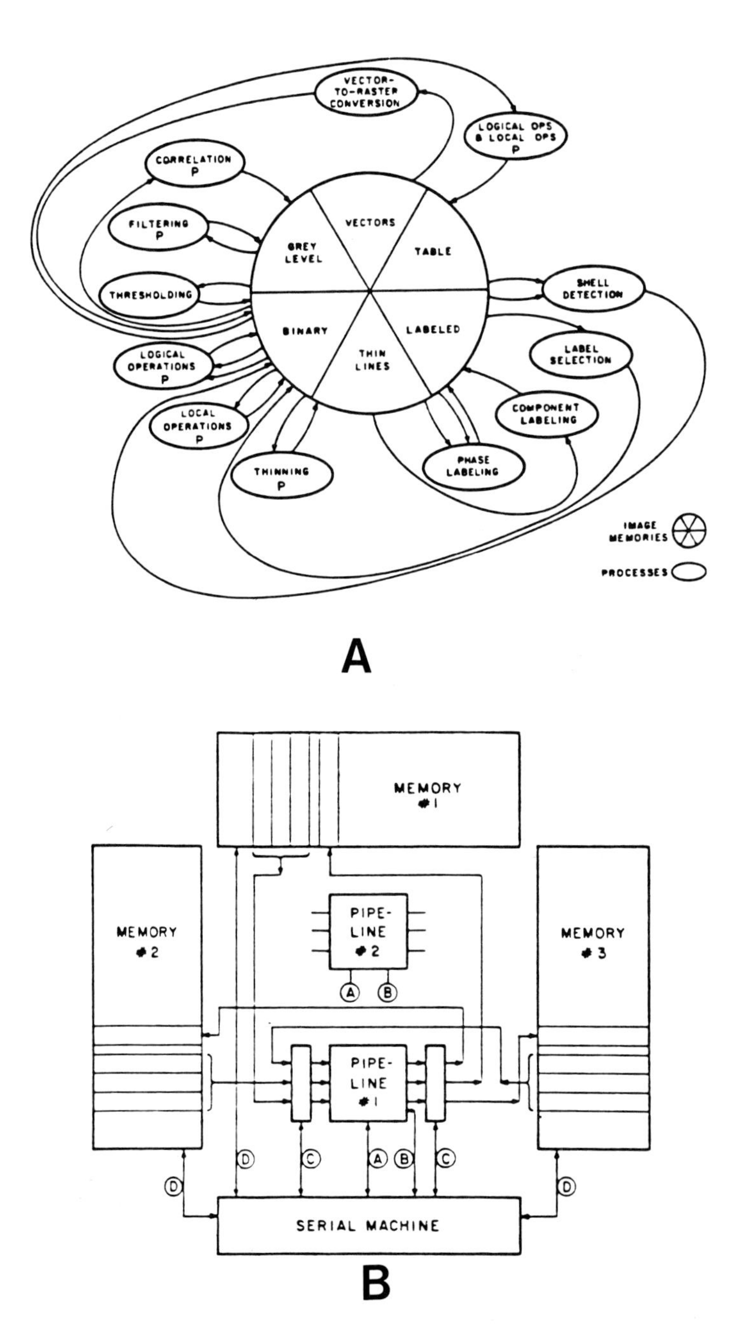

Figure 1. A, Computational aspects of image processing: (I) interactive, (S) sequential, and (P) "pipelineable; B, example of multimemory and multipipeline system (after Kasvand and Fabbri, 1984).

for data such as pixel neighborhoods that can be arranged and treated in a continuous flow. Figure 1B shows an example of a multimemory and multipipeline system envisaged for geological data processing.

The reason why many different solutions are being suggested to cope with the massive amount of pictorial information present in most real image-processing tasks, is the need for higher computational power. Figure 2A shows diagrammatically the basic structure of an "array processor". A single processor is repeated spatially along rows and columns to generate a grid where each node contains one processor, and each is interconnected to its immediate neighbors. An example of an array processor is the CLIP7 described by Fountain (1984) that uses a scanning mode along 8 lines of 256 processors.

Another significant class of computer architectures is the "pipeline" structure which enables a collection of independent processors to execute computations on a consecutive set of data similar to a hydraulic mode of liquid flow. A schematic diagram of part of a pipeline computer system termed CYTOCOMPUTER is shown in Figure 2B.

In order to combine the advantages of array processors and pipeline processors (Fig. 2C), the pyramidal architecture was proposed by Dyer (1980). In practice this architecture contains only 1/3 more hardware than an array processor, but is able also to process data vertically, and most of all, a new process of computation may take place, namely a weighted consensus taken over subsets of processors in the pyramid. The internal structure of the pyramid computer has been described by Cantoni and others (1984).

Some of these architectures are available computer systems, as for example the GENESIS 2000 and its smaller version or work station FLASHPAK built by Machine Vision International (Ann Arbor, Michigan), or the GOP (General Operator Processor) built at the Picture Processing Laboratory, Linkoeping University (Sweden) (Fig. 3).

Programming of Image-Processing Software

Special software packages for processing images have existed for sometime. For example Moore (1968) described the STRIP system (Standard Taped Routines for Image Processing) developed at the National Bureau of Standards in Gaithersburg, Maryland.

Recently a package for computing mathematical morphology transformations of binary images was proposed by Fabbri (1980, 1981, 1984). The package was termed GIAPP (Geological Image Analysis Program Package) and handles digital images of maximum size of 1024 x 1024 pixels. It is written in FORTRAN and was developed at the National Research Council of Canada (in collaboration with the Geological Survey of Canada and the University of Ottawa).

MORPHOLOG is a new package produced by the Centre de Geostatistique et de Morphologie Mathematique of Fontainebleau (France). It processes images of 256 x 256 pixels, and is written in FORTRAN. It can compute a large variety of morphological transformations.

An extensive general-purpose minicomputer system for processing images, also of considerable portability, is GIPSY (General Image Processing System; Haralick, Krusemark, and Nelkirk, 1981). It originated from an older one termed KANDIDATS (KANsas Digital Image DATa System) described by Haralick and others (1976). It can handle digital arrays of the size of remotely sensed images, that is several millions of pixels.

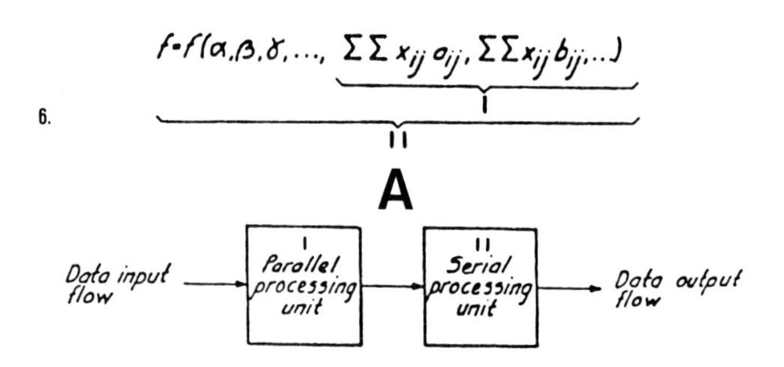

Simplified block diagram of system architecture.

B

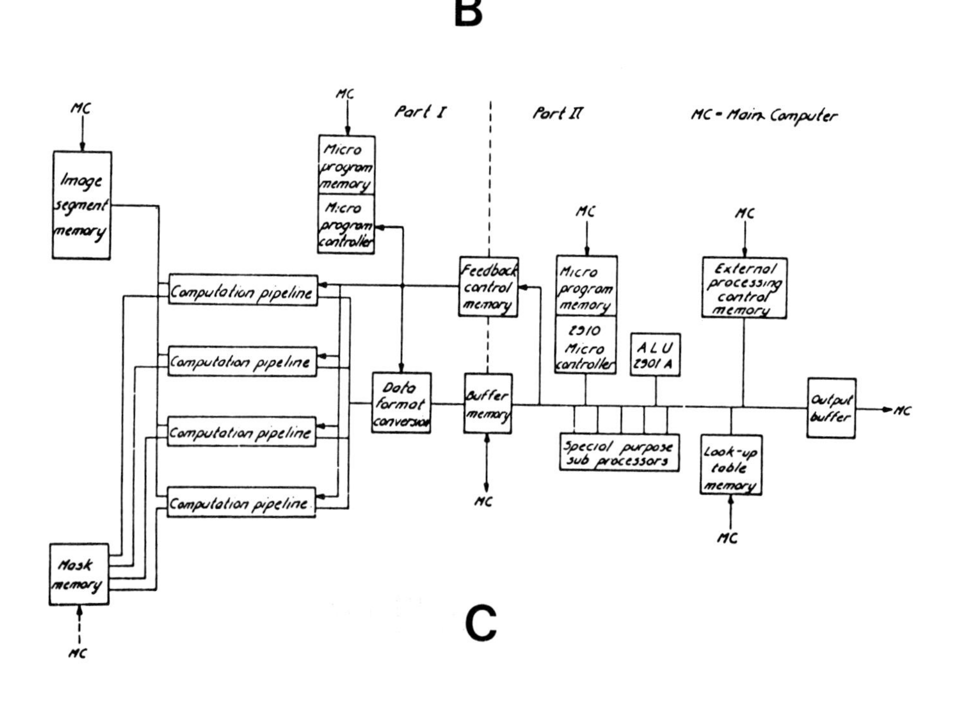

Figure 2. A, Basic structure of "array processor"; B, schematic diagram of pipeline computer system termed CYTOCOMPUTER (after Sternberg 1979, 1982); C, combination of advantages of array processor (A) and pipeline (B) (after Fabbri and Levialdi, 1984).

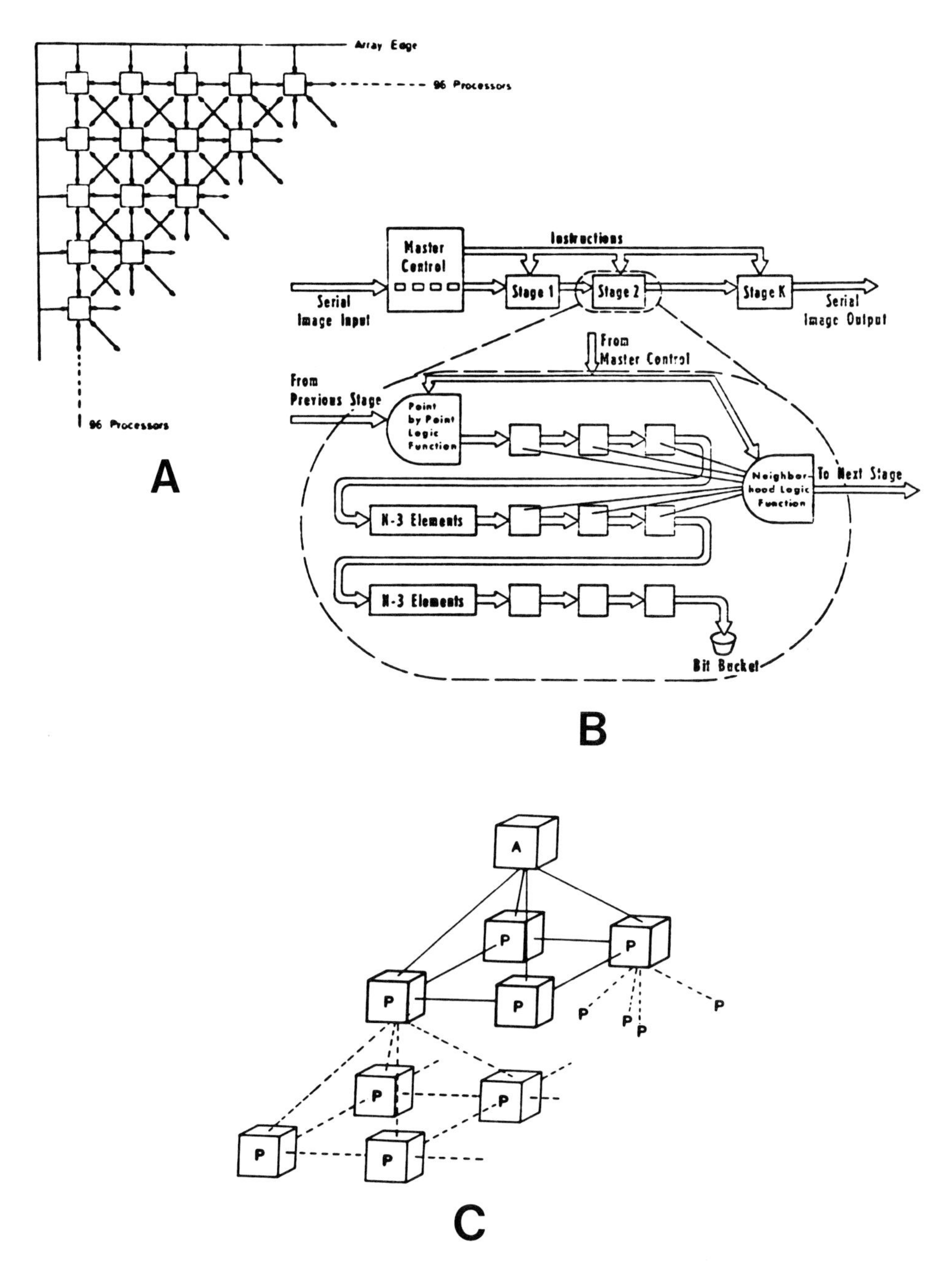

Figure 3. A, General form of operations that are computed on digital images (operation of arithmetical as well as logical type are included); B, simplified block diagram of GOP system and structure; C, block diagram of GOP image processor (after Lundgren and others, 1981).

Development of Languages and Image Algebra

For the AT4 prototype image analyzer described by Lantuejoul (1981), the high-level interpreted language MORPHAL (MORPHological Algorithmic Language) was created. It emphasizes the basic concept of quantitative image analysis, image transformations, and measurements of binary images.

The CYTOCOMPUTER uses a high-level language termed C3PL (Cytocomputer Parallel Picture Processing Language), that is an command interpreter more than a language, in the strictest sense.

GENESIS 2000 incorporates a mathematical morphology enhancement of the "C" programming language, termed BLIX (Basic Language for Image X-formation) for the Motorola 68000 computer and for the UNIX operating system.

GOP is written in FORTRAN and uses a high-level language, INTRAC, which is portable and interactive.

According to Levialdi (1983) there are nearly 50 languages (operating) for image processing. A classification of the main operations generally performed on images is as follows:

(1) utilities (I/O, memory management, etc.);
(2) representation (visual, etc.);
(3) arithmetic operations;
(4) geometrical transformations;
(5) enhancement;
(6) analysis; and
(7) classification.

No language seems to exist, however, that contains all seven types of operations. Levialdi (1983) analyzed four main languages for image processing, giving particular attention to PIXAL, an ALGOL-based language which allows parallellism, and to IPC, or Image Processing C, that is an expansion of the "C" language.

Giardina (1984) advocated a Universal Imaging Algebra that involves numerous new and conventional operators useful in image analysis and recognition. Operators of morphological type have been considered by various authors (Gillies, 1978; Haralick, 1979; Sternberg, 1980; Serra, 1982a, 1982b) for binary and also for gray-level images. Examples of basic operations defined are translation, thresholding, pixel counting, and covariance function. The syntactical nature of some operations can be visualized in the polyadic graph shown in Figure 4A, or in the block diagrams in Figure 4B (translation operation), or Figure 4C (erosion operation). According to Giardina (1984), several hundred operations exist in this algebra. The difficult task of determining the variety of this algebra is being conducted in a slow but systematic fashion. A universal image algebra leads toward a higher level of programming by using the concepts of fuzzy sets and a new more efficient man-machine interaction.

New Applications

Several new applications of image analysis in the geosciences represent promising fields of future development. Some representative examples are reviewed here.

Image Analysis. Fourier spectra for quantitative shape analysis in sedimentology were reviewed by Clark (1981) and used recently by Full and Ehrlich (1982) and by Lin (1982a, 1982b) as an alternative to processing digital images.

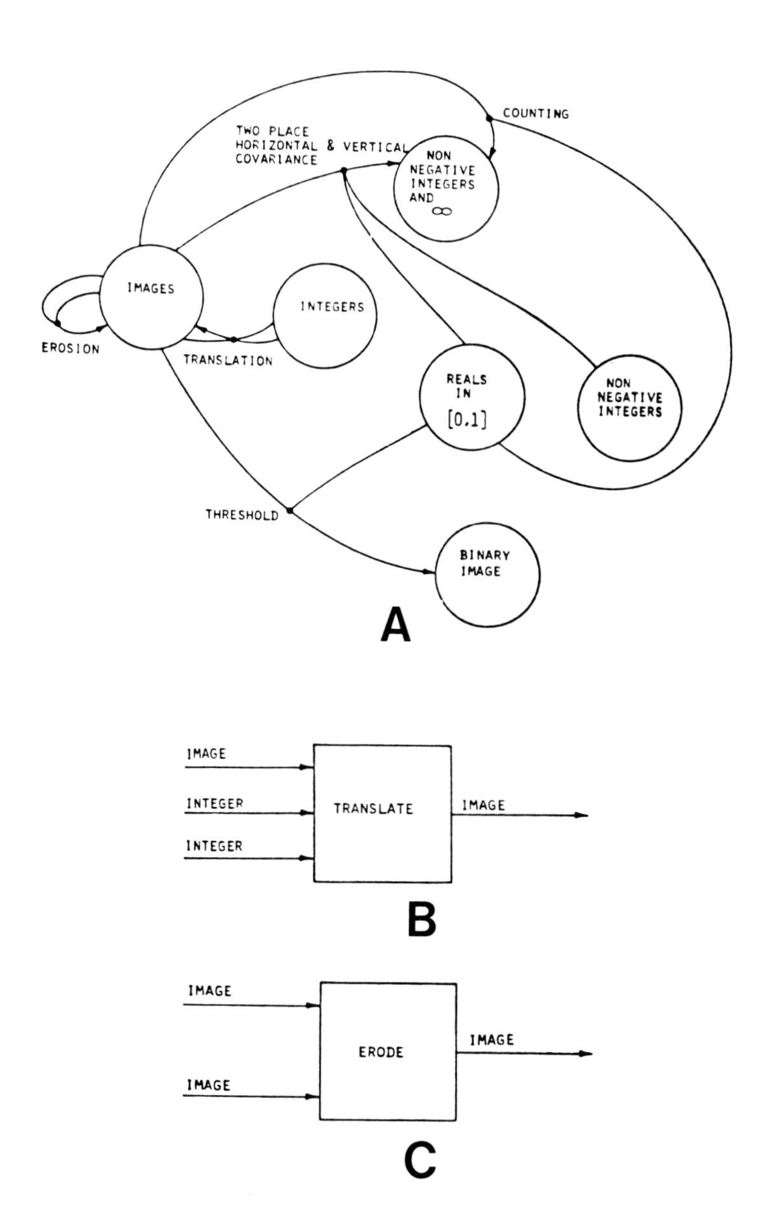

Figure 4. Syntactical nature of operations as shown in A, polyadic graph; B, translation operation; or C, erosion operation.

Textures. Pong and others (1983) described applications of image analysis to mineral processing in which the package GIPSY was employed. In a review by Fabbri and Kretz (1984) recent approaches to the analysis of crystalline textures by the study of two-dimensional grain sequences in rocks are described as applications with great potential in petrology and in the general study of polyphase crystalline aggregates.

The notion of the neighborhood of a point which is essential for analyzing dot patterns, has been studied by Ahuja (1982) by using Voronoi neighborhoods.

Cartographic Data. Convential image-processing techniques were applied to geological computer cartography by Coa and Debray (1984) for the extraction and tracing of line structures. The authors used an 8-bit microprocessor-based system: the Pericolor 1000.

Map Data Integration. Digital map data integration with lake-sediment geochemical data were used for evaluating mineral and energy resources by Chung (1983) and Bonham-Carter and Chung (1983). The authors combined methods of multivariate statistics and image processing for constructing statistical models for uranium exploration.

Image Processing. A formal treatment of image analysis by mathematical morphology was introduced by Serra (1982a) who also used several geological applications. He showed how set theory can be used to construct an image algebra by developing criteria and models for image analysis. In particular, Serra considered the morphology of gray-tone functions, also studied by Geotcherian (1980), and random sets.

New approaches to "morphological filtering" proposed by Serra (1982b, 1983) fully develop the general usage of mathematical morphology in image processing.

A general approach to image processing in geology was proposed by Fabbri (1984). He described applications that span the geological field from maps to microscopic images of thin sections. Geological reasoning was given for pattern recognition and texture analysis; also for processing binary images in parallel on a sequential computer were given.

Kasvand and Fabbri (1983) suggested the modeling of spatial computer architectures for the analysis of geoscience data.

Remote Sensing. Switzer, Kowalik, and Lyon (1982) studied smoothing techniques on LANDSAT spectral data to increase the accuracy in pixel classification by discriminant analysis. Carrere and others (1984) used LANDSAT data for characterization of geological map units by using spectral signature and morphological appearance (texture). They employed with success supervised classification methods to construct thematic maps, in a study area in Provence, France.

An interesting application of fractals to texture analysis for improving the classification of a satellite image in a morphological study was proposed by Nguyen and Quinqueton (1982).

The term quadtree (Samet, 1983) is used to describe a class of hierarchical data structures whose common property is that they are based on the principle of recursive decomposition. They are used currently for point data, regions, curves, surfaces, and volumes. An example of the quadtree concept is represented for a region shown in Figure 5. The quadtree is proposed as an alternative representation for binary images because it is hierarchical in nature and it facilitates the performance of a large number of operations. The concept of quadtree leads to a multiresolution representation of an image by an exponentially tapering "pyramid" of arrays, each of one-half the size of the preceding one (Rosenfeld, 1983). Therefore, it is well suited for a pyramidal computer architecture.

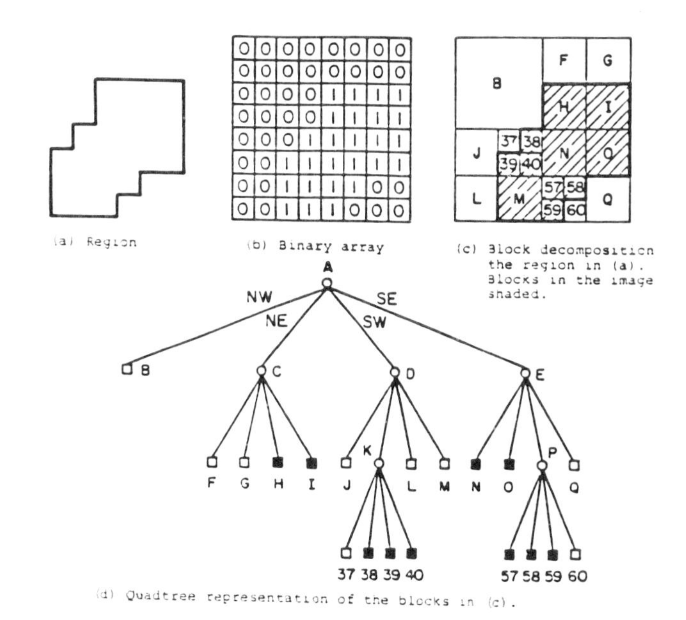

Figure 5. Quadtree concept (after Samet, 1983).

An application of quadtree hierarchical data structure to a geographical information system was made by Rosenfeld and others (1982). A database that was constructed, contained three maps that described the flood plain, elevation contours, and land-use classes of a region in California. The study is an enlightening exposition of advantages and disadvantages of applying the new concept of quadtrees. This method also has been evaluated computationally for processing raster and vector data in cartographical image processing by Bell and others (1983).

A complete bibliographical survey of image-processing activity in 1983 has been made by Rosenfeld (1984). It presents about 1100 references related to computer processing of pictorial information.

CONCLUDING REMARKS

This review has looked at the origin and the future trends of image processing in the geosciences. The fields of study and the applications listed represent new directions in quantitative geology.

A few years ago Griffiths (1974, p. 88) wrote this about quantification and the future of geoscience: "the increase in black box sensor system for gathering information is summarized in Figure 6 where tremendous surge in 'remote sensing' both in the 'very small' and the 'very large' is a self-evident feature of our observing systems."

His words were remarkable and not coincidental, when he said:

> "...the visual image is not a mere representation of 'reality' but a symbolic system." Gombrich, 1972..."Quantification may be defined as the assignment of symbols to observations...according to a rule. In its

broadest connotation this procedure includes the development of language as an assignment of symbols to observations and is therefore the basis of all communications (see quotation by Gombrich.)" (p. 83).

REFERENCES

Agterberg, F.P., 1967, Computer techniques in geology: Earth Sci. Rev., v. 3, p. 47-77.

Agterberg, F.P., and Fabbri, A.G., 1978a, Spatial correlation of stratigraphic units quantified from geological maps: Computers & Geosciences, v. 4, no. 3, p. 285-294.

Agterberg, F.P., and Fabbri, A.G., 1978b, Statistical treatment of tectonic and mineral deposit data: Global Tectonics and Metallogeny, v. 1, no. 1, p. 16-28.

Ahuja, N., 1982, Dot patterns using Voronoi neighborhoods: IEEE Trans. on Pattern analysis and machine intelligence, PAMI-4, p. 336-343.

Bell, S.B.M., Diaz, B.M., Horoyd, F., and Jackson, M.J., 1983, Spatially referenced methods of processing raster and vector data: Image and Vision Computing, v. 1, p. 211-220.

Bonham-Carter, G.F., and Chung, C.F., 1983, Integration of mineral resource data for Kasmere Lake area, Northwest Manitoba, with emphasis on uranium: Jour. Math. Geology, v. 15, no. 1, p. 25-45.

Cantoni, V., Ferretti, M., Levialdi, S., and Stefanelli, R., 1984, PAPIA: pyramidal architecture for parallel image analysis, in press.

Cao, T.T., and Debray, B., 1984, A simple algorithm for extracting and tracing line structures using a microprocessor-based system: Proc. intern. coll. on Computers in Earth sciences for natural resource characterization, Nancy, France (April 9-13, 1984), Sciences de la Terre, in press.

Carrere, V., Nguyen, P.T., Abrams, M., and Chorowicz, J., 1984, Le traitement numerique des donnees LANDSAT: une aide a la cartographie des unites lithologiques: Proc. intern. coll. on Computers in Earth sciences for natural resource characterization, Nancy, France (April 9-13, 1984): Sciences de la Terre, in press.

Chermant, J.L., and Coster, M., 1978, Fractal object in image analysis, in Intern. symp. on quantitative metallography, Florence, Italy (November 21-23, 1978): Associazione Italiana de Metallurgia Eds., p. 125-138.

Chung, C.F., 1983, SIMSAG: integrated computer system for use in evaluation of mineral and energy resources: Jour. Math. Geology, v. 15, no. 1, p. 47-58.

Clark, M.W., 1981, Quantitative shape analysis: a review: Jour. Math. Geology., v. 13, no. 4, p. 303-320.

Davis, J.C., 1971, Optical processing of microporous fabrics, in Cutbill, J.L., ed., Data processing in biology and geology: Systematics Assoc. Spec. Vol. No. 3, Academic Press, London, p. 69-87.

Digabel, H., 1976, Manuel d' utilization d' AT4: Internal report of the Centre de Morphologie Mathematique, Fontainebleau, France.

Dyer, C., 1980, Space efficiency of region representation by quadtrees, in Picture data description and management: Proc. of IEEE workshop, Asilomar, California, p. 31-36.

Fabbri, A.G., 1980, GIAPP: geological image analysis program package for estimating geometrical probabilities: Computers & Geosciences, v. 6, no. 2, p. 153-161.

Fabbri, A.G., 1981, Image processing of geological data: unpubl. doctoral dissertation, Univ. Ottawa, 514 p.

Fabbri, A.G., 1984, Image processing of geological data: Van Nostrand-Reinhold, Stroudsburg, Pennsylvania, 244 p.

Fabbri, A.G., and Kretz, R., 1984, A study of two-dimensional sequences in rocks: 27th Intern. Geol. Congress (Moscow) proc., in preparation.

Fabbri, A.G., and Levialdi, S., 1984, New computer architectures suitable for spatial analysis in the Earth sciences: Proc. intern. coll. on Computers in Earth sciences for natural resource characterization, Nancy, France (April 9-13, 1984): Sciences de la Terre, in press.

Fountain, T.J., 1984, Plans for the CLIP7 chip: Image Processing Group, Imperial College, London, Internal Report.

Full, W.E., and Ehrlich, R., 1982, Some approaches for location of centroids of quartz grain outlines to increase homology between Fourier amplitude spectra: Jour. Math. Geology, v. 14, no. 1, p. 43-55.

Giardina, C.R., 1984, The universal imaging algebra: Pattern Recognition Letters, v. 2, p. 165-172.

Gillies, A.W., 1978, An image processing computer which learns by example, in Nevatia, R., ed., Image understanding systems and industrial applications: Proc. Society of Photo-Optical Instrumentation Engineers, SPPIE, San Diego, California, v. 155, p. 120-126.

Geotcherian, V., 1980, From binary to grey tone image processing using fuzzy logic concepts: Pattern Recognition, v. 12, no. 1, p. 7-15.

Gombrich, E.H., 1972, The visual image: Scientific American, v. 227, no. 3, p. 82-96.

Griffiths, J.C., 1974, Quantification and future of geoscience, in Merriam, D.F., ed., The impact of quantification on geology: Syracuse Univ. Geology Contr. 2, p. 83-101.

Haralick, R.M., 1979, Statistical and structural approaches to texture: Proc. IEEE, v. 67, no. 5, p. 786-804.

Haralick, R.M., Bryant, W.F., Minden, G.J., Singh, A., Paul, C.A., and Johnson, D.R., 1976, KANDIDATS image processing system, in Proc. symp. on machine processing of remotely sensed data: Purdue Univ., West Lafayette, Indiana, p. 1A8-1A17.

Haralick, R.M., Krusemark, S., and Neikirk, K., 1981, GIPSY: introduction: Spatial Data Analysis Lab., Virginia Polytechnical Institute and State University, Report SDA 81-4, 74 p.

Kasvand, T., 1983, Computerized vision for the geologist: Jour. Math. Geology, v. 15, no. 1, p. 3-23.

Kasvand, T., and Fabbri, A.G., 1984, Considerations on pyramidal pipelines for spatial analysis of geoscience data, in Freeman, H., and Pieroni, G., eds., Computer architectures for spatially distributed data: Proc. NATO ASI, Cetraro, Italy, Springer-Verlag, New York, in press.

Lantuejoul, C., 1981, An image analyser, in Duff, M.J.B., and Levialdi, S., eds., Languages and architectures for image processing: Academic Press, New York, p. 165-177.

Levialdi, S., 1983, Languages for image processing, in Faugeras, O.D., ed., Fundamentals in computer vision: Cambridge Univ. Press, Cambridge, p. 460-478.

Lin, C., 1982a, Microgeometry I: autocorrelation and rock microstructure: Jour. Math. Geology, v. 14, no. 4, p. 343-360.

Lin, C., 1982b, Microgeometry II: testing for homogeneity in Berea Sandstone: Jour. Math. Geology, v. 14, no. 4, p. 361-370.

Lundgren, K., Antonsson, D., Arvidson, J., and Grandlund, G., 1981, GOP, a two stage microprogrammable pipelined image processor: Proc. of 2nd Scandinavian Conf. on image analysis, Helsinki, Finland (June 15-17, 1981), p. 408-414.

Mainwaring, P. R., and Petruk, W., 1984, SEM-based automatic image analysis and its use in mineral beneficiation-a review: Proc. of Scanning Electron Microscopy Symp. SEM'84 (Philadelphia), preprint 20 p.

Matheron, G., 1972, Elements pour une theorie des milieux poreux: Masson & Cie., Paris, 166 p.

Miller, P.R., Reid, A.F., and Zuiderwyk, M.A., 1982, QEM*SEM image analysis in the determination of modal essays, mineral association and mineral liberation: Proc. XIV Int. Mineral Processing Congr., Toronto (Oct. 17-23, 1982), Paper VIII-3.

Moore, G., 1968, Automatic scanning and computer processes for the quantitative analysis of micrographs and equivalent subjects, in Cheng, G.C., Ledley, R.S., Pollock, D.K.,and Rosenfeld, A., eds., Pictorial pattern recognition: Thompson Book Co., Washington, D.C., 521 p.

Muller, W., 1974, The Leitz-Texture Analysing System (Leitz-TAS): Leitz Scientific and Technical Information Suppl. 1, v. 4, p. 101-116.

Nguyen, P.T., and Quinqueton, J., 1982, Space filling curves and texture analysis: Proc. 6th Intern. Conf. on pattern recognition, Munich, Germany (Oct. 19-22, 1982), p. 282-285.

Pincus, H.J., and Dobrin, M.B., 1966, Geological applications of optical data processing: Jour. Geophys. Res., v. 71, no. 20, p. 4861-4869.

Pong, T.C., Haralick R.M., Carig, J.R., Yoon, R.H., and Choi, W.Z., 1983, The application of image analysis techniques to mineral processing: Pattern Recognition Letters, v. 2, p. 117-123.

Preston, F.W., and Davis, J.C., 1976, Sedimentary porous material as a realization of stochastic processes, in Merriam, D.F., ed., Random processes in geology: Springer-Verlag, New York, p. 63-86.

Reid, A.F., Gottlieb, P., MacDonald, K.J., and Miller, P.R., 1984, QEM*SEM image analysis of ore minerals: volume fraction, liberation, and observation variances: Proc. ICAM, Intern. Congr. on Applied Mineralogy in the Mineral Industry, (Los Angeles, California), in press.

Rosenfeld, A., 1983, Quadtrees and pyramids: hierarchical representation of images, in Haralick, R.M., ed., Pictorial data analysis: NATO ASI Series, v. F4, Springer-Verlag, Berlin, p. 29-42.

Rosenfeld, A., 1984, Picture processing 1983: Computer Vision, Graphics and Image Processing, v. 26, p. 347-393.

Rosenfeld, A., Samet, H., Shaffer, C., and Webber, R.E., 1982, Application of hierarchical data structures to geographical information systems: Computer Vision Laboratory, Univ. Maryland Tech. Rept. TR-1197, 160 p.

Samet, H., 1983, Using quadtrees to represent spatial data, in Freeman, H., and Pieroni, G., eds., Computer architectures for spatially distributed data: Proc. NATO ASI, Cetraro, Italy (June 6-17, 1983), Springer-Verlag, New York, in press.

Serra, J., 1966, Remarques sur une lame mince de minerai lorrain: B.R.G.M. Bull., v. 6, p. 1-36.

Serra, J., 1972, Morphologie mathematique, in Laffitte, P., ed., Traite' D' Informatique Geologique: Masson et Cie. Editeurs, Paris, p. 194-238.

Serra, J., 1974, Theoretical bases of the Leitz-Texture-Analysing-System: Leitz Scientific and Technical Information Suppl. 1, v. 4, p. 125-136.

Serra, J., 1982a, Image analysis and mathematical morphology: Academic Press, New York, 610 p.

Serra, J., 1982b, Les filtres morphologiques: Centre de Geostatistique et de Morphologie Matematique, Cahier N-744, 17 p.

Serra, J., 1983, Quelques semi-group des filtrages morphologiques: Centre de Geostatistique et de Morphologie Mathematique, Cahier N-807, 30 p.

Srivastava, G.S., 1977, Optical processing of structural contour maps: Jour. Math. Geology, v. 9, no. 1, p. 3-38.

Sternberg, S.R., 1979, Parallel architectures for image processing: Proc. IEEE Computer Society's 3rd Intern. Computer Software and Application Conference, p. 1-6.

Sternberg, S.R., 1980, Languages and architectures for parallel image processing, in Gelsema, E.S., and Kanal, L.N., eds., Pattern recognition in practice: North Holland, Amsterdam, p. 35-44.

Sternberg, S.R., 1982, Pipeline architectures for image processing, in Preston, K., Jr., and Uhr, L., eds., Multicomputers for image processing: Academic Press, London, p. 291-306.

Switzer, P., 1975, Estimation of the accuracy of qualitative maps, in Davis, J.C., and McCullach, M.J., eds., Display and analysis of spatial data: NATO ASI, Nottingham, England, John Wiley & Sons, London, p. 1-13.

Switzer, P., 1976, Applications of random process models to the description of spatial distributions of qualitative geologic variables, in Merriam, D.F., ed., Random processes in geology: Springer-Verlag, New York, p. 124-134.

Switzer, P., Kowalik, W.S., and Lyon, R.J.P., 1982, A prior probability method for smoothing discriminant analysis classification maps: Jour. Math. Geology, v. 14, no. 5, p. 433-443.

Mapping Applications for the Microcomputer

A. C. Olson

Magnum Computer Systems, Inc.

ABSTRACT

This paper discusses the hardware capabilities required for the implementation of a successful computer-generated mapping system and relates these requirements to the capabilities of presently available 16-bit microprocessors. The mapping software installed on the microcomputer is GeoMetriX. This application software is suitable for exploration, resource analysis, civil engineering, and mine planning. The hardware requirements addressed are color, screen resolution, memory, disk storage, 16-bit bus, and arithmetic coprocessor, as well as printers, plotters, and digitizers. In addition, the current state of microprocessor technology is described briefly. Common operating systems are discussed for their influence on software transportability. The resulting hardware configuration is presented as a solution for microcomputer-based mapping systems.

INTRODUCTION

The advent of the microcomputer during the mid-1970s led to the downloading of many computer applications to the personal computer or individual workstation level. Initially, these applications consisted of text processing, spread sheet, and small database-management programs. Attempts to implement mapping application software on these machines were impeded by the constraints of 8-bit architecture.

Mapping applications require large amounts of addressable memory for the storage of the large two-dimensional arrays needed to process cartographic data. This also imposes a requirement for sufficient processing power to perform the computations on these large data volumes and to display the map data in the most timely manner. The capabilities of 8-bit central processing units (CPU) were overwhelmed by these requirements. Not only are 8-bit microprocessors limited to an addressable memory space of 64K bytes, but performance-wise, they are slow.

It was not until 16-bit microprocessors were introduced that implementation of mapping software became feasible. These 16-bit microprocessors are faster, their computational speeds can be accelerated with the addition of an arithmetic coprocessor, and they are able to address 1024K bytes of memory. At this time, 16-bit machines provide sufficient computing power to allow the implementation of sophisticated mapping software.

This paper discusses the special problems posed by conversion of a large mapping program from the mainframe environment to microcomputer operation. It also addresses the microcomputer architecture and specifications necessary to satisfy mapping-application requirements.

BACKGROUND

To best understand the factors that influenced the selection of the microcomputer hardware and support software for mapping applications, it is useful to examine the features contained in a typical software package. The mapping software to be implemented on the microcomputer is termed GeoMetriX (GMX). GMX is a set of interactive graphics programs designed to aid geologists and engineers in performing volumetric calculations and mapping applications on a microcomputer. It uses the same time-proven algorithms that are contained in mainframe versions of this mapping system. Enhancements have been added to improve speed and accuracy in the microcomputer environment.

The GMX software supports the following interactive functions:

- Grid-value interpolation,
- Grid-value manipulation,
- Trend-surface generation,
- Point-data editing,
- Window (zoom) capability,
- Boundary definition,
- Logical boundary combination,
- Contour-map generation,
- Resource-map generation,
- Area calculations, and
- Volume calculations.

Grid-value interpolation creates an interpolated set of grid-node values based on the spatial distribution of measurements taken at irregularly spaced observation points. These measurements may include data such as structure elevations, bed thickness, chemical concentrations, or geophysical information associated with a pair of location coordinates. Because the measured-point data are spaced irregularly, it is difficult to use these points for resource estimates, engineering calculations, or mapping applications without further processing. This processing transforms the information into a format that is more manageable and more descriptive of the spatial characteristics that the point-data sample is intended to represent.

This data transformation is accomplished by superposing an imaginary square grid system over the map area. The origin of the map coordinate system must be identified and the grid interval must be specified. The horizontal and vertical dimensions of the map also must be provided. The system then processes the observation point data to derive an interpolated value for each of the grid nodes.

Two interpolation algorithms are available, one linear and the other quadratic. The density and the uniformity of the distribution of the observation points determines which algorithm is to be used. Both interpolation algorithms are simple computationally, thus minimizing the demand for computer resources. At the same time, the results are sufficiently accurate to produce a satisfactory map product. The influence of the observation data is weighted by the inverse exponential of the distance of the observation point from the grid node undergoing interpolation.

A gridded data set is required to perform the contouring and the volumetric functions discussed later.

Grid-value manipulation allows the modification of one set of gridded data by another set of gridded data. The four arithmetic functions (add, subtract, multiply, and divide) are used to operate on corresponding grid-node values from two different sets of measurements for the same area. This permits, for example, the subtraction of a structure elevation from a topographic elevation to obtain the distribution of overburden, the division of the overburden values by the corresponding thickness values to obtain overburden-to-thickness ratios, or the multiplication of the deposit thickness by the concentration of a mineral to obtain, through volumetric calculations, the total amount of that resource in the deposit.

Trend-surface generation allows the least-squares fit of a two-dimensional polynomial or trigonometric surface (or a combination of the two) to provide a surface that depicts the theoretical characteristics of the spatial distribution of the observed measurements. The residuals, that is the deviation of the measured values from the theoretical trend value, can be used to identify anomalous regions where the data deviates from the general trend. If the residuals exceed the theoretical values by a significant amount, for example twice the root-mean-square (RMS) error of the theoretical-surface fit, then there may be sufficient significance to warrant further investigation. The trend-surface option also is used effectively to check the entered data for accuracy. Those points with values that exceed three times the RMS error indeed may be values that were entered incorrectly into the data file.

Window (zoom) allows the user to enlarge a subarea of the map in the view area of the screen. The user may identify the rectangular area to be enlarged either through numeric entry of the corner coordinates or by designating diametrically opposed corner points with the cursor. The user may return to the original display scale at any time. If the user elects, the selected subarea may be directed to the pen plotter for hardcopy output.

Point-data editing permits the user to display, identify, and interactively change these data by the cursor. To delete, or change existing data values, the cursor first is placed over the location of the point to be edited and the "D" (for delete) or the "C" (for change) key is struck. A large X with a circle around it is placed over the deleted value, while a prompt appears in the left-hand display area for a new change value. Once the new value is entered, a circle appears about the point to show that it has been changed. New points also may be added with the screen cursor but the locational accuracy is not as precise as when points are added by a digitizer. GMX permits both a measured value and a label identifier to be input at the time that a point is added or changed.

Boundary definition is a flexible method of defining bounded areas by sets of short line segments that are known as chains. The algorithm allows boundaries to be "sliced" to reduce the size of the boundary sets for the purpose of testing other line data for intersections with the boundary, or to determine if a data point is contained within the bounded area. This technique allows large sets of boundary chains to be processed without placing an excessive computational burden on the system.

Outcrop, political, and ownership boundaries can be added to the database by a digitizer. Boundaries also may be created by using the cursor at the graphics terminal. These boundaries may be used to limit the extent of a volumetric calculation, or to suppress the plotting of contours and other graphics information occurring outside the boundary.

Logical boundary combination allows the Boolean combination (union, intersection, and not) of any pair of bounded areas to form new regions. For example, given the polygonal region A (Fig. 1A) and the polygonal region B (Fig. 1B), GMX can compute rapidly the union ("or") of these two regions (Fig. 1C) and the intersection ("and") (Fig. 1D), as well as region A and not region B (Fig. 1E) and region B and not region A (Fig. 1F).

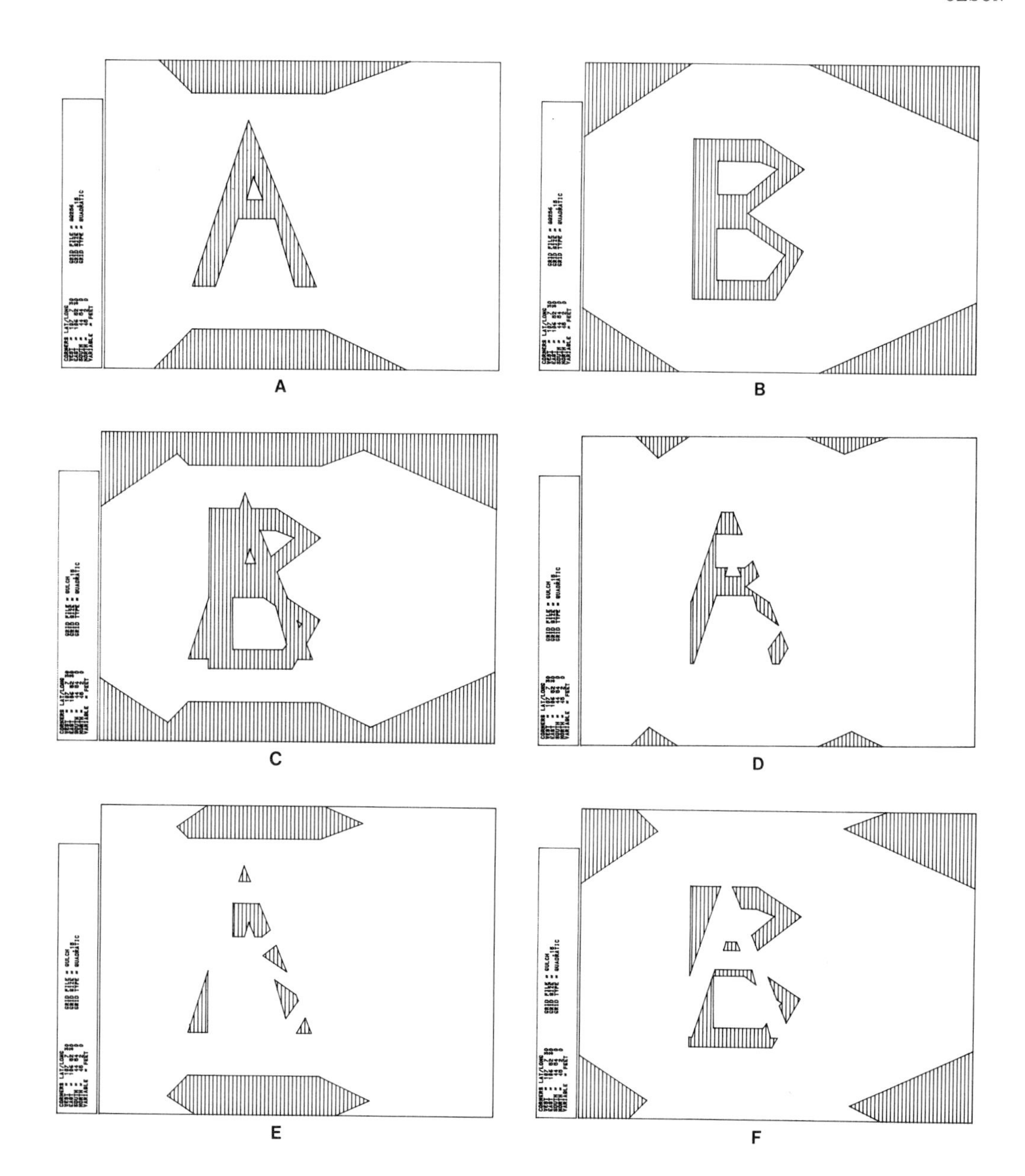

Figure 1. A Polygonal region A; B polygonal region B; C region A or region B; D region A and region B; E region A and not region B; and F region B and not region A.

A more complex example shows the delineation of a resource whose thickness lies between 5 and 10 feet (Fig. 2A) and a lease tract (Fig. 2B). The intersection of these two areas shows all of the resource contained within the lease tract and having a thickness between 5 and 10 feet (Fig. 2C). This resultant area is used to delineate the isopachous map of the resource (Fig. 3A) and the resource map itself (Fig. 3B).

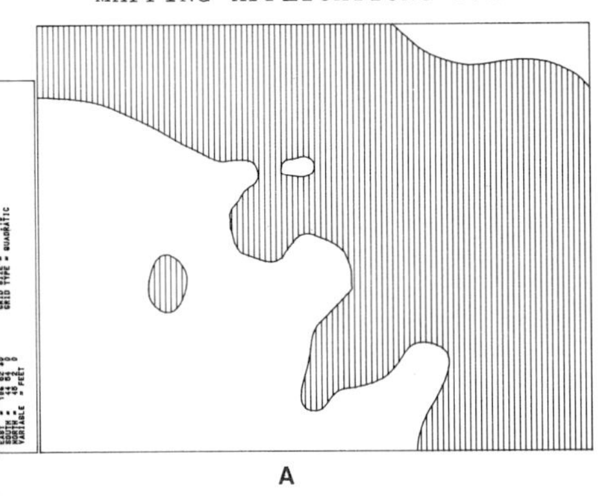

A

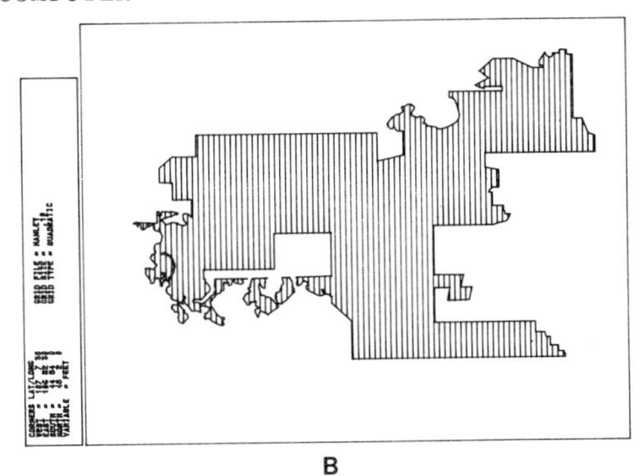

B

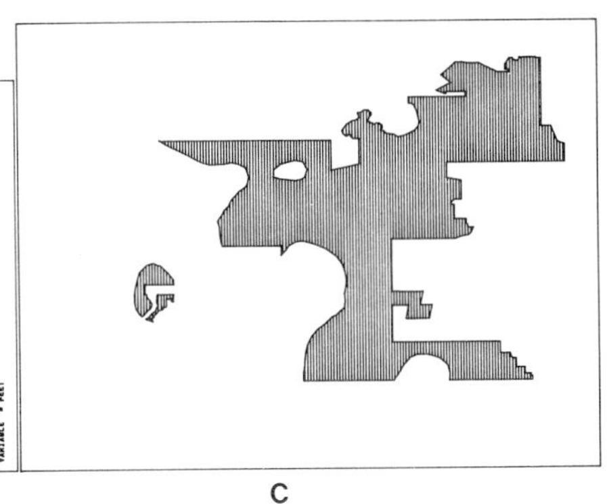

C

Figure 2. A Resource thickness greater than 14 inches; B lease tract boundary; and C delineation of resource greater than 14 inches and lying within lease tract.

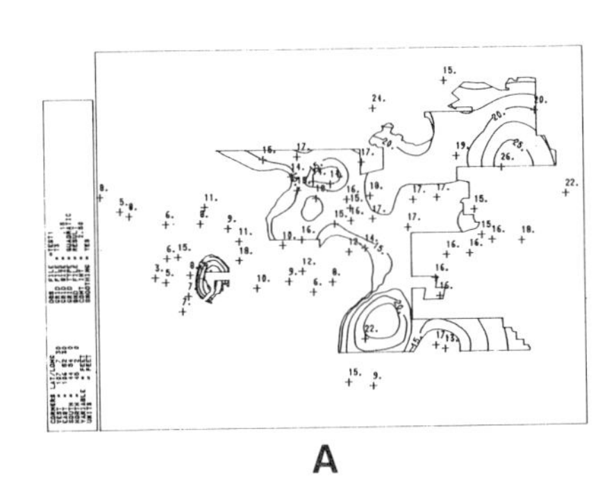

A

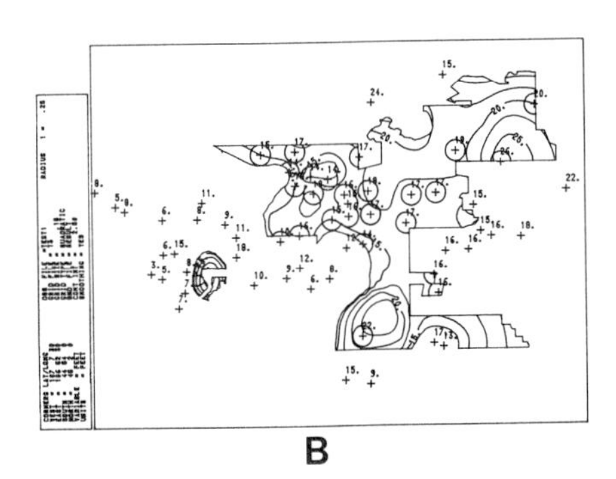

B

Figure 3. A Isopachous map of resource lying within boundary from Figure 1; and B resource map (with isopach) using boundary from Figure 1.

Contour-map generation produces a set of prompts to be filled in by the user. These prompts include name of the point-data file, if an overlay of the point data is desired. This includes the option of plotting the corresponding data values, the point labels, or both. The user then selects the gridded data file to be contoured, and is prompted for the contour spacing, frequency of bold contours, and any boundary files to be overlaid.

The first boundary file specified also may be used, at the user's discretion, to clip the contour lines at the boundary. If the dashing option is selected, the contours are dashed if they extend beyond the boundary. The user may elect to continue the contours as solid lines outside the boundary.

Resource-map generation produces a contour map, as discussed, plus an additional overlay with the resource circles plotted around the observation points. The radii of these resource circles may be designated by the user or he may select standard distances such as the U.S. Geological Survey reliability categories (Wood and others, 1983) for use in measuring bedded deposits such as coal.

Volume calculations are performed by using a gridded data file that contains the thickness of the stratigraphic deposit. The calculations may be constrained by a user-specified deposit. In addition, if the user has specified the radii of the reliability circles, the volumes are computed for each of the specified reliability categories. If the user supplies the density factor, the volumes may be converted to tonnages.

Area calculations similar to the volume calculations discussed also may be constrained by a user-specified boundary. The calculations may be performed separately for each of the reliability categories.

PURPOSE

The task was to convert this suite of mapping software onto a microcomputer system in order to offer an affordable alternative to those who do not have access to mainframes. In addition, to produce a system that could be transported into the field and that could be used at any location with an electrical power source. This would allow exploration teams, for example, to examine the results of their work before quitting the location.

The first step was to evaluate available microcomputer equipment to determine what best met our requirements. We were looking for an architecture that provided, as a minimum:

- Color graphics,
- Good resolution,
- A minimum of 640K bytes of memory,
- Arithmetic coprocessor,
- High-capacity floppy disk storage,
- An easy-to-use operating system, and
- Compatibility with other microcomputers.

The next step was to obtain the hardware and necessary support software (operating system, compiler, editor) and to implement the system.

MICROPROCESSOR TECHNOLOGY

The earlier 8-bit systems normally incorporated one of the following popular chips for the central processing unit (CPU):

- Zilog Z-80,
- Intel 8080,
- Intel 8085, or
- Motorola 6800.

Each of these chips uses a 16-bit (two byte) addressing scheme. This limits the addressable memory to 64K bytes (two raised to the power of sixteen). In addition, they use an 8-bit bus structure to transfer data internally between the memory and the CPU registers. Furthermore, the clock rate for the 8-bit chips is slower than the 16-bit chips, usually 2 to 2 1/2 MHz as compared to 4 to 8 MHz. This clock rate affects the speed at which the computer operations are performed in the CPU registers.

The currently popular 16-bit systems are based normally on one of the following CPU chips:

- Zilog Z8001,
- Zilog Z8002,
- DEC LSI-11,
- Texas Instruments 9900,
- Motorola 68000,
- National Semiconductor 16000,
- Intel 8088,
- Intel 8086, or
- Intel 80186.

Advantages of these 16-bit chips are: (1) they operate at a higher clock rate (4 MHz or greater) to gain almost twice the computing speed to the 8-bit chips; (2) except for the Intel 8088, they all have a 16-bit bus that doubles the internal data-transfer rate to that of the 8-bit bus; and (3) most use a 20-bit (or greater) addressing scheme to access 1M byte of memory. In addition, most of the 16-bit microprocessors now available are register oriented. Thus, it is easier to manipulate data when they are stored in registers, rather than when they are stored in memory.

This discussion will focus on the Intel 8086 and the Intel 8088 chips because these two CPU chips are used generally in the popular brands of personal computers and because of the enormous amount of support (both hardware and software) that Intel and its second sources provide for these chips. The 8086 was first introduced in early 1978. It has fourteen 16-bit registers, eight of which are general purpose. It also has a 16-bit internal bus.

The 8088 chip is identical to the 8086 internally, but uses an 8-bit data bus. IBM selected this chip for use in their PC, based on the premise that the 8-bit bus would allow it to execute much of the already existing 8-bit software. This decision also allows the use of existing, inexpensive 8-bit chips for memory and for I/O.

The first software for the 16-bit microcomputers, for the most part, was conversions of popular software already available on the 8-bit machines. Although register structures differ between the 8080/8085/Z-80 series and the 8086/8088 series chips, existing 8-bit software initially was converted by direct translation from the 8-bit to the 16-bit instruction sets. This was the most expedient method of converting existing software for immediate market availability. The 8086/8088 registers, however, are optimized for specific processing functions. Placing data in nonoptimal registers allows the function to be performed, but efficiency may be far less than that of the 8-bit devices.

Recent months have seen the recoding and enhancing of existing software packages so that they indeed take advantage of the technology utilized for the 8086/8088 chips. Care must be taken, though, to ensure that a commercially available software package is indeed the most efficient version for use with the 16-bit microprocessor.

The Intel 8086/8088 CPU chips allow the addition of the Intel 8087 arithmetic coprocessor to speed the arithmetic computational process. The 8087 monitors the 8086-instruction stream looking for its own instructions and executes them without any help from the 8086. This is not just a support chip, but it is a complete and powerful microprocessor designed to perform arithmetic operations up to 500 times as fast as they could be processed normally by the CPU alone. Because computer software is required to perform many other functions besides arithmetic operations, this does not indicate that there is an overall improvement of 500 in computing speed. Only the arithmetic operations are being performed at the higher rate. For mapping operations, experience has shown an improvement of 3 to 4 times that of the CPU alone.

It must be noted that although most 16-bit systems have an addressable memory space of 1024K bytes, usually only 640K bytes are reserved for the high-resolution, color-bit map that is used to store the digital image of the screen.

The microcomputer samples this bit map during each cycle and writes the pixels out on the screen display according to how the graphics program has set them in the bit map. Thus, graphics data can be erased and new data can be displayed without completely erasing and rewriting the screen each time a change is made. This capability allows the microcomputer to display graphics images in a dynamic manner, although the microprocessor cannot manipulate complex images as rapidly as more sophisticated graphics terminals.

EVALUATION RESULTS

The NEC APC was selected as the target system for the software conversion. The requirements are met by the following architecture:

- Eight-color graphics screen;
- Resolution is 640 by 480 pixels;
- Maximum memory of 640K bytes;
- CPU is the Intel 8086
  (16-bit registers, 16-bit bus);
- Intel 8087 arithmetic coprocessor;
- Two 8-inch floppy disk drives (1M byte each);
- MS-DOS operating system; and
- Compatible with other systems using
  MS-DOS or PC-DOS.

In addition, the following peripheral devices for the input and output of cartographic and textual data have been added:

- Dot-matrix printer,
- Pen plotter, and
- Digitizer.

GMX uses the dot-matrix printer to produce a printed copy of the data displayed on the graphics screen (Fig. 4). Although not of publication quality, the printer is useful for providing a rapid hardcopy record of the plotted data. A small plotter with a 10.6 inch by 15.75 inch plotting area is attached to the microcomputer. Larger plotters also are available. A digitizer tablet is provided for entry of point and line data. Digitizers are available in a variety of sizes.

In order to execute application software on a microcomputer, an operating system is required. The operating system performs the system-management functions such as assigning input and output to specific computer ports so that information may be transferred between the CPU and the peripheral devices that are connected to these ports. Operating systems also include a number of utility programs that assist in file management, set up communications protocols, perform line editing activities, and so on.

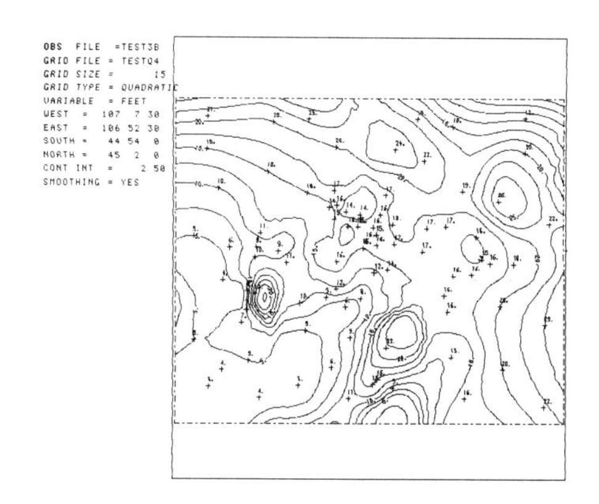

Figure 4. Example of screen display output from dot-matrix printer.

There are several excellent operating systems available for the current generation of microcomputers. These include:

- MS-DOS by Microsoft;
- CP/M and CP/M-86 by Digital Research; and
- UNIX by Bell Laboratories.

CP/M was one of the first operating systems available for the 8-bit machines and was popular on that generation of microcomputers. When the first 16-bit machines appeared, CP/M was translated to run on these new architectures. CP/M-86, however, was the first version to be optimized for the 16-bit CPUs. MS-DOS gained popularity when Microsoft adaped it to the 8088 for use with the IBM PC. The IBM version of MS-DOS is known as PC-DOS. With only slight difficulty, software written under either version will run on the sister systems.

CP/M and MS-DOS both have similar features, making it possible for the user to operate either system with minimal dislocation. UNIX, and some of the microcomputer subsets such as Zenix, are even more powerful operating systems, but they require more memory for their operation. As microcomputers continue to become more powerful, UNIX and its derivatives will gain in popularity.

MS-DOS was selected for the mapping system because of its immense popularity and because of the compatibility that it provides with the IBM PCs and the PC clones. MS-DOS is available for almost every 16-bit system on the market and allows the diskettes created on one system to be read on a different system. This does not guarantee universal compatibility of software, however, because each manufacturer has different hardware features built into the system, such as sound ports and graphics bit maps. The application programs that use these features must be adapted specially for each machine. Operating system compatibility, however, does allow most general purpose functions to be interchanged without special adaptation.

Subsequent to the selection of the NEC APC for a mapping computer, a number of other microcomputers meeting, or close to meeting, our selection criteria came on the market. Those significantly meeting our requirements include the Tandy 2000 and the Zenith/Heathkit Z-100. The Texas Instruments Professional Computer, the Columbia 1600, and the IBM PC and PC-XT can be configured to execute this mapping software, but have a lower display resolution.

The GRiD Compass is a small briefcase computer with monochrome resolution of 640 by 480 pixels. It, too, can operate under MS-DOS and can provide sufficient memory and graphics capability to operate mapping software and to drive the peripheral graphics devices.

The graphics-display capability of the IBM PC is only 640 by 200 pixels. However, there are third-party suppliers who sell a color graphics board with a resolution of 640 by 400 pixels. IBM has announced recently that they will be producing an enhanced graphics board with 640 by 400 pixel resolution color graphics.

OBSERVATIONS

The task of implementing the GMX software on this hardware system was more difficult to accomplish than first anticipated. The source code is written in FORTRAN. Until May 1984, there were no FORTRAN compilers available that allowed the use of more than 64K bytes of memory to be used to store the source code. GMX consists of nearly 15,000 lines of FORTRAN code, therefore, translating it into another high-level language was not a desirable alternative.

In addition, GMX contains some large arrays for internal storage of the gridded data. The arrays for storing boundary data also are large. The challenge was to reduce these array sizes to a point where the software would fit into the 640K bytes of memory while at the same time maintaining output. This meant reducing two large grid data arrays from 65,000 words each to 16,000 words each, and two 20,000 by 2 word-boundary arrays to two 8,000 by 2 word-boundary arrays. As you will note in the examples, this produces acceptable output.

The goal to produce a quality microcomputer mapping system that provides an acceptable alternative to a mainframe- or a mini-based system was achieved. Depending on the circumstances of the user and his willingness to trade off additional time to perform the mapping computations against the lower cost andthe convenience of having such a system at his fingertips, the microcomputer now provides a satisfactory alternative to the more expensive computer systems.

REFERENCE

Wood, G. H., Jr., Kehn, T. M., Carter, M. D., and Culbertson, W. C., 1983, Coal resource classification systems of the U. S. Geological Survey: U. S. Geological Survey Circ. 891, 65 p.

DIGITS: a Simple Digitizer System for Collecting and Processing Spatial Data using an Apple II

D. J. Unwin

University of Leicester

ABSTRACT

A simple data capture routine for the Apple II (TM) with Graphics Tablet Digitizer is listed and described. This allows geologists to collect location data on a variety of types of maps. It has proved useful in both teaching and research.

INTRODUCTION

Within the geosciences it may be necessary to collect pure locational data, such as the outline of a drainage basin, or similar data to which a third attribute, such as elevation, is added. Usually, but not always, these data will be collected as (x,y) pairs or (x,y,z) triplets in a plane Cartesian coordinate system and the number of items may be large.

Performed by hand, the collection of such data is both tedious and error prone but the task can be automated using some form of digitizer. Until recently, digitizers have been both expensive and difficult to use, giving as much work in editing and error correction as in the original data capture. They also have been used primarily in work, such as computer-assisted cartography, that necessarily involves high precision and resolution. However, a great amount of work in geoscience involves using similar data for which such precision, resolution, and complexity in use are not needed. Examples of this matter include collecting (x,y,z) triplets for input to contouring or trend-surface analyses, defining the two-dimensional shape of sedimentary particles, calculating down-river distances, determining drainage basin areas, and so on.

The advent of cheap microcomputers and robust, easy-to-use, tablet digitizers has meant that tasks such as these can be accomplished easily and quickly under direct computer control. This paper describes a simple digitizer computer program that has been useful in a wide variety of tasks. It is implemented in BASIC for the Apple II microcomputer equipped with a Graphics Tablet digitizer.

THE EQUIPMENT

The Apple Graphics Tablet is a robust tablet device (Fig. 1) that senses the (x,y) position of a switched, pen-like cursor on a relatively small digitizing surface. It communicates with the microprocessor by way of 'firmware' in the form of an interface board which is located in an input/output mapped

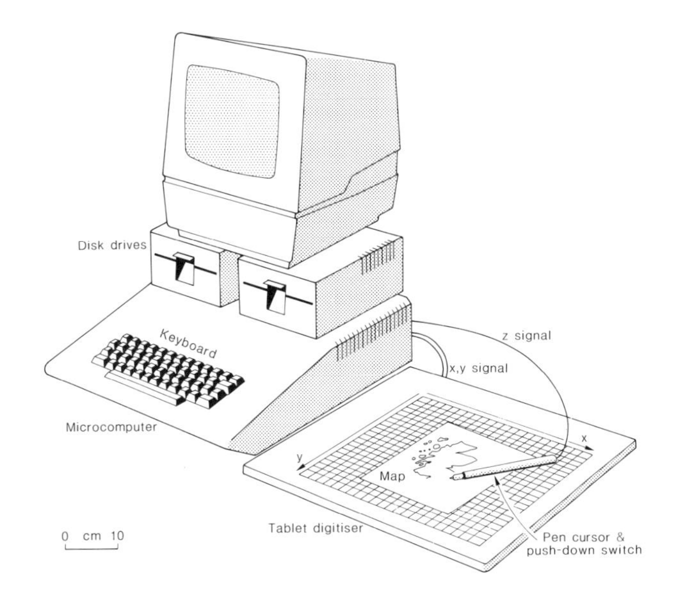

Figure 1. Apple Graphics Tablet configuration.

peripheral expansion slot. This makes the (x,y) data from the cursor available to the processor as triplets of the form

$$x, y, z$$

in which x and y are the locational coordinates in tablet units relative to an origin in the top left-hand and with y increasing down the tablet. The resolution is 0.1 mm in both x and y. The z value returned is a status flag that can be used to detect whether the pen has been pressed down.

Within Applesoft BASIC, these data are accessed using the pseudo-BASIC command

IN#n

(n = number of peripheral expansion slot)

after which any subsequent INPUT statement is assumed to be from the peripheral. The command PR#n has a similar function for outputs. The digitizer can be operated in one of several modes but in this application use is made of the simple point mode. Further details of the system is contained in the relevant manufacturer's manual and the system used also is described in some detail by Unwin and Dawson (1985, chapter 8).

THE PROGRAM

The BASIC program (listed in the Appendix) accesses the tablet and performs a limited number of operations. As listed, it is for a 48K machine running DOS 3.3, and it assumes that a printer is available for 'hard copy' together with one disk drive for data storage.

The program has the following features and facilities:

It is menu driven. Table 1 gives a list of the available commands and their effects. These are self-explanatory, and other than obvious restrictions that, for example, one cannot call DISTANCE if there are no data, items may be called in any order. A typical task might be to use these commands to determine the area of a drainage basin in km^2. This would involve calling DATA in order to input a string of (x,y) pairs defining the watershed, followed by ORIGIN to set an origin to the southwest of the basin and SCALE to indicate a known distance equivalent (say 10 km) on the tablet. Finally, the command AREA would return the required value.

Table 1. DIGITS: menu of available commands

Command	Effect
DATA	Captures x, y pairs or, optionally, x, y, z triplets from digitizer or digitizer and keyboard, or reads a data file from disk.
ORIGIN	Sets origin (0,0) at indicated point and reverses sense of y values.
TRANSLATE	Translates origin from (0,0) to any specified (x,y) values and modifies data accordingly.
SCALE	Scales both x and y axes to user-indicated values using known distance on tablet.
LIST	Lists entire data file either on monitor or as hard copy on printer. Note that printer is not essential to this function.
FILE	Stores current (x, y, z) data on disk using system conventions for naming.
AREA	Determines area of polygon enclosed by current set of x,y points considered as vertices of that polygon. Polygon is closed by software and result is delivered in whatever units are set by SCALE.
DISTANCE	Determines total distance along line given by current set of x, y points, again in whatever units are in force.
FINISH	Ends job.
TEST	Loads simple test data set for 1-inch square roughly in center of tablet.
COMMANDS	Lists available commands.

It should be noted that the captured data are held in the arrays A, B and C, dimensioned to consist of 500 elements and that the attribute data held in C are input from the keyboard. When collecting simple x,y pairs, these z coordinates are all set at unity.

It has been written paying particular attention to 'top down' modularity principles, so that its structure can be best appreciated by reference to Table 2. There are four distinct levels of subroutines involved with a rigorous functional separation between them. Level 1 (100-300) sets up initial parameters whereas Level 2 (1000-2750), which is called from Level 1, displays initial messages and allows the user to select peripheral slots appropriate to their system. Also within Level 2 are routines that create the menu options and capture individual user commands. Level 3 subroutines (6000-16020) form the heart of the program and are called from Level 2. As can be seen from an examination of Table 2, there is one subroutine at this level for each command in the menu. Finally, at Level 4 (20000-27040) there are a large number of short subroutines to perform standard tasks.

Table 2. DIGITS: program structures

Level 1 (100-300)	Level 2 (1000-2750)	Level 3 (6000-16020)	Level 4 (200000-27040)
Main	Message (1000-1140)	Data collection (6000-6740)	Yes/no routine (20000-20100)
	Set controls (1500-1580)	Set origin (7000-7210)	Collect data from disc (21000-21180)
	Set up menu (2000-2130)	Scale axes (8000-8320)	Initialize tablet (22000-22040)
	Get a command (2500-2750)	List data (9000-9290)	Capture on x,y pair (23000-23060)
		File data on disc (10000-10190)	Switch to keyboard for input (24000-24020)
		List commands (11000-11190)	Clear keyboard strobe (24500-24520)
		Find area (12000-12150)	Switch output to screen (25000-25010)
		Find distance (13000-13110)	Switch to printer for output (25500-25510)
		Load test data (14000-14150)	Pause while listing on screen (26000-26060)
		Translate origin (15000-15120)	'Beep' when a point is captured (27000-27040)
		Finish run (16000-16020)	

As presented, DIGITS clearly is dependent on both the hardware and software of the Apple II, but its overall structure would enable fairly easy conversion to other systems by changes to the disk-filing routines (1550,10000-10190, and 21000-21180) and a series of short routines in Level 4 (22000-25510, 27000-27040) that address the peripheral slots for both tablet and printer. Note that some of these facilities could be omitted in their entirety and that the attached printer is assumed to be of the Silentype variety. Other printers certainly will need some minor changes in 25500-25510.

DIGITS has an in-built capacity for expansion to include other functions in Levels 3 and 4. To add functions, for example, to perform statistical analyses on the collected data, all that is needed is to increase the number of commands (1570), add an appropriate name to the menu and driving routines (2120, 2740), and supply the desired code.

CONCLUSIONS: USES AND LIMITATIONS

DIGITS began life as a simpler program designed to give students 'hands on' experience with digitizer use and only subsequently was used in research for elementary data collection, and, area and distance measurement. It must be stressed that it is not, and never was intended to be, a complete digitizer system capable of collecting data for automated cartography where high resolution and repeatability are required. There is, for example, no provision to rotate coordinate axes to correct for distortion in the source document nor is it possible easily to link one set of data to another so that the small size of the tablet becomes a major constraint. It, however, has been of considerable research use.

REFERENCE

Unwin, D.J., and Dawson, J.A., 1985, Computer programming for geographers: Longman, London and New York, 246 p.

AVAILABILITY

Bona-fide research workers and teachers in public-sector-funded institutions can obtain a copy of DIGITS at no cost, but I would ask that they send a ready formatted disk together with a correctly addressed and stamped packet for its return. This note should serve as a suitable manual for its use and I do not offer any guarantees as to its suitability or accuracy in use. The address is: D. J. Unwin, Department of Geography, University of Leicester, University Road, Leicester LE1 7RH, U.K.

APPENDIX

Listing of Program

```
1  REM  VERSION OF 26.9.84
100  REM  *** DIGITS ***    LEVEL 1
110  REM  A DIGITISER PROGRAM FOR X,Y,Z
120  REM  DATA AND APPLE II WITH GRAPHICS
130  REM  TABLET .
140  REM  ----------------------------
150  REM  AUTHOR : D.J.UNWIN
160  REM  SOURCE : UNIVERSITY OF LEICESTER , UK.
170  REM  -----------------------------------
180  DIM P$(15),A(500),B(500),C(500)
190  DEF  FN A(X) =  INT (X * 1E3 + 0.5) / 1E3
200  GOSUB 1000: REM  INITIAL MESSAGE
210  GOSUB 1500: REM  INITIAL CONTROLS
220  PRINT "A LIST OF AVAILABLE COMMANDS"
230  PRINT "MAY BE OBTAINED BY TYPING"
240  PRINT "<COMMANDS"
250  LET D$ = " UNITS": REM  DEFAULT
260  PRINT "TO BEGIN ENTERING DATA"
270  PRINT : PRINT "TYPE <DATA"
280  GOSUB 2000: REM  SET UP MENU
290  GOSUB 2500: REM  GET COMMAND
300  END
1000  REM  *** INITIAL MESSAGE ***LEVEL 2
1010  HOME : PRINT : PRINT : PRINT
1020  PRINT "DIGITISER PROGRAM FOR APPLE"
1030  PRINT "==========================="
1040  PRINT
1050  PRINT "DIGITS ALLOWS YOU TO COLLECT"
1060  PRINT "X,Y OR X,Y,Z DATA USING THE"
1070  PRINT "GRAPHICS TABLET"
1080  PRINT
1090  PRINT "NOTE THAT THE TABLET GIVES"
1100  PRINT "CO-ORDINATES IN ITS OWN UNITS"
1110  PRINT "RELATIVE TO AN ORIGIN IN THE"
1120  PRINT "TOP LEFT HAND CORNER"
1130  PRINT
1140  RETURN
1500  REM  *** CONTROLS *** LEVEL 2
1510  PRINT " WHICH SLOT IS TABLET IN (0-6) ";
1520  INPUT N2
1525  PRINT
1530  PRINT " WHICH SLOT IS PRINTER IN"
1535  PRINT " TYPE 0 IF NO PRINTER .. ";
1540  INPUT N3
1550  LET F$ =  CHR$ (4): REM  APPLE DOS
1560  PRINT : PRINT : PRINT
1570  LET N5 = 11: REM  NO COMMANDS
1580  RETURN
2000  REM  *** COMMANDS *** LEVEL 2
2010  LET P$(1) = "DATA"
2020  LET P$(2) = "ORIGIN"
2030  LET P$(3) = "SCALE"
2040  LET P$(4) = "LIST"
2050  LET P$(5) = "FILE"
2060  LET P$(6) = "COMMANDS"
2070  LET P$(7) = "AREA"
2090  LET P$(8) = "DISTANCE"
2100  LET P$(9) = "TEST"
2110  LET P$(10) = "TRANSLATE"
2120  LET P$(11) = "FINISH"
2130  RETURN
```

```
2500  REM  ***CAPTURE A COMMAND ***
2510  REM                          LEVEL 2
2530  PRINT : PRINT : PRINT : PRINT
2540  PRINT "WHICH COMMAND";
2550  INPUT C$
2560  FOR I = 1 TO N5
2570  IF C$ = P$(I) THEN 2630
2580  NEXT I
2590  PRINT "COMMAND NOT RECOGNISED"
2600  GOSUB 11000: REM  LIST COMMANDS
2610  GOTO 2530
2620  REM  --- DECODE AND ACT
2630  REM   ---- THEN RETURN AT THIS LEVEL
2640  IF I = 1 THEN  GOSUB 6000: GOTO 2510
2650  IF I = 2 THEN  GOSUB 7000: GOTO 2510
2660  IF I = 3 THEN  GOSUB 8000: GOTO 2510
2670  IF I = 4 THEN  GOSUB 9000: GOTO 2510
2680  IF I = 5 THEN  GOSUB 10000: GOTO 2510
2690  IF I = 6 THEN  GOSUB 11000: GOTO 2510
2700  IF I = 7 THEN  GOSUB 12000: GOTO 2510
2710  IF I = 8 THEN  GOSUB 13000: GOTO 2510
2720  IF I = 9 THEN  GOSUB 14000: GOTO 2510
2730  IF I = 10 THEN  GOSUB 15000: GOTO 2510
2740  IF I = 11 THEN  GOSUB 16000
2750  RETURN
6000  REM  *** DATA COLLECTION *** LEVEL 3
6010  HOME : LET D$ = " UNITS"
6020  PRINT "DATA ENTRY ROUTINE"
6030  PRINT "=================="
6040  PRINT : PRINT "READS A DATA FILE FROM DISK OR CAPTURES DATA FROM DIGITISER
 "
6060  PRINT
6070  PRINT " ARE THE DATA ON DISK ";
6080  GOSUB 20000: REM  YES NO ROUTINE
6090  IF T = 0 THEN 6140
6100  GOSUB 21000
6110  PRINT : PRINT " DATA READ : CHECK USING LIST"
6130  GOTO 6740
6140  REM  ** ROUTINE TO COLLECT DATA FROM DIGITISER **
6150  PRINT
6160  PRINT " YOU MUST PROVIDE X,Y,Z DATA"
6170  PRINT " FROM THE DIGITISER AND KEYBOARD"
6180  PRINT
6190  LET Z1 = 0
6200  PRINT
6210  PRINT " DO YOU WANT TO TYPE IN"
6220  PRINT " A Z FOR EACH POINT ";
6230  GOSUB 20000
6240  PRINT
6250  PRINT " INDICATE YOUR POINTS USING THE PEN"
6255  PRINT "SIGNAL END BY PRESSING ANY KEY "
6260  IF T = 1 THEN Z1 = 1
6270  REM  ** INITIALISE TABLET **
6280  GOSUB 22000
6290  LET I = 0
```

```
6300  REM  ** CAPTURE A POINT **
6310  GOSUB 23000
6320  REM  ** REM VALIDITY CHECK **
6330  IF Z < 0 THEN 6670
6340  IF  ABS (Z) >  = 10 THEN  PRINT "OFF - SCALE "
6350  LET Z =  ABS (Z)
6360  IF Z >  = 10 THEN Z = Z - 10
6370  IF Z <  > 2 THEN 6310
6380  LET I = I + 1: REM  INCREASE COUNT
6390  LET A(I) = X
6400  LET B(I) = Y
6500  LET C(I) = 1: REM  SET Z AT ONE
6510  REM  ** CALL THE BEEP ROUTINE **
6520  GOSUB 27000
6530  REM  ** END OF BLEEP **
6560  PRINT "POINT NO ";I;" X = ";A(I);" Y = ";B(I)
6570  IF Z1 = 0 THEN 6640
6580  REM  ** KEYBOARD INPUT OF Z **
6590  GOSUB 24000
6600  PRINT " Z = ";
6610  INPUT C(I)
6620  REM  ** CLEAR KEYBOARD **
6630  GOSUB 24500
6640  IF I < 500 THEN 6310
6650  PRINT " TOO MANY POINTS "
6660  PRINT " FILE IS NOW FULL"
6670  GOSUB 24500: REM  CLEAR STROBE
6690  PRINT " END OF DATA -";I;" POINTS"
6710  LET N = I
6720  PRINT " NOTE THAT THESE DATA ARE "
6730  PRINT " IN TABLET CO-ORDINATES    "
6740  RETURN
7000  REM  *** NEW ORIGIN *** LEVEL 3
7010  HOME : REM  CLEAR SCREEN
7020  PRINT " SETTING THE ORIGIN"
7030  PRINT " =================="
7040  PRINT " X,Y WILL BE SET "
7050  PRINT " RELATIVE TO YOUR ORIGIN"
7060  PRINT " AND THE SENSE OF Y REVERSED"
7070  PRINT
7080  PRINT " PRESS PEN WHERE YOU WANT"
7090  PRINT " TO SET THE ORIGIN"
7110  GOSUB 23000: REM  GET POINT
7120  IF Z <  > 2 THEN 7110
7130  PRINT " GOOD ... DATA BEING TRANSLATED "
7140  PRINT " WITH THIS POINT AS 0,0"
7150  REM  ** TRANSLATE THE X,Y DATA ***
7160  FOR I = 1 TO N
7170  LET A(I) = A(I) - X
7180  LET B(I) = Y - B(I)
7190  NEXT I
7195  PRINT
7200  PRINT " JOB DONE ..."
7210  RETURN
```

```
8000  REM  *** SCALE AXES *** LEVEL 3
8010  HOME
8020  PRINT "SCALING OF THE X,Y DATA"
8030  PRINT "======================"
8040  PRINT
8050  PRINT " YOUR X,Y WILL BE SCALED BY"
8060  PRINT "INDICATING A KNOWN DISTANCEON"
8070  PRINT "THE TABLET USING THE CURSOR"
8080  PRINT
8090  PRINT "PRESS CURSOR AT START"
8100  GOSUB 23000
8120  IF Z <  > 2 THEN 8100
8124  LET X1 = X: LET Y1 = Y
8130  PRINT "GOOD! NOW PRESS CURSORON END"
8140  GOSUB 23000
8150  IF Z <  > 2 THEN 8140
8155  LET X2 = X: LET Y2 = Y
8160  PRINT "OK": PRINT : PRINT
8170  LET D =  SQR ((X1 - X2) ^ 2 + (Y1 - Y2) ^ 2)
8180  PRINT "THAT WAS ";D;" IN TABLET UNITS"
8190  PRINT "TYPE IN THE ACTUAL DISTANCE";
8200  INPUT D1
8210  PRINT
8220  PRINT "WHAT UNITS OF DISTANCE ARE YOU "
8230  PRINT "USING (EG MILES, KM ETC)";
8240  INPUT D$
8250  LET S3 = D / D1
8260  PRINT : PRINT "DATA BEING SCALED"
8270  FOR I = 1 TO N
8280  LET A(I) = A(I) / S3
8290  LET B(I) = B(I) / S3
8300  NEXT I
8310  PRINT : PRINT "JOB DONE": PRINT
8320  RETURN
9000  REM *** LISTING THE FILE***LEVEL3
9010  HOME
9020  PRINT "LIST OF THE CURRENT X,Y,Z"
9030  PRINT "========================"
9040  PRINT "DO YOU WANT A PRINTED LIST";
9050  GOSUB 20000
9060  REM  SWITCH ON SCREEN LIST
9070  GOSUB 25000
9080  IF T <  > 1 THEN 9100
9090  GOSUB 25500
9100  PRINT : PRINT "X","Y","Z"
9110  PRINT
9120  LET L = 8: REM  INITIAL LINE COUNT
9130  FOR I = 1 TO N
9140  PRINT  FN A(A(I)), FN A(B(I)), FN A(C(I))
9160  IF T = 0 THEN  GOSUB 26000
9170  NEXT I
9180  GOSUB 25000: REM  SWITCH BACK TO 0
9190  RETURN
10000  REM ***FILE DATA ON DISK***LEVEL3
10010  LET Q$ =  CHR$ (4): REM  OUTPUT BOARD
10020  PRINT "FILE DATA ON DISK"
10030  PRINT "================="
10040  PRINT "GIVE A <7 CHARACTER FILENAME"
10050  PRINT "BEGINNING WITH A LETTER AND"
10060  PRINT "NOTE THAT FILES OF THIS SAME"
10070  PRINT "NAME ALREADY ON DISK ARE"
10080  PRINT "DELETED !! THESE ARE"
10085  PRINT Q$;"CATALOG"
```

```
10090  PRINT : PRINT "YOUR NEW NAME =";
10100  INPUT E$
10110  PRINT Q$;"OPEN";E$
10120  PRINT Q$;"DELETE";E$
10130  PRINT Q$;"OPEN";E$
10140  PRINT Q$;"WRITE";E$
10150  PRINT N: REM  STORE NO OF XYZ
10160  FOR I = 1 TO N: PRINT A(I): PRINT B(I): PRINT C(I): NEXT I: PRINT
10170  PRINT Q$;"CLOSE";E$: REM  CLOSE FILE
10180  PRINT Q$;"CATALOG"
10190  RETURN
11000  REM  *** COMMANDS *** LEVEL 3
11020  PRINT "AVAILABLE COMMANDS"
11030  PRINT "=================="
11040  PRINT
11050  PRINT "DATA     = INPUT CO-ORDS"
11060  PRINT "ORIGIN   = DEFINE ORIGIN"
11070  PRINT "SCALE    = SCALE X,Y"
11080  PRINT "LIST     = LIST CURRENT DATA"
11090  PRINT "FILE     = STORE CURRENT X,Y,Z ON DISK"
11100  PRINT "COMMANDS= LIST OPTIONS"
11110  PRINT "AREA     = CALCULATE AREA"
11120  PRINT "DISTANCE= FIND DISTANCES"
11130  PRINT "TEST     = LOAD TEST DATA"
11140  PRINT "TRANSLATE = MOVE ORIGIN"
11150  PRINT "FINISH   = END JOB"
11160  PRINT
11170  PRINT "ONLY THE ABOVE ARE ALLOWED"
11180  PRINT "THEY MUST BE TYPED EXACTLY"
11190  RETURN
12000  REM  *** AREA CALCULATION***LEVEL3
12010  REM  CLOSE THE X,Y AND GIVES
12020  REM  AREA IN UNITS SQUARED
12030  HOME : PRINT : PRINT
12040  PRINT "AREA CALCULATION"
12050  PRINT "================"
12060  PRINT : PRINT "WAIT AWHILE ..."
12070  LET S = 0.0
12080  LET K = N - 1
12090  FOR I = 2 TO K
12100  LET S = S + B(I) * (A(I + 1) - A(I - 1))
12110  NEXT I
12120  LET S = S + B(1) * (A(2) - A(N)) + B(N) * (A(1) - A(N - 1))
12130  LET S =  ABS (0.5 * S)
12140  PRINT "ENCLOSED AREA = ";S;" SQ ";D$
12150  RETURN
13000  REM ***DISTANCES CALCULATION*** LEVEL 3
13010  HOME
13020  PRINT "DISTANCE ALONG THE LINE"
13030  PRINT "=======================": PRINT
13040  LET D3 = 0.0
13050  FOR I = 1 TO N - 1
13060  LET J = I + 1
13070  LET D1 =  SQR ((A(I) - A(J)) ^ 2 + (B(I) - B(J)) ^ 2)
13075  LET D3 = D3 + D1
13080  NEXT I
13090  PRINT "CALCULATION DONE"
13100  PRINT : PRINT "DISTANCE IS = ";D3;" ";D$
13110  RETURN
```

```
14000  REM  *** TEST DATA *** LEVEL 3
14010  HOME
14020  PRINT "TEST": PRINT "===="
14030  PRINT " READS DATA FOR THE CENTRE"
14040  PRINT " SQUARE OF THE TABLET"
14050  READ N
14060  FOR KK = 1 TO N
14070  READ A(KK),B(KK),C(KK)
14080  NEXT KK
14090  DATA  4
14100  DATA  3147,3599,1
14110  DATA  3660,3603,2
14120  DATA  3660,3093,3
14130  DATA  3153,3087,4
14140  PRINT : PRINT " TEST DATA OK"
14150  RETURN
15000  REM ***ORIGIN TRANSLATE***LEVEL 3
15010  HOME : PRINT
15020  PRINT "ORIGIN TRANSLATION"
15030  PRINT "=================="
15040  PRINT : PRINT "NEW VALUE AT X-ORIGIN TO BE = ";
15050  INPUT X8
15060  PRINT : PRINT "NEWVALUE AT Y-ORIGIN  TO BE = ";
15070  INPUT Y8
15080  FOR I = 1 TO N
15090  LET A(I) = A(I) + X8: LET B(I) = B(I) + Y8
15100  NEXT I: PRINT
15110  PRINT "JOB DONE"
15120  RETURN
16000  PRINT "**** END OF RUN ****"
16010  STOP
16020  RETURN
20000  REM  ** YES/NO ROUTINE ** LEVEL 4
20010  INPUT H$
20020  IF H$ <  > "YES" THEN 20050
20030  LET T = 1
20040  GOTO 20100
20050  IF H$ <  > "NO" THEN 20080
20060  LET T = 0
20070  GOTO 20100
20080  PRINT " ANSWER YES OR NO IN FULL"
20090  GOTO 20010
20100  RETURN
21000  REM *LOAD DISK FILE LEVEL 4
21010  REM  NOTE THAT NO PROTECTION
21020  REM  IS OFFERED IF NAME NOT
21030  REM  IN CATALOG.....!!!!!!!
21040  HOME : PRINT :
21050  PRINT "DATA FROM DISK"
21060  PRINT "=============="
21070  LET Q$ =  CHR$ (4): PRINT Q$;"CATALOG"
21080  PRINT : PRINT "DATA FILE NAME =";
21090  INPUT E$
21100  PRINT Q$;"OPEN";E$: REM  OPENFILE
21110  PRINT Q$;"READ";E$: REM  SET READY
21120  INPUT N: REM  NO POINTS
21130  FOR I = 1 TO N
21140  INPUT A(I): INPUT B(I): INPUT C(I)
21150  NEXT I
21160  PRINT Q$;"CLOSE";E$: REM  CLOSEFILE
21170  PRINT : PRINT "DATA READ ... RETURN"
21180  RETURN
```

```
22000  REM  ** INITIALISE TABLET ** LEVEL 4
22010  PRINT F$;"PR#";N2
22020  PRINT "T1,F,C,Q"
22030  PRINT F$;"PR#0"
22040  RETURN
23000  REM  ** CAPTURE AN X,Y PAIR ** LEVEL4
23010  PRINT F$;"IN#";N2
23020  INPUT X,Y,Z
23040  GOSUB 27000: REM  BLEEP
23050  PRINT F$;"IN#0"
23060  RETURN
24000  REM  ** SWITCH TO KEYBOARD ** LEVEL 4
24010  PRINT F$;"IN#0"
24020  RETURN
24500  REM  ** CLEAR KEYBOARD ** LEVEL 4
24505  REM  THIS IS VERY APPLE DEPENDENT
24510  POKE  - 16368,0
24520  RETURN
25000  REM *SWITCH TO SCREEN LIST LEVEL 4
25010  PRINT F$;"PR#0": RETURN
25500  REM *SWITCH TO PRINTER IN SLOT N LEVEL 4
25510  PRINT F$;"PR#";N3: RETURN
26000  REM *PAUSE WHILE LISTING LEVEL4
26010  LET L = L + 1
26020  IF L < 20 THEN 26070
26030  PRINT "PRESS ANY KEY FOR MORE"
26040  GET J$
26050  IF J$ = " " THEN 26040
26060  LET L = 3
26070  RETURN

27000  REM  ** BEEP ROUTINE ** LEVEL 4
27010  FOR KK = 1 TO 10
27020  LET X8 =  PEEK ( - 16336)
27030  NEXT KK
27040  RETURN
```

Multiarea Measurement from Maps and Soil Thin Sections Using a Microcomputer and Graphics Tablet

L. Robertson

The Macaulay Institute for Soil Research

ABSTRACT

Areas of two-dimensional irregular shapes can be measured using an Apple II microcomputer and Graphics Tablet. Additional software, and modification to the manufacturer's software is listed, to enable multiple areas on a percentage basis to be measured on source documents, such as maps and photomicrographs.

Hardware modification in the form of a flatbed cursor with illuminated disk improves greatly the ease with which measurements may be made, and permits micromorphological area measurements to be carried out on soil thin sections while simultaneously viewing the microscope image.

The efficiency of the method is evaluated by comparison with more traditional area-measurement techniques.

INTRODUCTION

Many investigations require single or multiarea measurement of two-dimensional irregular polygons. This paper describes a rapid method for making such measurements using an Apple II microcomputer and Graphics Tablet, and provides details of the necessary software modifications and optional hardware additions required.

The efficiency of the method is evaluated by comparison with other area-measurement techniques, and its use is described by reference to two specific applications:

a) multiarea measurement of part of a Soil Survey of Scotland soil map. Subsequent analysis of the data yields the percentage cover occupied by individual mapping units, and

b) area measurement of features observed during micromorphological examination of soil thin sections and polished surfaces of resin-impregnated soil blocks.

APPLICATION 1 (MAP AREA ANALYSIS)

As an example of how the method may be applied, multiarea analysis of data from a portion of a 1:25000 soil map (Fig. 1) will be considered, and comparison with other methods of analysis discussed. It should be appreciated that the procedures are applicable equally to the area analysis of plans, drawings, photomicrographs, and so on.

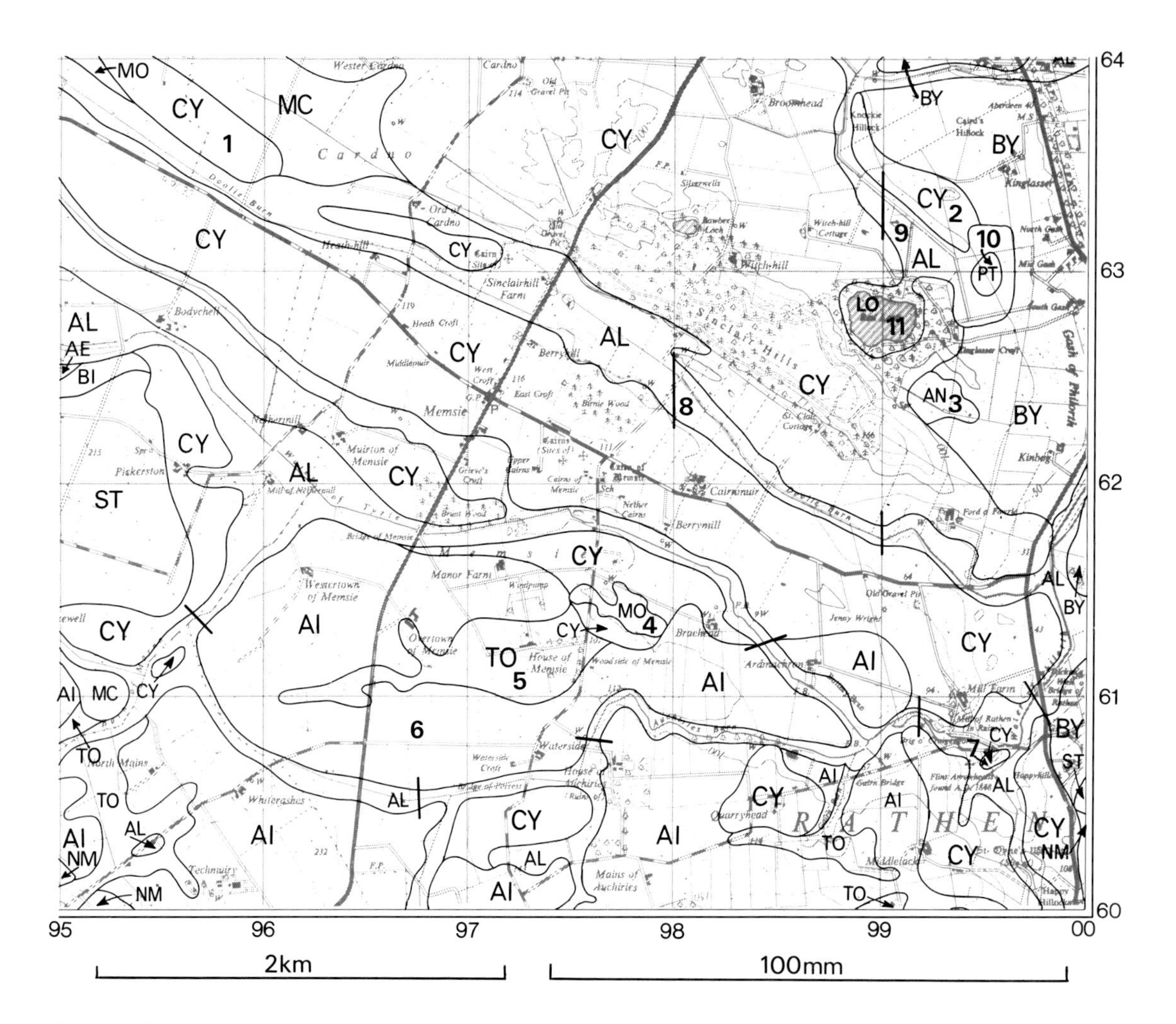

Figure 1. Southeastern corner of 1:25000 Soils Map (NJ 96), used as example for multiarea measurement and analyses. Significance of symbols is discussed in text.

The map consists of a series of irregularly shaped units (each representing an area of a particular Soil Series). Each mapping unit is identified by the inclusion of alphabetic symbols (Series code). Although the map shows the spatial distribution of the various units, the actual areas occupied by each Soil Series can be obtained only by measurement.

Several methods may be used and these will be reviewed briefly.

CUTTING AND WEIGHING. Each individual area can be cut out, and an estimate of the areas obtained by weighing. This method is possible only with simple documents (Boul and Hole, 1961) printed on constant density paper, and of course is destructive. This method has not been assessed, because its application is limited.

GRID-SQUARE OVERLAY MEASUREMENTS. A transparent overlay with a square grid pattern may be placed over the map, and the number of complete and partial squares occupied by each mapping unit counted manually (Robinson, 1963; Raisz, 1962). A fine grid is required to give reasonably precise results and the procedure is time consuming. However, if more sophisticated methods are not available this technique is used frequently.

PLANIMETRIC ANALYSIS. This method, involving the use of a polar planimeter (Robinson, 1963), has been used extensively in the past by the Soil Survey, but also is time consuming and is prone to errors, due mainly to 'wheel slip.'

IMAGE ANALYSIS. The speed and efficiency of area analysis using image analyzers cannot be overlooked. However, this equipment is expensive, and normally is available only in specialized laboratories. Although used extensively in micromorphological investigations (Delgado and Dorronsoro, 1983), my experience with its use for area analysis of colored soil maps has been disappointing, as it is difficult to achieve sufficient tonal separation between different units.

DIGITIZATION AND COMPUTER ANALYSIS. Sophisticated digitizing tables and associated computer hardware (Margerison, 1976) are expensive and not widely available. However, the relatively cheap microcomputer with peripherals is becoming readily available, and its application in multiarea analysis as described here has considerable potential.

EQUIPMENT

The equipment used (Fig. 2) consists of an Apple II Europlus computer (48K RAM) with monitor, two disk drive units with controller card, Epson MX/82 dot-matrix printer with Apple II Interface Kit Type 3 and an Apple Graphics Tablet (A2M0029).

The Graphics Tablet consists of a flat digitizing table, with a maximum active area of approximately 28 X 28 cm. Data are entered by drawing on the Tablet using the electromagnetic pen. Basic operating procedures are controlled by the Graphics Tablet Software, a suite of machine code and BASIC programs, which are stored on a floppy disk supplied with the Tablet. However, to improve the accuracy of area and distance measurements, the 'primary program changes' discussed by Roberts (1983), should be made to the Graphics Tablet Software.

A plastic overlay covers the active area of the Tablet, and a range of functions may be selected from a menu printed on the overlay by touching the pen to the required menu box. The most relevant functions for this application are DELTA (the value of which determines the number of points digitized for a given distance of pen travel), CALIBRATE which allows the Tablet to be calibrated in user-selected units for area and distance measurements and AREA/DISTANCE used for the calculation of single area or distance measurements.

Figure 2. Apple microcomputer with Graphics Tablet and dot-matrix printer. Printer illustrated is not suitable if 1:1 dumping of screen images is required, and should be replaced with an Epson MX/82.

A general review of the Graphics Tablet has been given by Bell (1982), and full details regarding the various functions, their operation and application, are contained in the manufacturer's Operation and Reference Manual (Apple Computer Inc., 1979).

SOFTWARE

The main operating program for the Graphics Tablet named TABLET-CODE APPLESOFT, is written in Applesoft II BASIC, and performs all the commands and functions of the Tablet.

To carry out a simple area calculation, the Tablet is calibrated, a suitable DELTA setting is used, the AREA function is selected from the menu, and the polygon to be measured is traced on the Tablet using the pen. The defined area is reproduced on the monitor (Fig. 2), and after a slight delay, its area value is displayed.

However, in many applications this simple facility is limited, because only occasionally are single area measurements required. To cater to multiarea measurement applications, the TABLET-CODE APPLESOFT program has been modified by the author, so that the value of each area measured together with an associated identifying code, is saved on disk for subsequent analysis, if required.

The supplied TABLET-CODE APPLESOFT program is fairly large (12121 bytes), and occupies nearly all available free RAM space. If increased greatly it will not function correctly, and therefore it is necessary to delete some functions to make room for the modifications discussed here. A suggested procedure for modifying the program, together with the new program lines required for this application is given in Appendix 1A.

USE OF THE MODIFIED GRAPHICS TABLET SOFTWARE

Using a Graphics Tablet Software disk, modified as described in Appendix 1A, the Graphics Tablet Software option is selected by keyboard entry, from the displayed menu. The modified TABLET-CODE APPLESOFT program is loaded automatically into memory, and a filename for data storage, map identification code, calibration unit type, and user's initials are entered from the keyboard when requested.

A reminder to set up and switch on the printer is displayed, and the relevant header information is output to it. A further video reminder to select a DELTA setting, AUDIO FEEDBACK, and to CALIBRATE the Tablet is displayed.

To obtain the highest possible accuracy, a DELTA setting of 1 should be used. However, with this setting, areas with a perimeter exceeding about 770 mm are not measurable, and a higher DELTA setting must be used if such areas are encountered. The DELTA setting may be changed between area measurements, if required. The AUDIO FEEDBACK facility should be switched on, as this gives a useful audio response during the tracing process.

The CALIBRATE function is selected from the Tablet Menu, and using the map scale, two points are entered on the Tablet when requested. The number of scale units represented by the distance between these points is entered via the keyboard, as is the unit type. The Tablet now is placed in the default DRAW mode ready for data input.

Taking care not to dislodge the overlay, the part of the map or diagram to be analyzed is placed over the active area of the Tablet, preferably below the overlay. The active area of the Tablet may be ascertained readily by moving the pen lightly (without depression) over the Tablet surface, and noting the pen cursor position as shown on the monitor screen.

Area measurements are made by selecting the AREA function from the Tablet Menu and carefully tracing round the entire perimeter of the polygon to be measured. Care should be taken that the pen is not lifted prematurely. On lifting the pen from the Tablet, the area of the feature traced will be calculated, and an identifying code will be requested (Series Code in this example), which should be entered via the keyboard. The value of the area calculated and its identifying code will be displayed momentarily on the video screen, saved on disk and also printed out, after which the Tablet will return to the DRAW mode.

The video image may be annotated at this time using the pen, or a further area may be measured by selecting the AREA function as outlined. This process is repeated as often as required. An identifying code of zero should be entered if the pen is lifted inadvertently from the Tablet before drawing is completed. or if a manual error is made during digitization. During subsequent analysis of the data, software controls may be used to delete or ignore such data records.

With fairly simple source documents such as the map example discussed here, the entire area measurement process may be completed quickly, but with larger applications it may be necessary to interrupt the work. In this situation, the video image may be saved, by selecting the SAVE function from the Tablet menu, and retrieved later with the LOAD function.

When data entry from the Tablet is completed, ESC from the keyboard allows the user to exit from the Graphics Tablet Software, with the information assembled stored on the disk data file for subsequent analysis.

If further area measurements are to be carried out on the same source document at a later date, rerunning the Graphics Tablet Software as outlined, using the same disk data filename, will result in new area data being appended to the already existing data file. This has been done to cater for large data volumes which cannot be assembled at one time. A simple program termed 'READER'

(Appendix 1B) may be used to examine the contents of any data file assembled during the operation of the modified Graphics Tablet Software.

The data stored in the disk files normally would be subjected to subsequent analyses, the type of analyses depending on the particular requirements of the user. By way of example, data from two data files assembled during two digitizations of the map shown in Figure 1 have been used as input to the BASIC 'Multiarea analysis' (Appendix 1C), and a compilation of the relevant output data is given in Table 1.

This program uses the data stored in the disk data file to prepare an analysis of the percentage area of the map occupied by each individual Soil Series. The user may calculate these percentages in two ways. They may be expressed as a percentage of the total area actually measured (Table 1, Columns 5 and 7), or as a percentage of the true total area of the map (Table 1, Columns 6 and 8) if this is known. The latter is a useful check on the degree of error involved during digitization and rounding for maps whose true total area can be assessed readily, whereas the former option is designed for maps or parts of maps with irregular boundaries whose total true area is not available readily. Prior to leaving the program, the user is given the option of deleting the data file, or retaining it for future use.

Table 1. Composite area values for unit types delineated on Figure 2, as measured using Apple Graphics Tablet. Scales used for Graphics Tablet calibration are shown on Figure 2.

UNIT CODE	No. of areas	Area covered by each unit type:		Area occupied by each unit type expressed as a percentage of			
		sq mm	sq km	Tot. area measured (31427.7 sq mm)	Actual area of map (32000 sq mm)	Tot. area measured (19.56 sq km)	Actual area of map (20 sq km)
MO	2	104.3	0.06	0.3	0.3	0.3	0.3
CY	15	13671.4	8.57	43.5	42.7	43.8	42.9
MC	2	698.4	0.44	2.2	2.2	2.3	2.2
BY	4	2312.1	1.41	7.4	7.2	7.2	7.1
AL	12	5648.7	3.51	18.0	17.7	17.9	17.6
PT	1	34.2	0.02	0.1	0.1	0.1	0.1
LO	1	100.5	0.06	0.3	0.3	0.3	0.3
AN	1	87.6	0.05	0.3	0.3	0.3	0.3
AI	8	6557.0	4.11	20.9	20.5	21.0	20.6
ST	2	930.5	0.55	3.0	2.9	2.8	2.8
NM	3	50.5	0.02	0.2	0.2	0.1	0.1
TO	5	1163.0	0.72	3.7	3.6	3.7	3.6
AE	1	4.0	0.00	0.0	0.0	0.0	0.0
BI	1	65.5	0.04	0.2	0.2	0.2	0.2
TOTALS	58	31427.7	19.56	100.1	98.2	100.0	98.1

COMPARISON WITH OTHER METHODS OF AREA MEASUREMENT

This method of multiarea measurement and analysis, as with many other methods available, is prone to both hardware limitations and operator error. In an attempt to evaluate these, and give an assessment of the degree of error likely to be involved, the test map was examined using other available area measurement techniques.

The areas designated 1 to 11 on Figure 1 were measured three times using the Graphics Tablet, to show the degree of variability likely to occur (Table 2). The inconsistency of results is due to minor variations which occur during repeated tracings of the same area, and gives some indication of the degree of operator error likely to be involved. A hardware modification which helps to minimize this error is discussed later.

The same eleven areas also were measured using a Ferranti Cetec System 4 Digitizer, manually using a 1 mm square-grid overlay, and planimeter. The area values obtained by these methods are also presented in Table 2.

The areas measured ranged from about thirteen to about 3000 sq mm, and because the data were not distributed normally they were transformed. The eleven mean values from each method were plotted mean against standard deviation, and mean against variance to establish the best transformation to use. Because the latter gave a better straight line fit to the data, a square-root transformation of the area data was applied (Quenouille, 1950), and an analysis of variance performed to determine the within-areas mean square values for the two methods (Table 3). A variance-ratio test affirmed that the variances were not different significantly, and the application of a Fisher-Behrens test (Fisher and Yates, 1982), ascertained that the area measurements could be drawn from the same population (Table 4).

The time involved in carrying out the area measurement and analysis using the various methods also was assessed. The grid-overlay method was by far the slowest, taking many hours to carry out the area measurements of the entire map and associated clerical procedures. The planimetric analysis also was slow, and from the figures available, seems to be the most unreliable method. Digitization of the entire test map, together with area measurements, using the System 4 method would take about three hours (P.F.S. Ritchie, 1984, pers. comm.). Use of the Graphics Tablet with its modified software compares favorably with the latter method for simple area measurement, is simple to use, and is efficient in terms of operator time. For example, the digitization of the test map shown in Figure 1 and the subsequent multiarea analysis took just over one hour to complete.

Table 2. Comparative area values derived from Figure 2 using four different area measurement techniques. Planimetric area values reflect limited degree of accuracy possible with method.

AREA CODE	APPLE GRAPHICS TABLET. Area value mm sq.	APPLE GRAPHICS TABLET. Mean value mm sq.	FERRANTI/CETEC SYST. 4 Area value mm sq.	FERRANTI/CETEC SYST. 4 Mean value mm sq.	GRID O/L. Area value mm sq.	PLANIMETER Area value mm sq.
1	344.42	347	340.36	341	318	330
	346.96		339.92			
	348.29		341.35			
2	223.76	221	224.51	223	220	210
	219.00		225.72			
	220.03		219.47			
3	96.78	96	88.92	90	88	90
	95.96		90.66			
	95.63		90.00			
4	84.98	86	86.92	87	82	60
	85.31		87.15			
	86.12		87.35			
5	552.83	557	544.64	545	514	540
	558.49		547.46			
	559.46		543.78			
6	3062.35	3059	3080.74	3090	2975	3240
	3050.38		3096.81			
	3063.01		3093.58			
7	12.33	12	13.91	14	14	10
	11.71		14.42			
	12.07		13.44			
8	224.26	224	234.44	230	227	280
	225.22		226.29			
	222.00		229.99			
9	385.05	384	393.59	391	371	420
	382.82		389.92			
	383.40		390.45			
10	36.01	37	36.09	36	38	40
	37.30		36.13			
	36.13		36.45			
11	99.46	102	97.20	100	91	90
	103.08		102.65			
	104.18		101.01			

Table 3. Results of analysis of variance to determine the within-areas mean square values for the square roots of the areas computed by the Graphics Tablet and Ferranti Cetec Digitizer, respectively.

Area Code	Square roots of Graphics Tablet derived area values (mm sq)		
	1	2	3
1	18.56	18.63	18.66
2	14.96	14.80	14.83
3	9.84	9.80	9.78
4	9.22	9.24	9.28
5	23.51	23.63	23.65
6	55.34	55.23	55.34
7	3.51	3.42	3.47
8	14.98	15.01	14.90
9	19.62	19.57	19.58
10	6.00	6.11	6.01
11	9.97	10.15	10.21

Source	Degs. of freedom	Sum of Squares	Mean Square
Between Areas	10	5973.6942	
Within Areas	22	0.0922	0.00419
Total	32	5973.7864	

Mean = 16.87
Standard Devn. = 0.065

At the 5% level, the mean square term = 0.0180.
Thus the variance within areas is not significant at the 5% level.

Area Code	Square roots of Ferranti Cetec derived area values (mm sq)		
	1	2	3
1	18.45	18.44	18.48
2	14.98	15.02	14.81
3	9.43	9.52	9.49
4	9.32	9.34	9.35
5	23.34	23.40	23.32
6	55.50	55.65	55.62
7	3.73	3.80	3.67
8	15.31	15.04	15.17
9	19.84	19.75	19.76
10	6.01	6.01	6.04
11	9.86	10.13	10.05

Source	Degs. of freedom	Sum of Squares	Mean Square
Between Areas	10	6023.1297	
Within Areas	22	0.1346	0.00612
Total	32	6023.2643	

Mean = 16.90
Standard Devn. = 0.078

At the 5% level, the mean square term = 0.0263
Thus the variance within areas is not significant at the 5% level.

Table 4. Results of Fisher-Behrens test carried out on data derived from Apple Graphics Tablet and Ferranti Cetec System area measurement.

The ratio of the variances = $0.078^2/0.065^2 = 1.44$

For (32,32) degrees of freedom the 5% limit of the Variance Ratio is 1.80

With the given variance ratio being within that limit, the variances can be taken as not being significantly different.

In assessing whether both samples could be drawn from the same population, a Fisher - Behrens test was used where the errors may be due to different causes.

$$t = \frac{\bar{X}_1 - \bar{X}_2}{\sqrt{EMS_1/n_1 + EMS_2/n_2}}$$

$$= \frac{(16.90 - 16.87)}{\sqrt{0.0061/33 + 0.0042/33}} = 1.695$$

where $\bar{X}_1$ and $\bar{X}_2$ are the mean values

n_1 and n_2 are the numbers in samples, and

EMS_1 and EMS_2 are the error mean squares.

The 5% t value for θ with 22 degrees of freedom is 2.056 where

$$\theta = \tan^{-1}(EMS_1/EMS_2)$$

$$= 0.8796 \text{ radians} = 50.2 \text{ degrees}$$

The area measurements derived from both the Apple Graphics Tablet and the Ferranti Cetec methods, may therefore be considered as drawn from the same population.

APPLICATION 2 (MICROMORPHOLOGICAL)

The techniques described for simple maps are equally applicable for measuring the areas of features in soil thin sections as seen through the microscope. They have been used extensively in a soil micromorphological context to measure the cross-sectional areas of soil pores seen on the surfaces of serially ground resin-impregnated soil blocks, and directly from soil thin sections. Details of the results of these investigations and comparison with image-analysis techniques are reported on elsewhere (Darbyshire, Robertson, and Mackie, 1984).

For such applications, several related methods may be used, the one selected being dependent on the equipment available:

(A) Previously prepared photomicrographs may be placed directly on the Graphics Tablet, and selected features traced with the pen as described.

(B) Using a simple mirror system or camera lucida attached to the microscope, drawings of the required features may be prepared, and these drawings used as source documents for area analyses.

(C) Using a camera lucida mounted on the microscope, it also is possible to set up the equipment as shown in Figure 3, and carry out the analyses directly, while viewing the drawing area of the Graphics Tablet, and the microscope image simultaneously. Because this method obviates the necessity to prepare intermediate source documents, as required in the two previous methods, it has obvious advantages.

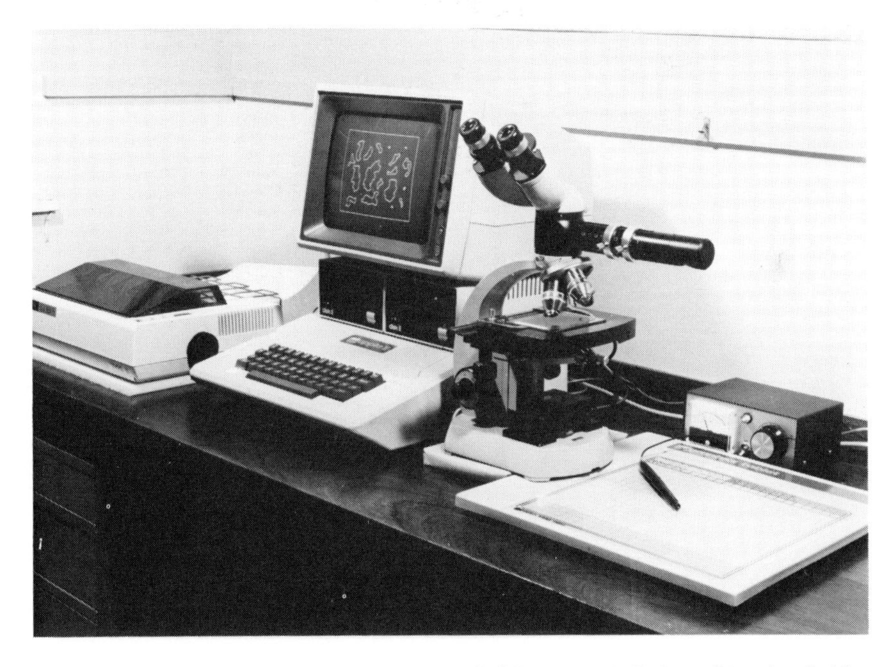

Figure 3. Apple microcomputer, Graphics Tablet, and Zeiss Standard 18 microscope fitted with camera lucida, set up for micromorphological multiarea measurements.

However, with all these methods it is difficult to measure small or irregularly shaped features (for example, many soil pores). The main problem is that the electromagnetic pen supplied with the Graphics Tablet is rather large in diameter, and tends to obstruct the operator's field of view, particularly if method C is being used. It also is difficult to hold the pen in a vertical position, while tracing irregular shapes.

To help overcome these problems, a flatbed cursor (Fig. 4) was designed and manufactured at the Macaulay Institute. The cursor is of simple construction, being manufactured essentially from two slabs of perspex, channelled to accept the necessary wiring, and screwed together. The field coil consists of 22 turns of ca 85 micron diameter (Imperial Standard Wire Gauge 44) lacquered copper wire wound onto a narrow 10 mm diameter bobbin, inset into a clear perspex disk. The cursor is activated by a small push-for-on switch mounted on the cursor body. This cursor has been used extensively, and has proved to be more satisfactory for applications A and B than the pen supplied. However, as no alteration is required to the rest of the hardware, pen and flatbed cursor may be interchanged, although it is necessary to run the Graphics Tablet MENU ALIGNMENT program if an exchange is made.

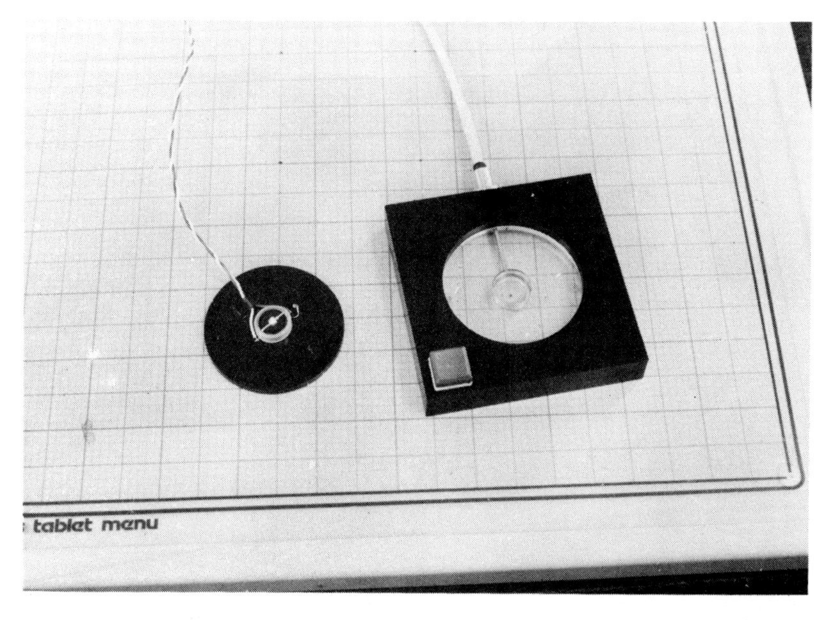

Figure 4. Flatbed cursor and LED disk designed for use with Apple Graphics Tablet. Cursor is compatible directly with Apple hardware and may be in place of electomagnetic pen as and when required.

In the situation of method C, it is difficult to trace round complex features accurately when using the camera lucida configuration, even with the flatbed cursor, because of the difficulty of matching the illumination of the Graphics Tablet and the microscope image. To remove this problem, a further modification is necessary.

A black perspex disk, which fits neatly into the recess on the upper surface of the flatbed cursor was prepared, and a miniature red light emitting diode (LED) was mounted centrally on it. This was wired into the paddle socket on the mother-board of the computer (Annunciator 1). The LED may be switched on or off from the keyboard or under software control, by referencing the relevant controlling softswitch (49242 or 49243).

The cursor, with LED disk installed and switched on is placed on the Graphics Tablet. The microscope image is focussed normally, and the camera lucida adjusted so that the illuminated LED appears as a sharply focussed red dot superimposed on the microscope image. After calibrating the Graphics Tablet, using a micrometer slide on the microscope stage, area measurements may be made by selecting the AREA function from the menu, moving the image of the LED onto the periphery of the feature to be measured as seen in the microscope image, pressing the cursor switch and tracing around the feature. Termination of digitization is completed when the cursor switch is released. After coding the feature from the keyboard, the calculated area is displayed momentarily on the monitor, printed out and saved to disk as previously described. The monitor image may be saved to disk after all tracing is completed, if required, for subsequent 1:1 hardcopy printout.

To reduce to a minimum curvature errors, which can be a problem towards the edge of the microscope image field, it is useful if a rectangular graticule mask can be fitted in the microscope eyepiece, to limit the field of view to the central portion of the microscope image. The total area under examination should be calculated using a micrometer slide, and it is useful if the active area of the Graphics Tablet outlined by the mask, is delineated on the monitor screen. Area measurements of individual or multiple feature types now may be carried out, as described in APPLICATION 1, and if the total area of thin section under examination has been determined the area occupied by each feature type expressed as a percentage of the total area under examination, may be determined by running the MULTIAREAS program already described, in conjunction with the disk data file just prepared.

CONCLUSIONS

It has been demonstrated that using an Apple II microcomputer and Graphics Tablet, a simple method for carrying out single or multiarea, distance or perimeter measurements, from original source documents such as maps, plans, or drawings, or, in the situation of micromorphological investigations from secondary documents, such as photomicrographs, can be made available. The method involves some modification to the software normally supplied, but this is not difficult, and no great computer programming experience is required to carry out these changes. Further sophistication of the software to produce more detailed shape or form parameters to suit particular applications is possible, but consideration should be given to relating the programming and development time required, to the cost of purchasing commercially available software. Inexpensive hardware modifications, the implementation of which will greatly improve the ease with which measurements may be made, and also enable, in the situation of micromorphological investigations, area measurements to be carried out directly from, and <u>while simultaneously viewing</u> the microscopy image, also are possible.

Many small laboratories possess, or have access to the basic type of equipment described, and it is believed that the techniques advocated here, will to some extent bridge the gap between the more traditional but time expensive facilities afforded by the more sophisticated image analyzers available only in specialized laboratories.

ACKNOWLEDGMENTS

The author would like to thank Mr. A. Lilly for carrying out the grid square and planimetric measurements, Mr. K. W. M. Brown for statistical analysis and helpful discussion, Mr. P. F. S. Ritchie and Mrs. J. G. Cairns for undertaking the comparative area measurements using the Ferranti Cetec System 4 Digitizer, and Mr. J. Mitchell and Mr. W. S. Shirreffs for preparation of the Figures.

Particular thanks are due to Mr. G. J. Gaskin and his colleagues in the Technical Services Section for advice and construction of the flatbed cursor and LED disk.

REFERENCES

Apple Computer Inc., 1979, Graphics tablet - operation and reference manual: Ref. A2L0033, 123 p.

Bell, D., 1982, Microcomputer color graphics system: Pitman Press, Bath, England, 178 p.

Boul, S.W., and Hole, F.D., 1961, Clay skin genesis in Wisconsin soils: Am. Soil Sci. Soc. Proc., v. 25, no. 5, p. 377-379.

Darbyshire, J.F., Robertson, L., and Mackie, L.A., 1984, A comparison of two methods of estimating the soil pore network available to protozoa: Soil Biology and Biochemistry, in press.

Delgado, M., and Dorronsoro, C., 1983, Image analysis, in Bullock, P., and Murphy, C.P., eds., Soil micromorphology 1: techniques and applications: Academic Press, Berkhamstead, England, p. 71-86.

Fisher, R.A., and Yates, F., 1982, Statistical tables for biological, agricultural and medical research: Longman Group Ltd., New York, p. 146

Margerison, T.A., 1976, Computers and the renaissance of cartography: Natural Environment Research Council, Experimental Cartography Unit, London, 20 p.

Quenouille, M.H., 1950, Introductory statistics: Butterworth, London, 248 p.

Raisz, E., 1962, Principles of cartography: McGraw-Hill Book Co., New York, 315 p.

Roberts, D. Mcl., 1983, Apple graphics tablet software modification: Laboratory Microcomputer, v. 2, no. 1, p. 15-18.

Robinson, A.H., 1963, Elements of cartography: John Wiley & Sons, New York, 343 p.

APPENDIX 1A

MODIFYING THE GRAPHICS TABLET SOFTWARE

A back-up copy of the Graphics Tablet Software disk supplied with the Tablet should be made, and the following procedures carried out.

With the back-up copy of the Graphics Tablet Software inserted in the disk drive proceed as follows:

LOAD TABLET-CODE APPLESOFT

DEL 1850,1930

DEL 2340,2480

DEL 2500,2580

Change the following lines:

```
192  TEXT : HOME : VTAB 12: HTAB 13: INPUT "QUIT? (Y OR N) ";A$: IF A$ = "Y" THEN  HOME :
     PRINT D$;"CLOSE";F$: VTAB 12: HTAB 10: PRINT "LOADING HELLO PROGRAM": POKE 104,8:
     POKE 103,1: PRINT D$"RUN HELLO,D1": STOP
1840  GOSUB 1130: GOTO 170: REM  BOX FN DELETED
2010 AR = AR + (X%(1) - X%(N% - 1)) * ((Y%(1) + Y%(N% - 1)) / 2):AR =  ABS (AR)      / WM
     ^ 2: IF AR < 999999999 THEN AR = ( INT (AR * 1000)) / 1000
2020  GOSUB 2300: PRINT D$;"IN£0": POKE 41, PEEK (41) + 4: INPUT "ENTER AREA          CODE
     ";NO$: GOSUB 2300: VTAB 23:B$ = "AREA IS ": POKE 41, PEEK (41) + 4: GOSUB 2030:
     RETURN
2030 B$ = B$ +  STR$ (AR) + " SQ." + UN$ + " FOR AREA " + NO$: PRINT B$
2330  GOSUB 1130: GOTO 170: REM  SLIDE FN DELETED
2490  GOSUB 1130: GOTO 170: REM  SEPERATE FN DELETED
```

Add the following lines:

```
15  REM   * MODIFIED FOR MULTIPLE AREAS MEASUREMENT.  L. ROBERTSON, 1983 *

65  GOSUB 3000

2031  PRINT D$;"PR£";PS: PRINT AR,,NO$
2032  PRINT D$;"PR£0"
2040  PRINT D$;"WRITE";F$: PRINT ID$: PRINT NO$: PRINT AR: PRINT D$: RETURN

3000 PS = 1: REM  PRINTER SLOT
3001 D$ =  CHR$ (4)
3010  TEXT : HOME : INVERSE : PRINT "MULTIPLE AREAS MEASUREMENT   L.ROBERTSON": POKE 34,2:
     NORMAL
3100  PRINT : VTAB 5: PRINT "PLEASE ENTER THE FOLLOWING INFORMATION": PRINT : PRINT
3105  PRINT : INPUT "DISC FILENAME FOR DATA ";F$: PRINT
3110  INPUT "MAP IDENTIFICATION CODE ";ID$
3120  PRINT : INPUT "CALIBRATION UNIT TYPE (FORMAT XX) ";UN$
```

The modifications advocated by Roberts (1983) also may be incorporated into the software at this point, and the modified version of the TABLET-CODE APPLESOFT program resaved on disk under the same name.

This new version of the TABLET-CODE APPLESOFT program is 12220 ($2F40) bytes long as compared with 12121 ($2F61) bytes for the original version.

```
3130  IF  LEN (UN$) <  > 2 GOTO 3120
3140  PRINT : INPUT "YOUR INITIALS, FOR IDENTIFICATION ";NM$
3142  PRINT D$;"OPEN";F$: PRINT : PRINT D$;"WRITE";F$: PRINT ID$: PRINT UN$: PRINT NM$:
     PRINT D$;"CLOSE";F$: PRINT : PRINT D$;"APPEND";F$: PRINT : PRINT D$
3150  HOME : VTAB 6: PRINT "SET PRINTER---PRESS ANY KEY TO CONTINUE";: GET R$
3160  PRINT : PRINT D$;"PR£";PS: PRINT "MULTIPLE AREAS MEASUREMENT   L.ROBERTSON": PRINT :
     PRINT : PRINT "MAP IDENTIFICATION CODE IS ";ID$: PRINT : PRINT "OPERATOR I.D. CODE
     IS..... ";NM$
3164  PRINT : PRINT "DISC FILENAME FOR DATA IS ";F$
3165  PRINT : PRINT : PRINT "AREAS (SQ.";: PRINT UN$;: PRINT ")              AREA CODE"
3170  PRINT : PRINT D$;"PR£0": POKE 1656 + PS,40: HOME : VTAB 8: INVERSE : PRINT "SET
     DELTA, AUDIO FEEDBACK & CALIBRATION ": PRINT "BEFORE OPERATING AREA FUNCTION ON
     TABLET": FOR P = 1 TO 1500: NEXT P
3180  NORMAL : RETURN
```

APPENDIX 1B

DISK DATA FILE READER

This short program also should be stored on the new GRAPHICS TABLET SOFTWARE disk. It may be used whenever it is necessary to examine the contents of any disk data file previously assembled during use of the modified Graphics Tablet Software. It should be saved under the name 'READER'.

```
10  REM   **** READER ****
15 SLOT = 1: REM  PRINTER SLOT
20  ONERR  GOTO 120
30  TEXT : HOME : VTAB 6: INPUT "ENTER FILE NAME TO READ ";F$
40  INPUT "PRINTER ON Y/N ?";P$
50  IF P$ <  > "Y" GOTO 70
60  PRINT : PRINT  CHR$ (4);"PR£1": POKE 1656 + SLOT,72
70  PRINT  CHR$ (4);"OPEN";F$
80  PRINT  CHR$ (4);"READ";F$
90  INPUT A$,B$,C$
100  PRINT A$,B$,C$
110  GOTO 80
120  PRINT : PRINT  CHR$ (4);"PR£0": POKE 1656 + SLOT,40: PRINT "END OF DATA...."
130  PRINT  CHR$ (4);"CLOSE"
140  END
```

APPENDIX 1C

MULTIAREA ANALYSES

This is a sample program showing how the data assembled in the disk data file by the modified Graphics Tablet Software may be used. The information shown in Table 1 is a compilation of data produced by running this program twice, using data previously assembled during two separate digitizations of the sample soil map (Fig. 1). It may be used as it stands in many situations, or may be modified by the user to suit particular requirements. As listed, it caters for a maximum of 500 individual area values derived from the relevant disk data file.

```
10  REM   **** MULTI-AREAS ****
20 D$ =  CHR$ (4)
25  TEXT : HOME : VTAB 6: INVERSE : PRINT "MULTI-AREA ANALYSIS FROM DISC-FILE DATA.":
     NORMAL : PRINT : PRINT : INPUT "ENTER DATA FILENAME REQD. ";F$
30  GOSUB 730
40  DIM SS$(500),A(500)
50  GOSUB 630
60 WRITE = 49312 + 256 * 1
70  TEXT : HOME : VTAB 6: INVERSE : PRINT "MULTI-AREA ANALYSES FROM DISC-FILE DATA.":
     NORMAL : PRINT : PRINT : PRINT "  HARD-COPY OF ANALYSIS REQD.  Y/N ? ";: GET R$
80  IF R$ <  > "Y" AND R$ <  > "N" GOTO 70
90  IF R$ = "N" GOTO 120
100  PRINT : PRINT : FLASH : PRINT : PRINT "         * SWITCH PRINTER ON *          ":
     NORMAL : PRINT : PRINT : PRINT "PRESS ANY KEY TO CONTINUE ";: GET R$
110  PRINT : PRINT D$;"PR£1"
120  REM   ** READ DATA FROM DISC-FILE INTO ARRAYS **
130 J = 0: ONERR  GOTO 220
140  PRINT : PRINT D$;"READ";F$
150  INPUT A$,NO$,A
160  IF  PEEK (112) -  PEEK (110) < 2 THEN  PRINT "   STAND BY    "
170  IF NO$ = "0" OR A = 0 GOTO 150
180 SS$(J) = NO$:A(J) = A
190 TT = TT + A
200 J = J + 1
210  GOTO 150
220  REM  ALL DATA READ INTO ARRAYS
230  FOR P = 1 TO 500: NEXT P
240  TEXT : HOME : PRINT "           ";: INVERSE : PRINT "MULTI-AREA ANALYSIS"
245  NORMAL : PRINT : PRINT "DISC DATA FILENAME IS ";F$
250  PRINT : PRINT "MAP I.D. CODE IS....";A$: PRINT : INVERSE
260  IF A$ <  > "" GOTO 280
270  PRINT : PRINT "NO DATA FILE NAMED ";F$;" EXISTS !": FOR J = 1 TO 3: PRINT  CHR$ (7):
     NEXT J: POKE 216,0: GOTO 620
280  PRINT : PRINT "AREA          TOT AREA        % AREA": PRINT "CODES..(NO)     SQ.
     ";W$;"          ......": PRINT
290  NORMAL
```

```
300 J = 0
310 TEMP$ =  LEFT$ (SS$(J),2)
320  IF SS$(J) = "" GOTO 400
330  IF  LEFT$ (SS$(J),2) = "**" GOTO 380
340  IF  LEFT$ (SS$(J),2) <  > TEMP$ GOTO 380
350 C% = C% + 1
360 SA = SA + A(J)
370 SS$(J) = "**"
380 J = J + 1
390  GOTO 320
400  IF SA = 0 GOTO 520
410  IF TRA = 0 GOTO 430
420 TT = TRA
430 PA =  INT (((SA / TT) * 100) * 10 + 0.5) / 10
440 PT = PT + PA
450  CALL WRITE:TEMP$;" (";C%;I4,")","  ",SA;F8.2,"        ",PA;F3.1, CHR$ (13):
460 J = 0
470 TEMP$ =  LEFT$ (SS$(J),2)
480  IF TEMP$ = "**" GOTO 500
490 T = T + SA:SA = 0:C% = 0: GOTO 310
500 J = J + 1
510  GOTO 470
520  PRINT : HTAB 30: PRINT "-------"
530  CALL WRITE:"                              TOTAL",PT;F4.1:
540  PRINT : HTAB 30: PRINT "-------"
550  PRINT : PRINT : PRINT "TOTAL AREA MEASURED =>  ";T;: PRINT " SQ.";W$
560  IF TRA = 0 GOTO 580
570  PRINT "TRUE AREA OF MAP     =>  ";TRA;: PRINT " SQ.";W$
580  PRINT "NUMBER OF AREAS      =>  ";J
590  GOSUB 690
600  PRINT : PRINT  CHR$ (4);"PR£0"
610  GOSUB 760
620  PRINT : PRINT "BYE. NICE WORKING WITH YOU!": END
630  TEXT : HOME : VTAB 6: INVERSE : PRINT "MULTI-AREA ANALYSIS FROM DISC-FILE DATA.":
     NORMAL : PRINT : PRINT
635  PRINT "THE ANALYSIS MAY BE CARRIED OUT USING   THE TRUE TOTAL AREA OF THE MAP, OR THE
     TOTAL AREA AS MEASURED VIA THE GRAPHICS TABLET DATA. "
640  PRINT : PRINT : PRINT "IS TRUE TOTAL AREA OF MAP KNOWN Y/N ? ";: GET R$
650  IF R$ <  > "Y" AND R$ <  > "N" GOTO 630
660  IF R$ = "N" THEN  RETURN
670  PRINT : PRINT : PRINT "ENTER TRUE AREA OF MAP (SQ.";: PRINT W$;: PRINT ") ";: INPUT
     TRA
680  RETURN
690  PRINT : PRINT D$;"PR£0": PRINT "RE-RUN Y/N ?";: GET R$
700  IF R$ <  > "Y" AND R$ <  > "N" GOTO 690
710  IF R$ = "N" THEN  RETURN
720  TEXT : HOME : VTAB 6: HTAB 9: PRINT "LOADING - PLEASE WAIT.": CLEAR : PRINT : PRINT
     CHR$ (4);"RUN MULTI-AREAS"
730  ONERR  GOTO 270
740  PRINT : PRINT D$;"OPEN";F$: PRINT : PRINT D$;"READ";F$: INPUT ID$: INPUT W$: INPUT
     NUL$: PRINT  CHR$ (4): RETURN
750  REM   RENAME/DELETE DATA FILE OPTION
760  TEXT : HOME : VTAB 6: PRINT "THE DATA USED IN THIS ANALYSIS IS STORED": PRINT "ON
     DISC UNDER THE NAME ";F$: PRINT : PRINT "DO YOU WISH TO RETAIN THIS DATA...Y/N ?";:
     GET R$
770  IF R$ <  > "Y" AND R$ <  > "N" GOTO 760
780  IF R$ = "N" GOTO 830
820  GOTO 850
830  PRINT : PRINT D$;"DELETE";F$
840  PRINT : PRINT "THE DISC DATA FILE NAMED ";F$: INVERSE : PRINT : PRINT "HAS BEEN
     DELETED !!"
850  NORMAL : RETURN
```

INDEX